Advances in Intelligent and Soft Computing 170

Editor-in-Chief

Prof. Janusz Kacprzyk
Systems Research Institute
Polish Academy of Sciences
ul. Newelska 6
01-447 Warsaw
Poland
E-mail: kacprzyk@ibspan.waw.pl

For further volumes:
http://www.springer.com/series/4240

Wojciech Zamojski, Jacek Mazurkiewicz,
Jarosław Sugier, Tomasz Walkowiak,
and Janusz Kacprzyk (Eds.)

Complex Systems and Dependability

Editors

Wojciech Zamojski
Institute of Computer Engineering,
Control and Robotics
Wrocław University of Technology
Wrocław
Poland

Tomasz Walkowiak
Institute of Computer Engineering,
Control and Robotics
Wrocław University of Technology
Wrocław
Poland

Jacek Mazurkiewicz
Institute of Computer Engineering,
Control and Robotics
Wrocław University of Technology
Wrocław
Poland

Janusz Kacprzyk
Polish Academy of Sciences
Systems Research Institute
Warszawa
Poland

Jarosław Sugier
Institute of Computer Engineering,
Control and Robotics
Wrocław University of Technology
Wrocław
Poland

ISSN 1867-5662
e-ISSN 1867-5670
ISBN 978-3-642-30661-7
e-ISBN 978-3-642-30662-4
DOI 10.1007/978-3-642-30662-4
Springer Heidelberg New York Dordrecht London

Library of Congress Control Number: 2012938746

Printed on acid-free paper

Springer is part of Springer Science+Business Media (www.springer.com)

Preface

We would like to present monographic studies on selected problems of complex systems and their dependability which are included in this volume of "Advances in Intelligent and Soft Computing" series.

Today's complex systems are integrated unities of technical, information, organization, software and human (users, administrators and management) resources. Complexity of modern systems stems not only from their complex technical and organization structures (hardware and software resources) but mainly from complexity of system information processes (processing, monitoring, management, etc.) realized in their defined environment. System resources are dynamically allocated to ongoing tasks. A rhythm of system events flow (incoming and/or ongoing tasks, decisions of a management system, system faults, "defense" system reactions, etc.) may be considered as deterministic or/and probabilistic event streams. This complexity and multiplicity of processes, their concurrency and their reliance on embedded intelligence (human and artificial) significantly impedes the construction of mathematical models and limits evaluation of adequate system measures. In many cases, analysis of modern complex systems is confined to quantitative studies (Monte Carlo simulations) which prevents development of appropriate methods of system design and selection of policies for system exploitation. Security and confidentiality of information processing introduce further complications into the system models and evaluation methods.

Dependability is the modern approach to reliability problems of contemporary complex systems. It is worth to underline the difference among the two terms of system dependability and systems reliability. Dependability of systems, especially computer systems and networks, is based on multi-disciplinary approach to theory, technology, and maintenance of systems working in a real (and very often unfriendly) environment. Dependability of systems concentrates on efficient realization of tasks, services and jobs by a system considered as a unity of technical, information and human resources, while "classical" reliability is restrained to analysis of technical system resources (components and structures built form them).

In the following few points we will briefly present main subjects of our monograph.

Problems of complex system modelling can be found in many chapters. Modelling of a system as a services network is investigated in chapter 11. Mathematical models of computer systems and networks and applied computation methods are presented in chapters 4, 6 and 19. Optimization of the traveller salesman problem modelled by a genetic algorithm is considered in chapter 13, while in 21 specific mechanism – service renaming – is proposed as a method for elimination of unexpected behaviour in component-based systems.

A statistical methodology called "nonparametric predictive inference" applied to reliability analysis for systems and networks is presented in chapter 8. Some specific view on several aspects on the interrelation between statistics and software testing and reliability is discussed in chapter 7. In chapter 17 a method of event monitoring in a cluster system is proposed. A functional testing toolset and its application to development of dependable avionics software are the topic of chapter 2.

The ISO/IEC standard 15408 "Common Criteria for IT security evaluation" deals with problems of IT security features and behaviour. Chapter 3 proposes an ontology-based approach to definition of specification means used in a process compliant with this standard, whereas building development environments for secure and reliable IT products based on design patterns of the "CC" project is the topic of chapter 12. Chapter 1 presents specific tool supporting both business continuity and information security management trying to integrate those two aspects in the most efficient way.

Encryption and security tools create a foundation of secure exploitation of computer systems and networks. An encryption method useful for oblivious information transfer is proposed in chapter 9, while specialized cipher devices implemented in FPGA devices are discussed in chapter 18. Chapter 10, in turn, is devoted to security issues in the process of hardware design and simulation.

The final subject deals with specific controlling and design issues in specialized complex systems applied for street lighting. In chapter 14 mathematical models of the lighting systems are introduced and then formal methods improving agent-aided smart lighting system design are presented in chapter 15. Computational support for optimizing systems of this kind is discussed in chapter 16 while in 20 specific rule-based approach to the problems of their control is proposed.

In the final words of this introduction we would like to express our sincere appreciation for all authors who have contributed their works as well as to all reviewers who has helped to refine the contents of this monograph. We believe that it will be interesting to all scientists, researchers, practitioners and students who work on problems of dependability.

The Editors

Wojciech Zamojski
Jacek Mazurkiewicz
Jarosław Sugier
Tomasz Walkowiak
Janusz Kacprzyk

List of Reviewers

Salem Abdel-Badeeh
Ali Al-Dahoud
Manuel Gil Perez
Janusz Górski
Zbigniew Huzar
Adrian Kapczyński
Jan Magott
Istvan Majzik
Grzegorz J. Nalepa
Sergey Orlov
Yiannis Papadopoulos
Oksana Pomorova
Maciej Rostański
Krzysztof Sacha
Marek Skomorowski
Barbara Strug
Stanisław Wrycza
Wojciech Zamojski

Contents

Validation of the Software Supporting Information Security and Business Continuity Management Processes

Jacek Baginski and Andrzej Białas

Abstract. The chapter presents the OSCAD tool supporting the business continuity (according to BS 25999) and information security management (according to ISO/IEC 27001) processes in organizations. First, the subject of the validation, i.e. the OSCAD software is presented, next the goal and range of the validation are briefly described. The validation is focused on the key management process related to risk analyses. A business-oriented, two-stage risk analysis method implemented in the tool assumes a business processes criticality assessment at the first stage and detailed analysis of threats and vulnerabilities for most critical processes at the second stage of the risk analysis. The main objective of the validation is to answer how to integrate those two management systems in the most efficient way.

1 Introduction

The chapter concerns a joint software implementation of two management systems very important for business processes of any modern organization. The first deals with business continuity, while the second concerns information security.

Business continuity [1], [2] is understood as a strategic ability of the organization to:

- plan reactions and react to incidents and disturbances in its business operations with a view to continue them on an acceptable, previously determined level,
- diminish possible losses if such harmful factors occur.

Different kinds of assets are engaged to run business processes – technical production infrastructures, ICT infrastructures with information assets, services, energy, media, materials, human resources, etc. All of them are needed by an organization to continue its operations and, as a result, to achieve its business objectives. Business continuity is very important for organizations:

Jacek Baginski · Andrzej Białas
Institute of Innovative Technologies EMAG, 40-189 Katowice, Leopolda 31, Poland
e-mail: jbaginski@emag.pl, a.bialas@emag.pl

W. Zamojski et al. (Eds.): Complex Systems and Dependability, AISC 170, pp. 1–17.
springerlink.com

- which have expanded co-operation links,
- which are part of complex supply chains,
- which work according to the Just-in-time strategy.

There is a specific group of organizations providing e-services that are totally dependent on ICT infrastructures. Their business processes depend strongly on efficient operations of IT systems and business continuity is a key issue for them.

There are many other factors which can disturb business continuity, such as: technical damages; random disruption events; catastrophes; disturbances in the provision of services, materials and information from the outside; organizational errors; human errors, deliberate human actions, etc. These factors should be identified, mitigated and controlled. The BS 25999 group of standards [1], [2] plays the key role in business continuity management (BCM), specifying the requirements for Business Continuity Management Systems (BCMS).

Information security [3], [4] is identified as the protection of integrity, confidentiality and availability with respect to information assets of any organization. Business processes of modern organizations depend on reliable and secure information processing. This processing influences the achievement of business objectives of these organizations, despite of their size and character. It is important for public or health organizations, educational institutions, and commercial companies. The information related to business processes should be protected while it is generated, processed, stored or transferred, despite of its representation (electronic, paper document, voice, etc.). Different threats exploiting vulnerabilities cause information security breaches. Any factors that may negatively influence information assets should be mitigated and controlled. The key role for information security management (ISM) belongs to the ISO/IEC 2700x family of standards, especially ISO/IEC 27001, specifying the requirements for Information Security Managements Systems (ISMS).

Both BCM and ISM systems are based on the Deming cycle (Plan-Do-Check-Act) and broadly use the risk analysis. They gave foundation to the OSCAD system, presented in this chapter. OSCAD is developed at the Institute of Innovative Technologies EMAG within a project co-financed by the Polish National Research & Development Centre (Narodowe Centrum Badań i Rozwoju) [5].

The BS 25999 standard plays the key role in the BCM domain. To improve the effectiveness of the BCM processes, a number of software tools were developed, e.g.: SunGuard LDRPS (Living Disaster Recovery Planning System) [6], ErLogix BCMS [7], ROBUST [8], RPX Recovery Planner [9]. These tools offer very similar functions, more or less detailed activities within the BCMS phases. Some of them require purchasing several separate modules for the full BCMS implementation (like risk analysis or incident management). OSCAD tries to gather those functions within one complex solution and offers additional elements, like:

- integration with ISMS and its specific elements (e.g. statement of applicability, information groups risk analysis, more emphasis on information confidentiality and integrity during the risk analysis);

- incidents database with incident registration (manual registration or automatic coming from external sensors) and management (with simplified task management for the activities related to the incident investigation and handling);
- simplified task management for any task with automated generation of most typical tasks (risk analysis, business continuity plans, incident management task) and users' notification about task generated for them.

Risk management is a significant element of BCM and ISM systems. There are a number of tools (like Cobra [10], Cora [11], Coras [12], Ebios [13], Ezrisk [14], Mehari [15], Risicare [16], Octave [17], Lancelot [18]) that specialize in risk management. More advanced risk methods and their ontologies are discussed in [19].

Risk is an important issue for different management systems in many domains. For example, the decision making process requires inclusion of a risk level related to the considered decision. That is why the OSCAD tool presented in the chapter is considered (among other tools) a potential element of the decision-making supporting tool, whose implementation is a subject of the ValueSec project [20].

The chapter presents the OSCAD software supporting business continuity and information security management processes, validation program for the OSCAD system, especially the risk management processes validation, and the conclusions for further development and improvements.

2 Computer-Supported Integrated Management System OSCAD

The OSCAD project is focused not only on the software development but also on providing an organizational and procedural layer for the joined BCMS/ISMS implementation. It is assumed that management activities contain purely human activities and human activities supported by the OSCAD tool. The OSCAD project provides software, a set of patterns of BCMS/ISMS documents, implementation methodology and know-how. Apart from the OSCAD core application, supporting management processes, two additional software components have been developed:

- the OSCAD-STAT web-based application gathering incident records, analyzing them and, on this basis, elaborating and providing statistics,
- the OSCAD-REDUNDANCY module, being a failover facility for the basic OSCAD system.

The BCMS and ISMS parts of OSCAD are integrated according to the Integrated Security Platform (ISP) whose concept was presented in [21]. The ISP assumes that different management systems (ISM-, BCM-, IT services-, quality-, business processes management systems, etc.) are integrated with the use of a common Configuration Management Database (CMDB). The ISP platform ensures the management systems integration from the logical point of view, while BS PAS 99 [22] provides recommendations for the organizational and procedural aspects of this integration. The ISP concept assumes the separation of common parts of the organization's management systems and their uniform, integrated implementation. Specific elements of particular management systems are implemented for each system separately.

The OSCAD software components were discussed in details in [23], [24], therefore in this chapter only concise information will be provided. The OSCAD system consists of a certain number of modules presented below (Fig. 1).

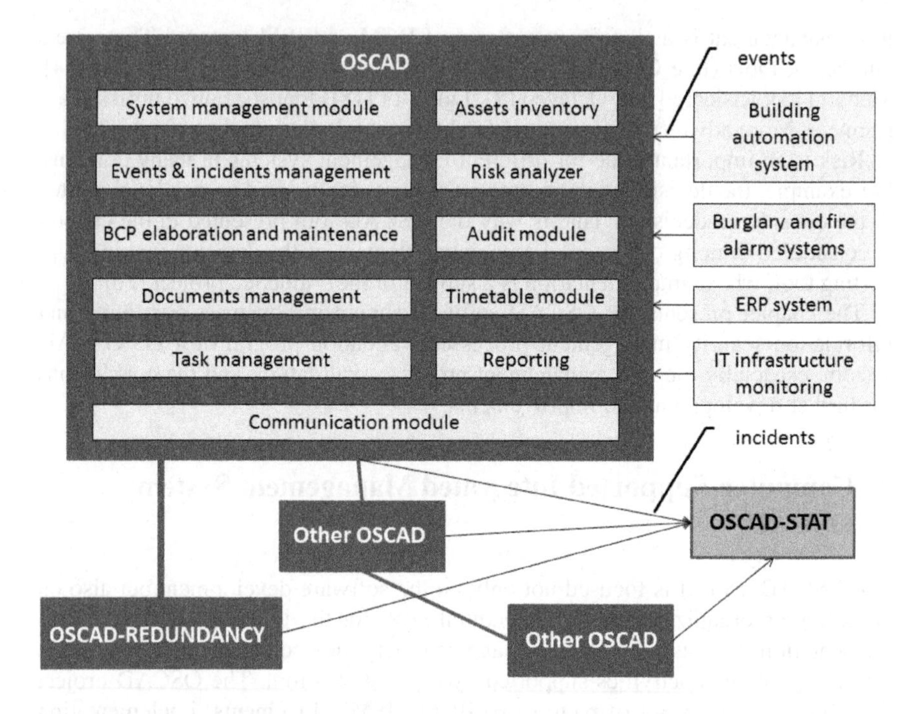

Fig. 1 OSCAD – block diagram of software modules

The assets inventory module stores information on assets and asset groups. Some assets are common for both systems and some are related either to the business continuity- or information security management system.

The system management module supports the execution of tasks by other modules and concerns proper configuration of the OSCAD system, preparation of pre-defined data, including dictionaries, basic information about the organization, risk analysis parameters (business loss matrix, acceptable risk level), and the roles of people who take part in business continuity and information security management processes.

The risk analyzer plays a key role in the OSCAD system. Thanks to this module it is possible to orientate the implementation of the system in the organization's operational environment. The module supports the execution of activities related to risk management, such as:

- Business Impact Analysis (BIA) of the loss of such parameters as availability, integrity of a process, service or information as well as information confidentiality (important in the case of an information security management system),
- determining criticality of processes based on the BIA analysis results,
- determining parameter value of Maximum Tolerable Period of Disruption (MTPD) and Recovery Time Objective (RTO),
- collecting information about threats and vulnerabilities for processes, information groups (possibility to select threats and vulnerabilities defined in bases/dictionaries or to enter new ones),
- risk level assessment with respect to existing security measures,
- preparing a risk treatment plan: risk acceptance or selection of security measures with respect to implementation costs.

The events and incidents management module performs the following tasks in the BCM system:

- event registration, incident classification and registration,
- preliminary evaluation and selection of proper proceedings,
- initiation of rescue operations to protect lives and health of employees and clients,
- problem analysis and start-up of a proper Business Continuity Plan (BCP),
- communication with interested parties and co-operation with contractors,
- providing accessibility of the basic and stand-by locations of incident management,
- providing all required documents,
- closing the incident,
- reporting,
- lessons learnt.

The BCP elaboration and maintenance module is responsible for supporting three basic activity groups: elaboration, start-up and testing of the BCP plan. The assumption was that plans are prepared for processes with the "critical" attribute assigned as a result of the conducted risk analysis. The BCP plan points at:

- assets necessary to begin and fulfill the plan,
- plan execution environment – basic and stand-by locations in the organization,
- contact list of people who execute the plan,
- operations to be performed.

The kinds of operations undertaken within the plan determine the kinds of tests. Tests planning is supported by the timetable module.

The task management module enables to define tasks which are to be performed by particular users of the systems and to control this performance. The module has the following basic functions: planning, ordering, supervising the performance and reviewing tasks. The tasks can refer not only to the management of business continuity and information security in the organization but also to any other aspects of its operations.

The audit module supports audit management in a sense that it collects information about the execution of each audit, supports the preparation of reports, assists in audit approval by people who supervise it. Audit planning as such is handled in the timetable module.

The timetable module supports to plan management operations undertaken at a certain time horizon within different modules.

The reports module is presents information about the BCM/ISM system, collected during its implementation and kept during exploitation, in the form of text or graphics. Additionally the information can be exported with the use of popular formats. The following reports are expected as minimum:

- information about the organization and its business processes,
- information about the organization's assets involved in the execution of the OSCAD system,
- summary of the risk analysis process,
- summary of planned tasks which are part of timetables,
- information about BCP plans,
- summary of audit results,
- summary of preventive and corrective actions,
- values of measures and indicators used for continual improvement of the BCM/ISM system,
- summary of different tasks execution,
- statistics related to incidents.

The documents management module supports registration, version control, circulation, confirmation, search, etc. of BCM and ISM systems documents in compliance with the requirements of standards. Documents and their templates are attached to the system and stored in the data repository. Sample documents are: "Business continuity management system policy", "Description of risk assessment methodology and risk acceptance criteria", or "Records supervision procedure".

The communication module enables information exchange between external systems and appointed persons. The phases of event detection and notification are performed automatically by the communication interface for the incident management module. Events registration can be performed either "manually" (entered by a person who gives information about the event) or automatically with the use of an XML template generated and sent by the system or by external devices, including other OSCAD-type systems.

Currently, the following external systems are taken into account. They can be divided with respect to the functions they perform:

- Enterprise Resource Planning (ERP) systems,
- IT infrastructure monitoring systems,
- building automation systems,
- burglary and fire alarm systems.

The communication interface enables connection with external systems by means of the monitoring station module which will also contain the implementation of all recognized communication protocols.

The efficiency parameters of the management system are stored in the measures and indicators module. This way it is possible to have periodical analyses and make decisions how to improve the system.

3 Risk Management Methodology of the OSCAD System

The key element of the OSCAD BCMS-ISMS joined implementation is the risk analysis facility. It was conceived as a common module for both management systems. The risk analysis method adopted in OSCAD is a business-oriented method, which means that its key aspect and starting point is proper identification and description of business processes functioning in the organization. Having those processes identified, the risk analysis is performed within a two-stage process. First, the criticality (weight) of each business process for the organization must be assessed. Next, a detailed risk analysis should be performed to identify threats, vulnerabilities and to select appropriate security measures for the most important vulnerabilities. Such approach complies with the recommendation for risk analysis described in [25] and is a simplified version of the method presented in [26].

3.1 Business Impact Analysis

Criticality of business processes is performed in the OSCAD tool with the use of the so called Business Impact Analysis (BIA) as described in [27]. During this analysis it is necessary to assess potential losses for the organization resulting from breaching its main security attributes. These attributes are: confidentiality of data, integrity of data and process activities, availability of data and processes.

The main tool for BIA is the business impact matrix. The table requires to define impact categories (rows). The most common categories, adequate for the most of organizations, are financial losses and law violation. Other categories may vary for each organization. These can be human (personnel/clients) related losses, environmental and other, depending on business operations of the organization.

For each category, a loss level should be assessed. To keep the unified grading scale, the same range of possible losses (levels of losses) should be defined to assess the whole of the organization's business processes. To make things simpler, the loss levels may be defined as a descriptive note or as a range of values, if possible (for example for financial losses). In case of descriptive values, they should be assigned to numerical values to enable their calculation in the software tool.

An example of descriptive loss levels interpretation is shown in Table 1. The N value is the maximum loss level and, just like impact categories, the number of possible loss levels should be adjusted to the organization's specific needs.

Table 1 Loss levels table [27]

Loss level	Value and description
1 – Lowest	No or insignificant losses in case of confidentiality, integrity or availability incident; Confidentiality example: data publicly available
…	…
…	…
N – Highest	Major losses, threat to further organization functioning Confidentiality example: Top secret data requiring maximum security

Business impact categories and loss levels with their descriptions for each categories together make the business impact matrix which is the core element of the proposed Business Impact Analysis. An example of the impact matrix is shown in Table 2. Using a prepared matrix, possible impacts resulting from the loss of security attributes (data/process availability, integrity and confidentiality) should be assessed. Each attribute should be assessed separately. The worst-case assessment from all categories is the final value for the particular attribute. This means that for each attribute (integrity, confidentiality, availability) all categories from the left column of Table 2 are assessed. The most pessimistic value for the given parameter is chosen as the final value.

Table 2 Business impact matrix example [27]

Business impact categories	Loss level
Human (personnel/clients) related losses	1 - No effect on personnel/clients 2 - Possibilities of injuries 3 - May cause human death or serious injury
Law violation	1 - No infringement 2 - Organization may have limited possibilities of activity caused by lawsuits 3 - May be a cause to stop all activities of the organization
Financial Losses	1 - Up to 50,000$ 2 - Up to 500,000$ 3 - Over 500,000$
Effectiveness drop	1 - No difficulties 2 - Break in process continuity for a couple of hours 3 - No further service possible

After carrying out the assessment of all categories and loss levels for each attribute, an overall weight of the process is calculated. The current OSCAD version offers three possible methods of weight computation:

- *worst case scenario – the process weight is the highest loss value in any attribute and category;*
- *arithmetical sum of all loss levels for availability/integrity/confidentiality attributes;*
- *product of all loss levels for security attributes.*

The implementation of other methods of weight computation, like square sum percentage or process weight matrix, is taken into consideration in other versions of OSCAD. These methods of computation were described in [27].

3.2 Detailed Risk Analysis

Based on the Business Impact Analysis results it is possible to calculate the weight of all processes – the importance, criticality of each process for the organization.

The next stage of the analysis – detailed risk analysis is performed in the first place for the most important (critical) processes.

At that stage it is necessary to identify the threats for the data, business processes and vulnerabilities which may cause these threats to materialize. Groups of threats and vulnerabilities can be defined in the OSCAD database – the so called 'dictionaries'. For all threat-vulnerability pairs identified for the organization the risk level (risk value) must be assessed next.

In the OSCAD tool, the risk level assessment is performed with the following formula:

$$R = \frac{Ts * Vs * Pc}{Ci * Cta} \quad (1)$$

R – risk value,
Ts – threat severity (level of impacts),
Vs – vulnerability severity (probability of occurrence),
Pc – process criticality (process weight),
Ci – existing controls implementation level,
Cta – existing controls technical advancement,

The current version of OSCAD allows to change the multiplication operator into arithmetic sum (in configuration options), depending on the organization's needs.

The vulnerability severity can be assessed based on the probability of its exploitation and with consideration of existing security measures. The information about incidents related to this vulnerability should be also taken into consideration. For example, possible values of severity can be defined as follows:

Once per 5 years -> *vulnerability severity = 1*
Once per 3 years -> *vulnerability severity = 2*
Once per year -> *vulnerability severity = 3*
...

The threat severity can be assessed based on possible financial or other losses that may appear in case of the threat materialization. For example:

Less than 10,000 -> *threat severity = 1*
10,000 – 100,000 -> *threat severity = 2*
100,000 – 500,000 -> *threat severity = 3*
...

The process criticality is the weight value assessed during the Business Impact Analysis of the process.

The existing controls implementation level can be assessed based on the percentage of implementation or as a descriptive value, for example:

Lack of implementation -> *implementation level = 1*
Partially implemented -> *implementation level = 2*
Fully implemented -> *implementation level = 3*

The technical advancement can be assessed taking into consideration the level of technical means used by the implemented controls, for example:

Procedural controls -> *technical advancement = 1*
Technical controls with manual handling -> *technical advancement = 2*
Technical, automated controls -> *technical advancement = 3*

All above values must be adjusted to the organization's needs and its mission.

Depending on the calculated risk level, a decision about the risk treatment should be made:

- when the risk is higher than the assumed acceptable level for the organization, risk reduction should be done – new controls (security measures) should be proposed for implementation,
- when the risk is lower than the level acceptable for the organization or it is not possible (in the technical or financial aspect) to reduce the risk further – the organization's management board can accept the risk and no further risk reduction activities are required; in this case risk should be monitored and organizational controls should be applied (e.g. crisis management procedures).

OSCAD supports both these risk treatment possibilities. In the first case, for each threat-vulnerability pair with a high risk level identified, new security measures can be added and taken into account, and then the risk level can be reassessed. Both risk values (before and after new controls addition) can be easily compared. In figure 2, the 'Risk' column presents risk levels for each pair. The values in brackets relate to the current situation. The numbers outside brackets are estimated values expected after the security measures implementation.

If the risk value is still higher than the acceptable level, additional security measures can be considered. The level of acceptable risk can be configured in the software depending on the organization's needs. Security measures selected and agreed during the risk analysis are then merged into the security treatment plan which should be accepted by the organization's management board.

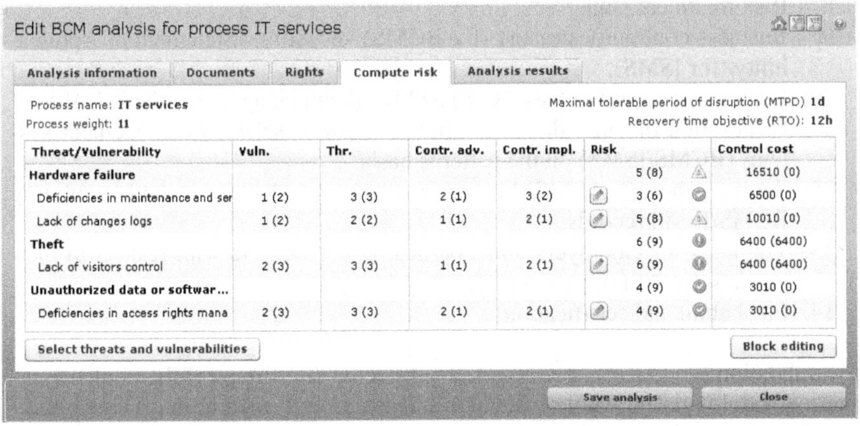

Fig. 2 OSCAD – Threat-vulnerability risk assessment main window

In the second case – risk acceptance, the organization's management board should be aware of threats that may occur if the risk level is not reduced. Based on such information from the risk management team, the management board should formally approve the remaining risk. It is also possible within the OSCAD tool – the risk analysis can be closed (justification required) even if the risk level exceeds the acceptable risk level.

Other typical risk treatment strategies, like risk avoidance or risk transfer are treated in OSCAD as security measures.

4 Validation Plan of the OSCAD System

The OSCAD implementation methodology assumes the implementation of the BCMS part, ISMS part, or both. The implementation range may encompass the entire organization, its part, selected parts of a few organizations, e.g. co-operating together and involved in a common project. The BCMS/ISMS implementation can run without or with the OSCAD supporting tool. The software-based implementation gives extra advantages comparable to the advantages of CAD/CAM systems.

The OSCAD implementation methodology encompasses the following steps:

1. Kick-off meeting – initializing the deployment process by the top management, planning and organizing the BCMS/ISMS implementation.
2. Training of the deployment team.
3. Business processes analysis.
4. Gap analysis (zero-audit) and elaboration of the strategy how to achieve standards/legal compliance.

5. Building the assets inventory for the organization.
6. Business Impact Analysis (high-level risk analysis), identification of critical processes.
7. Detailed (low-level) risk analysis.
8. Risk treatment plan.
9. Business continuity strategy (for BCMS) and SoA – Statement of Applicability (for ISMS).
10. Business continuity plans (BCP) and Incident management plans (IMP).
11. Elaboration of other documentation of both or selected management systems (BCMS/ISMS) on the patterns basis.
12. Setting the BCMS/ISMS in motion (a decision) and sampling BCMS/ISMS records.
13. Trainings for employers, third parties, OSCAD users with respect to their business/system roles.
14. Final audit and certification.

The validation process may encompass the selected or all steps of the implementation methodology. The whole validation has been performed at the EMAG Institute. The chapter focuses on validation issues concerning risk management.

5 Validation of the Risk Management Processes

5.1 Gathering Information about Business Processes

Before the risk analyses process starts, the significant issue is to properly identify the organization's processes and their sub-processes for which risk analyses will be performed.

For the validation purposes in EMAG, four main and ten supportive processes were identified. The OSCAD tool allows to register these processes in the database. Once a process has been registered in the OSCAD database, a new task is automatically registered in the task management module and the responsible person gets information that the new process has occurred and it needs the risk analysis. Additionally, each change in the process description causes a new task generation with information that the risk analysis of the this process should be revised.

The first issues appear as early as at the processes description stage. They are caused by the fact that the organization's management processes are partially handled by persons from several different organizational units. In that case, persons from each unit must be selected to participate in the business process description and assessment.

The sub-processes must be distinguished for some processes due to their complexity and activities diversity For example, one of the complex supportive processes was the 'IT support' process. Because of a wide scope of activities, it was divided into several sub-processes (related to the LAN, Internet, applications, telephones, and other aspects).

Processes should be divided in such a way to neither over- nor under-estimate any of them. For example, a supportive process like 'clean-up and facilities maintenance' may include less important activities for business continuity, like cleaning the offices, but on the other hand such activities like 'clearing (the streets or roofs) of snow' may have significant influence on business continuity and the personnel safety. In the first iteration of the system implementation 'clean-up and facilities maintenance' was assumed as one process. During the risk analysis it appears that equal treatment of these activities and performing their analysis as a whole will lead to over-estimation of offices cleaning or under-estimation of 'clearing the streets/roofs of snow' (the streets not cleared of snow may reduce access to buildings, while snow remaining on the roofs may cause injuries by falling icicles or even roof collapse).

The validation of risk analysis helps then to identify gaps in the process map previously prepared for the organization.

5.2 Validation of BIA Method

The presented risk analysis method requires to adjust the business impact matrix. Possible impact (loss) levels should be matched to the organization's needs. These levels will depend on the organization's business activities, legal and procedural requirements, annual income (it will have influence on the level of acceptable/unacceptable financial losses).

Taking into account the scope of activities of the organization where OSCAD was validated, the following impact categories were selected: financial losses, delay or discontinuity of business operations, law violation, employees' satisfaction, influence on employees' health and on the natural environment, loss of reputation.

During the validation, a matrix was prepared and a few first processes were analyzed. To keep the calculation of the processes weight as simple as possible, the arithmetic sum of impact levels (for availability, integrity and confidentiality loss) was used in the validation process.

For all of these categories four levels of possible losses were established. Numerical values (from 1 to 4) with description were assigned. For such assumption, the process weight values will be in the range from 3 to 12. For such values, for the initial risk analysis it was assumed that processes with the weight value higher than 7 will be marked as critical.

The first business impact analyses have shown that the matrix statements were not fitted properly and too many processes were assessed as critical for the organization. Even supportive processes, activities that are not really important for the organization's mission, had very high scores during the analysis. Some statements from the 3rd level of possible losses were moved to the 2nd level. After that operation, reassessed BIAs gave more appropriate results.

A proper analysis of each process requires participation of crucial organizational units involved in the process operation. Risk analyses were performed in the form of workshops with all owners of the processes or persons having appropriate knowledge appointed by the process owners, and with other process stakeholders (from other related processes). This helps to keep the objectivity during the analyses, because one process is always more valuable for its owner, but not necessarily must be important for the overall business activity of the organization. Additionally, a loss level may be strictly dependent on loss levels of other related processes. The IT process, for example, should be assessed with respect to the highest loss (the worst case) which may appear in other processes as a result of the IT process unavailability.

5.3 Validation of Detailed Risk Analysis

Once the business impact analysis of the process has been approved in OSCAD, the next step of the risk analysis can be performed, which requires more time and organization's resources. During this stage as many as possible threats and vulnerabilities should be identified, their severity and risk levels assessed, while security measures and controls should be selected for the vulnerabilities with the highest risk levels, as described in section 3.2. The identification of vulnerabilities requires more involvement of technical/IT staff. That is why the proposed risk analysis method assumes that the second stage – detailed risk analysis, should be performed only for the most critical processes.

During the first stage of the analysis (BIA) the risk management team asked the interviewees about the possible impacts (consequences) and the potential sources of those impacts. This way, already during the business impact analysis, they received information about the most important threats. Having threats and vulnerabilities identified, the risk parameters value assessment can be easily performed with the OSCAD tool support (Fig. 3). Incidents registered in the database through the incident management module can be taken into account during the threat and vulnerability severity (impact and probability) assessment.

For the highest risks the security measures (controls) can be selected from the database or defined to mitigate these risks.

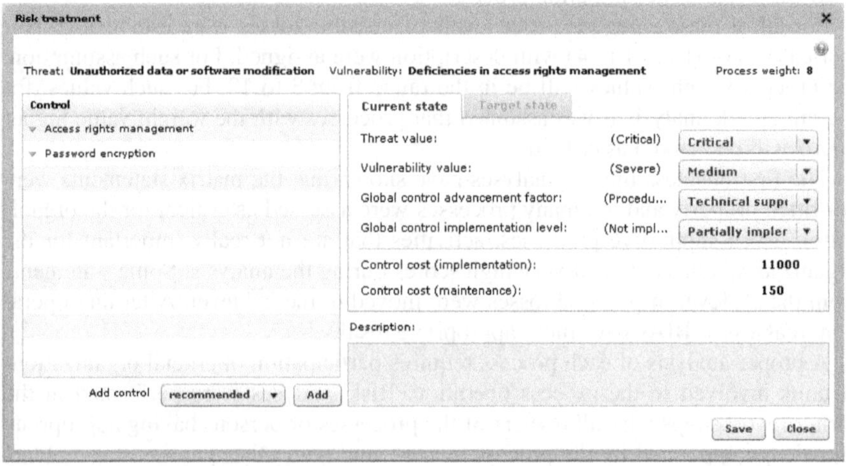

Fig. 3 Risk treatment window – threat-vulnerability assessment, controls selection

The results of the detailed risk analysis are the input to the next activities within the business continuity and information security management process. Based on the risk level values, a decision can be made which vulnerabilities should be treated as priority during the risk treatment plan elaboration and during the implementation of security measures. Business continuity plans should be prepared for the most critical processes and the most severe threats.

When the risk analysis is completed, the task is sent to the responsible person (using the task management module) to approve the results. Each risk analysis must be approved by the authorized person (selected in the software configuration). Information about all performed risk analyses, business processes weight, risk levels can be listed in the risk analysis module and in the reports module.

For the risk analysis that was approved, a written report and a risk treatment plan can be generated. Additionally, the tasks are created in the task management module for people responsible for the control implementation.

6 Conclusions

The chapter presents the computer-supported, integrated business continuity- and information security management system and its validation on the risk issues examples.

The OSCAD tool validation in the scope of the risk analysis allows to find gaps in the organization's business processes map. The validation enabled to identify those phases of the risk analysis process implementation, to which more attention should be paid. It helped to identify typical issues which may occur during the implementation, such as appropriate selection of impact matrix values and appropriate description of those values. During the validation of the BIA and detailed risk analysis processes, the risk value calculation method was tested. The validation showed the points of the OSCAD tool that require improvement. For example, inability to perform the analysis and assessment of possible impacts of security measure implementation or the impossibility to assess a few variations of security measures and to select the most appropriate from the assessed variants (taking into consideration the risk level expected after its implementation). The identified gaps and required improvements will be taken into consideration in further work on the software development.

The OSCAD project has the following innovative elements:

- possibility to support the BCMS, ISMS, or both management systems;
- possibility to gather, analyze and provide statistical data about incidents and use them in the lessons learnt, risk management, and improvement processes;
- open character; possibility to customize provided patterns for the needs and expectations of different kinds of organizations;
- advanced automation of management processes;
- possibility to gather information on incidents from different sources: monitoring the IT infrastructure or technical-production infrastructure, physical security infrastructure (burglary, fire), business processes through the ERP systems;

- implementation of failover technology for the BCM/ISM computer-supported systems,
- implementation of mobile technology (web access, SMS-notification);
- possibility to conduct a statistical analysis of events and reliability analysis of IT systems functioning (an option);
- possibility to exchange information between different organizations, especially those working within supply chains;
- easy integration with other management systems co-existing in the organization (PAS99).

The project results are dedicated mainly for organizations:

- which are elements of the critical infrastructure of the country (power engineering, production and distribution of fuels, telecommunications, etc.), providing media for people (water, gas, energy, etc.),
- operating in a financial sector (insurance, banking),
- offering e-services,
- operating as part of the public administration (government- or local government level),
- representing the sectors of health services and higher education,
- involved in the protection of groups of people,
- commercial and industrial companies.

The results of the conducted validation of the OSCAD software are used to improve the developed prototypes and to prepare dedicated versions for different application domains. The OSCAD project is currently in progress and its completion is planned by the end of 2012.

References

[1] BS 25999-1:2006 Business Continuity Management – Code of Practice
[2] BS 25999-2:2007 Business Continuity Management – Specification for Business Continuity Management
[3] ISO/IEC 27001:2005 – Information technology – Security techniques – Information security management systems – Requirements
[4] ISO/IEC 27002:2005 – Information technology - Security techniques - Code of practice for information security management (formerly ISO/IEC 17799)
[5] Institute EMAG (2010-2011) Reports of a specific-targeted project "Computer-supported business continuity management system – OSCAD"
[6] LDRPS, http://www.availability.sungard.com (accessed January 05, 2012)
[7] ErLogix, http://www.erlogix.com/disaster_recovery_plan_example.asp (accessed January 05, 2012)
[8] Resilient Business Software Toolkit ROBUST, https://robust.riscauthority.co.uk (accessed January 05, 2012)
[9] RPX Recovery planner, http://www.recoveryplanner.com (accessed January 05, 2012)
[10] Cobra, http://www.riskworld.net (accessed January 09, 2012)

[11] Cora, http://www.ist-usa.com (accessed January 09, 2012)
[12] Coras, http://coras.sourceforge.net (accessed January 09, 2012)
[13] Ebios, http://www.ssi.gouv.fr (accessed January 09, 2012)
[14] Ezrisk, http://www.25999continuity.com/ezrisk.htm (accessed January 09, 2012)
[15] Mehari, http://www.clusif.asso.fr (accessed January 09, 2012)
[16] Risicare, http://www.risicare.fr (accessed January 09, 2012)
[17] Octave, http://www.sei.cmu.edu (accessed January 09, 2012)
[18] Lancelot, http://www.wck-grc.com (accessed January 09, 2012)
[19] Bialas, A.: Security Trade-off – Ontological Approach. In: Akbar Hussain, D.M. (ed.) Advances in Computer Science and IT, pp. 39–64. In-Tech, Vienna – Austria (2009) ISBN 978-953-7619-51-0, http://sciyo.com/articles/show/title/security-trade-off-ontological-approach?PHPSESSID=kkl5c72ntlg3qc4t98de5shhc2 (accessed January 10, 2012)
[20] ValueSec Project, http://www.valuesec.eu (accessed January 10, 2012)
[21] Białas, A.: Development of an Integrated, Risk-Based Platform for Information and E-Services Security. In: Górski, J. (ed.) SAFECOMP 2006. LNCS, vol. 4166, pp. 316–329. Springer, Heidelberg (2006)
[22] BS PAS 99:2006 Specification of common management system requirements as a framework for integration
[23] Białas, A.: Integrated system for business continuity and information security management – summary of the project results oriented towards of the construction of system models. In: Mechanizacja i Automatyzacja Górnictwa, vol. 11(489), pp. 18–38. Instytut Technik Innowacyjnych "EMAG", Katowice (2011)
[24] Bialas, A.: Computer Support in Business Continuity and Information Security Management. In: Kapczyński, A., Tkacz, E., Rostanski, M. (eds.) Internet - Technical Developments and Applications 2. AISC, vol. 118, pp. 161–176. Springer, Heidelberg (2012)
[25] Stoneburner, G., Goguen, A., Feringa, A.: Risk Management Guide for Information Technology Systems. Recommendations of the National Institute of Standards and Technology. NIST Special Publication 800-30 (July 2002)
[26] Białas, A., Lisek, K.: Integrated, business-oriented, two-stage risk analysis. Journal of Information Assurance and Security 2(3) (September 2007) ISSN 1554-10
[27] Bagiński, J., Rostański, M.: The modeling of Business Impact Analysis for the loss of integrity, confidentiality and availability in business processes and data. Theoretical and Applied Informatics 23(1), 73–82 (2011) ISSN 1896-5334

A Functional Testing Toolset and Its Application to Development of Dependable Avionics Software

Vasily Balashov, Alexander Baranov, Maxim Chistolinov, Dmitry Gribov, and Ruslan Smeliansky

Abstract. Significant part of requirements to avionics software can only be tested on target hardware running non-instrumented software. General purpose functional testing toolsets require loading auxiliary software to target avionics devices to perform target-based testing. This work presents a toolset for functional testing of avionics software, aimed at testing without target instrumentation. The toolset supports automatic and human-assisted testing of software running on target avionics device(s) by providing input data through a variety of onboard interface channels and analyzing devices' responses and inter-device communication. Architecture of software and hardware counterparts of the toolset is described. A family of avionics testbenches based on the toolset is considered as an industrial case study. Special attention is paid to toolset application for testing of dependability requirements to avionics systems.

1 Introduction

Modern avionics systems (AS) are subject to strict requirements to dependability, functionality and real-time operation. To ensure that an AS fulfills these requirements, systematic testing of avionics software is performed on several phases of AS development.

Important specifics of avionics software testing is that significant part of requirements can only be tested on target hardware. These are requirements to

Vasily Balashov · Maxim Chistolinov · Ruslan Smeliansky
Department of Computational Mathematics and Cybernetics, Lomonosov Moscow State University, Leninskie Gory, MSU, 1, Bldg. 52, Room 764, Moscow, Russia
e-mail: hbd@cs.msu.su, mike@cs.msu.su, smel@cs.msu.su

Alexander Baranov · Dmitry Gribov
Sukhoi Design Bureau (JSC),
23A, Polikarpov str., Moscow, Russia
e-mail: abaranov@okb.sukhoi.org, dgribov@okb.sukhoi.org

W. Zamojski et al. (Eds.): Complex Systems and Dependability, AISC 170, pp. 19–35.
springerlink.com

real-time operation, to data exchange through onboard channels, to co-operation of multiple software layers (OS, system software, application software). Many dependability requirements belong to these classes. For instance, switching between primary and hot-spare devices requires real-time actions of several devices coordinated through onboard channels.

General purpose functional testing toolsets, such as Rational Test RealTime and VectorCAST, require instrumentation of target avionics devices to perform target-based testing. Such instrumentation involves loading auxiliary software to the devices and thus is prohibited for acceptance and field testing. These toolsets do not also support analysis of data communication through onboard channels, requiring deep customization for AS testing.

In this work we present a toolset for functional testing of avionics software and hardware/software avionics systems. This toolset is aimed at testing without target instrumentation, and supports testing of dependability requirements to AS. The toolset is developed in the Computer Systems Laboratory (CS Lab) of Computational Mathematics and Cybernetics Department of Lomonosov Moscow State University. The toolset is utilized by Sukhoi Design Bureau for testing of modern avionics systems. These systems consist of multiple devices (computational nodes, sensors, actuators, indicators) connected by dozens of data transfer channels.

The rest of this work is organized as follows. Section 2 lists several essential cases which require target-based avionics software testing. Section 3 estimates applicability of two existing general purpose functional testing toolsets to the task of target-based avionics software testing and emphasizes significant issues undermining this applicability. Section 4 presents the functional testing toolset utilized by Sukhoi, describes its architecture and capabilities. Section 5 describes the common architecture of testbenches based on the toolset. Section 6 considers an industrial case study, including a set of testbenches based on the presented toolset, each testbench aimed at a specific stage of AS development. The section also outlines the architecture of the target AS for this case study. Section 7 describes several approaches to testing of dependability requirements to AS, involving the presented testing technology. In the last section, future directions for technology development are proposed.

2 Cases for Target-Based Avionics Software Testing

There are following essential cases of avionics software development and support which require testing of avionics software running on target hardware.

1. Integration of software components with target hardware, debugging of software on target platform;
2. Integration of application software components, including components delivered by partner organizations responsible for specific software subsystems;
3. Integration of avionics subsystems consisting of several devices, or integration of the whole AS as a hardware/software system;
4. Functional and acceptance testing of AS software;

5. Acceptance testing of series-produced avionics devices and complete AS instances;
6. Diagnostics of avionics devices subject to reclamations;
7. Diagnostics of avionics devices onboard the aircraft.

Cases 1 and 2 typically involve operations with an individual (isolated) avionics device.

Specialized hardware/software environments (avionics testbenches) are involved to support testing activities related to cases 1-7. Testing toolset is the key component of testbench software.

Testing technology, including the used toolset and testbench architecture, should be unified for tasks 1-7 in order to minimize technological complexity of the testing process, as well as to enable reusing test suites between AS development stages.

3 Applicability of General Purpose Functional Testing Toolsets to Target-Based Avionics Software Testing

In this section we consider two existing functional testing toolsets and estimate their applicability to the task of testing avionics software on target device(s) without target instrumentation. The toolsets, namely Rational Test RealTime and VectorCAST, are not "purely" general purpose. They are designed to support testing of distributed real-time systems, however there are several common issues that substantially complicate their application to the above mentioned task.

Rational Test RealTime (RTRT) [1] from IBM is a cross-platform solution for testing real-time applications, including networked and embedded software. RTRT provides facilities for automated target-based software testing, memory leaks detection and code coverage analysis. RTRT is integrated with revision control and configuration management tools from IBM (ClearCase, ClearQuest).

RTRT was initially developed for unit testing. System testing support for distributed applications was added later. Target-based testing requires loading instrumental modules (system testing agents) to every device (computer) of the target system which runs software to be tested. Agents are responsible for supplying test data to the software under test, reading resulting data and communicating with the testing control computer.

RTRT supports two test description languages: Component Testing Script Language for unit tests and System Testing Script Language for system tests.

Conventional (recommended by vendor) technology of RTRT application to target-based testing of avionics software requires loading instrumental software modules to AS devices. This approach is not suitable for acceptance and field testing. It is also questionable whether the instrumental modules significantly affect real-time characteristics of the target software operation.

RTRT was applied in Sukhoi company for testing of Sukhoi Superjet civilian aircraft avionics. This application involved custom "adaptation layer" software running on dedicated computers and translating test data from RTRT to onboard channel messages sent to the (non-instrumented) target system. Development of such intermediate software is a complex task which requires access to internal details of RTRT implementation.

Testing solution based on RTRT with adaptation layer provides communication with the target system on "parameter" level, where parameter is a variable taken by the target application software as an input or generated as an output. This level is not suitable for testing devices that take or generate "binary" data (e.g. video frames, digital maps etc). It is also hardly suitable for dependability testing which requires low-level stimulation of the target system (e.g. injection of communication faults into channels, simulation of interface adapter breakdown or power failures).

VectorCAST [2] from Vector Software is an integrated toolset for unit and functional testing. It supports test generation, execution, code coverage analysis and regression testing. VectorCAST can be integrated with third-party requirement management tools through VectorCAST / Requirement Gateway module.

VectorCAST provides special features for avionics software testing. In particular, it implements recommendations of DO-178B standard which describes avionics software development processes. VectorCAST was applied to development of avionics software for several aircraft including JSF, A380, Boeing 777, A400.

VectorCAST supports unit and integration testing of avionics software on target platform. Similarly to RTRT, VectorCAST requires loading instrumental software modules (VectorCAST/RSP, Runtime Support Package) to target devices. This raises the same issues as mentioned above for RTRT.

There is no open information available on existence of "adaptation layer"-like software for VectorCAST which could allow application of VectorCAST to testing of non-instrumented AS through onboard channels. Development of such software layer is problematic as VectorCAST is a closed commercial product, and even if such layer existed, the resulting solution would have same limitations as noted above for RTRT-based solution.

Both testing toolsets reviewed in this section provide a rich feature set, but have similarly limited applicability to the task of testing of avionics software running on a non-instrumented target system. These toolsets also lack support for direct control of exchange through onboard channels, which additionally restricts their capabilities for testing AS dependability features.

4 Functional Testing Toolset Utilized by Sukhoi

This section describes the avionics testing tools (ATT) utilized by Sukhoi Design Bureau for functional testing of avionics software of several modern Sukhoi aircraft. The ATT toolset is developed in Computer Systems Laboratory (CS Lab) at Computational Mathematics and Cybernetics Department of Lomonosov Moscow State University.

4.1 Toolset Overview

In contrast to RTRT and VectorCast toolsets considered above, ATT toolset is aimed at testing of AS through onboard interface channels, without loading any instrumental modules to AS devices. ATT supports MIL STD-1553B, ARINC 429, Fibre Channel and several other standards of onboard interface channels required for testing of modern avionics. Onboard channel adapters are installed on instrumental computers that perform tests execution.

For every supported type of channel, following activities are supported:

- Preparation and sending data to AS through onboard channels (both packing of parameters into messages and low-level forming of "binary" messages are supported).
- Receiving of data from AS through onboard channels for subsequent analysis (both unpacking parameters from messages and low-level access to "binary" messages are supported).
- Monitoring of data exchange between AS devices through onboard channels and processing of monitoring results by tests (on same two levels as for data receiving).

ATT can be extended to support new standards of onboard channels. Testbenches based on ATT cover all cases for target-based testing of AS software listed in Section 2.

The toolset supports distributed execution of tests on multiple-computer testbench configurations, necessary for testing complex AS with large (up to several hundred) number of interfaces. Instrumental computers operate in synchronized time and perform time-coordinated sending and processing of test data.

To operate specialized channels requiring highly optimized hardware solutions for signal simulation, e.g. Fibre Channel-based video channels, ATT supports integration with signal simulation hardware (SSH) systems which are integral parts of ATT-based testbenches (typical testbench structure is described in Section 5). SSH systems operate under control of tests which run on instrumental computers.

ATT provides facilities for testing requirements to real-time operation of the AS. All operations performed by tests are synchronized with physical time. Timings of test data sending to AS, and timings of receiving the responses are measured with high precision (dozens of microseconds), enabling tests to verify AS reaction time to test stimuli.

ATT supports both fully-automatic test execution (including batch mode) and interactive testing. Interactive features include:

- Generation of requests from the test to the user:
  - For positive/negative confirmation (yes/no), necessary if testing requires visual assessment of AS reaction, e.g. correctness of image shown on avionics display device;
  - For text comments, e.g. rationale for positive or negative confirmation;

- Manual input of test data values by the user;
- Manual selection of tests execution order (within a single testing script), useful for debugging of AS software on the testbench.

Each testing session produces a log which contains status of all executed tests (pass/fail), user replies to test requests, etc. Testing log is subsequently processed to determine which requirements to AS have successfully passed testing (correspondence between tests and requirements is specified in test description). Testing logs and test data values (in numeric or graphical form) are displayed to the user in online visualization tool, enabling one to track the testing progress. Custom parameter visualization and input tools are supported, such as dials, sliders etc; plugin interface is provided for adding new custom tools.

An auxiliary log can be generated which records all user-entered data, including parameter values, replies to test requests, manual selection of tests execution order, etc. Such log can later be "replayed" to automatically reproduce an interactive testing session, which is useful for debugging of tests.

To supports development and debugging of tests "in advance", without actual avionics hardware available, ATT implements the following features:

- Support for software-simulated "virtual" onboard channels, namely MIL STD-1553B and ARINC 429;
- Execution of tests in user input expectation mode, in which the tests ask the user for data instead of getting it from avionics devices (this mode requires no modification of tests).

ATT toolset supports creation of avionics device simulation models using the hardware-in-the-loop simulation technology described in [3]. This technology is an integral part of the testing toolset. Simulation models utilize the resources of instrumental computers, including onboard channel adapter cards, to reproduce the modeled device's activity on the channels. Simulation models can be involved in testing reconfiguration features of AS subsystems in case of incomplete availability of hardware avionics devices participating in reconfiguration.

Hardware resources of an ATT-based testbench can be shared between tests and monitoring tools of "Channel analyzer" family [4]. Sharing is supported for MIL STD-1553B adapters with multiple terminal support, with monitoring support (including concurrent operation of bus controller and monitor functions), for multiple-channel ARINC 429 and Fibre Channel adapters. Operation of testing tools (both runtime an user interface subsystems) and monitoring tools in a common hardware/software environment based on Linux OS with real-time extensions enables implementation of a compact mobile workstation for testing and monitoring of avionics systems. This solution can be used for field diagnostics of AS, see Section 6 for details.

Test development subsystem of ATT supports automatic generation of interface part of tests' description (i.e. specification of data words and messages structure) from data contained in Interface control document database. Such database is maintained by Sukhoi software development teams for each version of avionics software, see [5] for details.

SUBVERSION-based revision control for tests and test logs is implemented in ATT, integration with ClearCase is on the roadmap.

4.2 Test Description Language Features

Test description language (TDL) of ATT toolset is intended for description of avionics devices testing scripts. Tests written in TDL are executed on instrumental computers of a testbench and communicate with tested devices through onboard channels. Tests also control the operation of SSH systems, including data exchange between SSH and avionics devices.

TDL is an extension of C language providing statements for defining test structure, binding test execution to physical time, control of data exchange through onboard channels, and control of testing process.

Basic TDL unit for test coupling is test component (TC). TC source code consists of *header* and *body*.

TC header specifies:

- Set, structure and hierarchical naming of test cases;
- Correspondence between test cases and requirements to AS;
- Set and types of interfaces to onboard channels through which communication with AS is performed during testing;
- Structure of data words and messages transferred and received through interfaces;
- Set and types of auxiliary variables (parameters) intended for data exchange between different TCs and between TC and SSH systems;
- Set of testing logs recorded by the TC.

  TC body specified testing scripts' activity on:

- Preparing test data;
- Sending test data into channels;
- Receiving data from avionics devices through channels and checking tested conditions (access to channel monitoring results is provided in a similar way);
- Direct control of channel adapters, including turning on/off, setting service flag values, fault injection;
- Interaction with the user during human-assisted (e.g. visual) checking of tested conditions;
- Control for generation of testing logs.

TDL provides following functionality for automatic checking of timing constraints on responses from avionics devices (i.e. for testing AS real-time characteristics):

- Waiting for a specified duration and then checking the condition on received data;
- Waiting for the condition on received data to become true, until a specified timeout expires;
- Constantly checking that the condition on received data remains true during a specified duration.

A TDL project may include several TCs, in particular intended for execution on different instrumental computers. Even if only one TC is used, it can access channel adapters located on different instrumental computers. It is useful in case a single TC implements scripts for comprehensive testing of AS or its subsystem through a large set of interfaces attached to different instrumental computers.

4.3 Toolset Software Structure

ATT toolset contains following major subsystems:

1. Test development subsystem, responsible for creation and editing of tests' source code in TDL;
2. Facilities for setup of testbench configuration, including:
   - Distribution of test components to instrumental computers;
   - Allocation of channel adapters to test components' interfaces;
   - Specification of data links from tests to SSH systems;
   - Level of detail for testing events recording;
3. Real-time test execution environment, responsible for:
   - Distributed real-time execution of tests on multiple instrumental computers;
   - Data communication and time synchronization between instrumental computers;
   - Remote access of test components to channel adapters located on different instrumental computers (via internal Ethernet network of the testbench);
   - Data communication between tests and SSH systems;
   - Data exchange through onboard channels between tests and tested devices;
   - Access of tests to results of channel monitoring;
   - Communication with experiment control tools to provide interactive testing control and online visualization of testing process;
   - Recording of testing results as logs and event traces;
4. Experiment control subsystem, responsible for interaction with the user, in particular for online visualization and interactive testing control;
5. Testing results processing subsystem, responsible for:
   - Visualization of testing results in form of testing logs, time diagrams, parameter graphs, channel monitoring logs;
   - Generation of reports on results of testing;
6. Server subsystem, responsible for:
   - Revision control of test source code, testbench configuration and testing logs;
   - Interaction with Interface control document database;
   - Sharing of testbench repository over network between all testbench computers.

The user accesses tools from groups 1-6 via Integrated development environment.

5 Architecture of a Testbench Based on the Presented Toolset

ATT toolset described in Section 4 is practically applied as a central software facility of avionics testbenches. Architecture of such testbenches, including typical components and scheme of their composition, evolved through years of toolset industrial application. The present section describes this architecture.

ATT-based testbench typically includes following components:

- Instrumental computers intended for running tests and performing data exchange with AS through onboard channels (currently, MIL STD-1553B, ARINC 429 and Fibre Channel);
- Specialized signal simulation hardware (SSH) systems operating under control of instrumental computers;
- Workstations for test engineers and AS software developers;
- Server (or a dedicated workstation) responsible for centralized functions such as testbench repository keeping and sharing, user authentication, etc;
- Network of onboard interface channels connecting avionics devices with instrumental machines and SSH systems, and avionics devices with each other;
- Auxiliary equipment, including systems for power supply and cooling, technological networks, racks, etc.

Structural scheme of a typical testbench is shown on Fig.1.

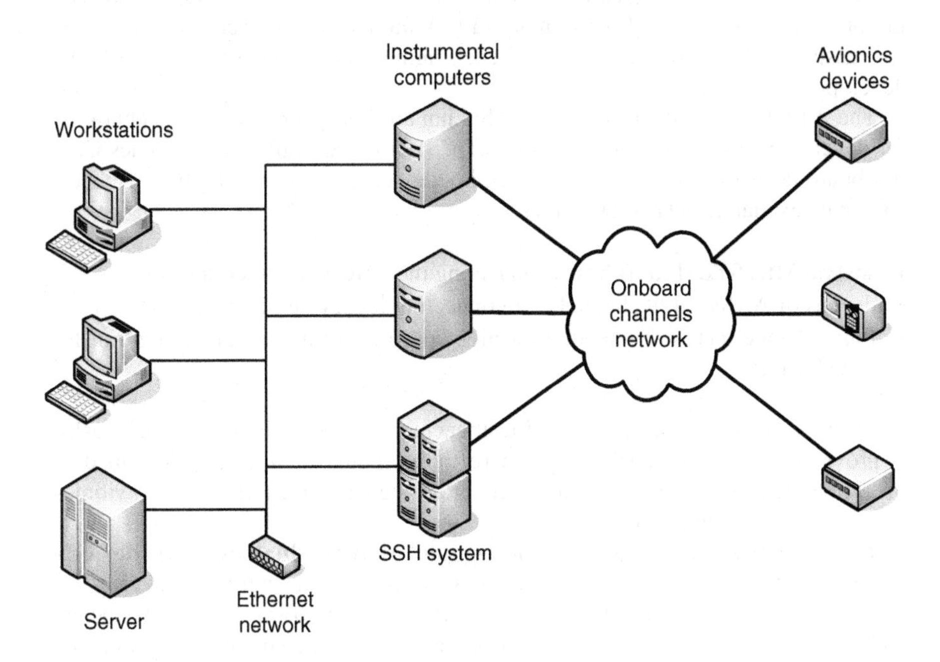

Fig. 1 Structural scheme of an ATT-based testbench

The number of instrumental computers, workstations, number and types of on-board channels depends on specifics of the tested AS and tasks of the testbench.

All testbench computers (workstations, server, instrumental computers, SSH system computers) use Linux as operating system. Instrumental computers and computers of SSH systems use Linux with real-time extensions.

In addition to Ethernet network, instrumental computers are connected by a dedicated time synchronization network. It connects their LPT ports and serves for transfer of precise time signals from a single "master" computer to other computers. Time synchronization error is within bound of 10 us.

The most advanced type of SSH system implemented to date is SSH TV, a multi-computer system supporting real-time generation and recording of multiple high speed video streams. It serves as test data source and feedback recorder for testing of avionics display and video processing devices. Static (picture) and dynamic (motion picture) test video streams are prepared in advance, before start of testing. Selection of a video stream to be transferred to a given channel at specific moment of time is performed by the test executed on an instrumental computer.

6 Industrial Case Study

This section describes the industrial application of ATT testing toolset presented in Section 4.

The ATT toolset is utilized for over six years by Sukhoi Design Bureau for testing of avionics software of modern aircraft. A family of testbenches is created on the base of this toolset, each testbench aimed at a specific stage of AS development.

The ATT-based testbenches used by Sukhoi are integral parts of the technology for development of Data control system (DCS) of the aircraft. DCS includes Central board computer (CBC), a set of display devices, and some auxiliary devices. DCS data exchange network includes:

- several MIL STD-1553B buses connecting the CBC with other devices;
- dozens of ARINC 429 channels for connecting legacy devices;
- Fibre channel network intended for high speed data transfer, including transfer of video data.

Description of particular ATT-based testbenches used by Sukhoi, and their tasks, is provided below. All testbenches share the common architecture described in Section 5. Together these testbenches cover all the cases for target-based avionics software testing listed in Section 2.

1. Testbench for testing and debugging of individual avionics device's software. This testbench is intended for working with an individual device, for instance the most complex DCS device – the CBC. Software of this testbench includes ATT toolset, tools for target-based software debugging and "Channel analyzer" tools for monitoring of data exchange through channels. Main purpose of this testbench is to support debugging of DCS device software and preparation of the device for integration with other devices. For final checks of the device's readiness for integration with other devices, an approved set of tests should be used.

Data exchange monitoring tools are used in the testbench for resolving "boundary" issues in case it is unclear whether the target device itself operates incorrectly or the tests provide improper input data.

As it is sufficient to support only those external interfaces that are provided by the selected DCS device, the number of instrumental computers in the testbench is usually small (not greater than 3). SSH systems are implemented in single-computer configurations. This enables composition of all testbench hardware, together with the DCS device, into a single rack. Sharing of hardware resources between testing and monitoring tools helps to keep hardware requirements low.

"Unified" testbench for working with different DCS devices from a specified set is organized in a similar way. Instrumental computers of such testbench cover all types of external interfaces provided by devices from the set, in quantities sufficient for working with every *single* device from this set.

Testbench of the described type can be used for testing and debugging of application software subsystems on a partially equipped avionics device, for instance a CBC with only some of processor and communication modules installed. This approach is practical for co-developer organizations responsible for specific application software subsystems.

2. Testbench for testing and debugging the software of a group of several connected avionics devices. This testbench is intended for workout of communication between two or several directly connected DCS devices, for instance CBC and a display device. The testbench can also be used for performing some activities of debugging the software of these devices.

The testbench hardware structure and configuration is similar to those of the testbench for an individual device (described above), except that a separate place outside the rack is needed for at least one DCS device, usually the display device.

A set of tests for this type of testbench is aimed at checking of co-operation of the bunch of connected DCS devices composing a DCS subsystem.

As in the previous case, a testbench can be created which supports different combinations of DCS devices, for instance the CBC and one of several display devices.

This type of testbench supports the activities necessary for preparation to integration of the whole DCS. Use of this testbench relieves the "main" DCS integration testbench from tasks of detecting and fixing the issues of co-operation between most intensely communicating DCS devices.

3. Testbench for DCS integration and acceptance testing. This testbench is intended for stepwise integration of DCS, and for DCS software acceptance testing. The testbench supports delivery of input data to all DCS input interfaces and receiving resulting data from all DCS output interfaces. Furthermore, monitoring of internal DCS channels is supported.

Due to need for complete workout of data communication protocols for all DCS interfaces and internal channels, a large number of instrumental computers

(5-6 or more) is included in the testbench hardware, along with full-featured SSH systems in multiple-computer configurations.

Testbench for DCS integration and acceptance testing represents several racks containing instrumental computers and DCS devices. The testbench includes several workstations, enabling simultaneous development and execution of tests.

Tools for monitoring of data exchange through channels are used in the testbench for analysis of inter-device communication issues and for resolving "boundary" issues.

4. Testbench for development of DCS functional tests. This testbench plays an auxiliary role and is dedicated to development of tests. It contains several workstations and a server connected by Ethernet network. All computers have ATT toolset installed.

Debugging of tests is performed on workstations using such features of the ATT toolset as virtual onboard channels and execution of tests in user input expectation mode.

5. Testbench for acceptance testing of series-produced DCS. This testbench is intended for performing acceptance tests for DCS device sets before their installation onboard the aircraft. Second purpose of this testbench is diagnostics of devices which are subject to reclamations.

Instrumental hardware of this testbench is essentially the same as of the testbench for DCS integration and acceptance testing, except for lack of multiple workstations as the testbench is not intended for development of tests.

The testbench includes a "reference" set of DCS devices which has passed full-scale testing. On arrival of a DCS device subject to reclamations, this device replaces the corresponding reference device, and a bunch of tests is automatically executed which is aimed at examination of the device under question. Decision on farther operations with the device is made on the base of testing results.

If a newly produced set of DCS devices arrives for testing, it completely replaces the reference set. In case of issues in DCS operation, new devices can be selectively replaced by reference devices to locate the improperly operating device.

6. Mobile workstation for monitoring and testing DCS devices onboard the aircraft. Mobile workstation is the most compact installation of ATT toolset and "Channel analyzer" tools for data exchange monitoring. Workstation hardware includes:

- A rugged industrial notebook with extension chassis providing PCI and/or PCI Express slots;
- Adapters of onboard interface channels (e.g. MIL STD-1553B, ARINC 429, Fibre Channel).

Testing toolset and monitoring tools named above use the same software environment, share the hardware resources (including adapters) and can be used concurrently.

In case the number of available extension slots is not sufficient for installation of adapters for all required types of interface channels, a replaceable extension chassis can be used. External adapters attached to USB ports can also be used, e.g. for monitoring of data exchange on MIL STD-1553B channels.

Mobile testing and monitoring workstation enables primary diagnostics of DCS devices operation without their transportation to the location of stationary test-benches. This possibility is essential in case of multiple geographically separated locations in which the aircraft containing the DCS is operated or field tested. If there are issues in operation of DCS "as a whole", use of mobile test and monitoring workstations allows localization of the problem and its attribution to a specific DCS device. This single device becomes the subject for farther examination on a stationary testbench.

Testing of DCS devices operation using the mobile workstation can be performed:

- On a grounded aircraft: the workstation is attached to control plugs of the onboard network and operates in "passive" mode, performing following activities:
  - data exchange analysis by monitoring tools;
  - execution of tests which automatically check the correctness of data exchange sequences and transferred data values;
- On a device (temporarily) disconnected from the other DCS devices: the workstation is attached to the device's inputs and outputs and executes the tests for checking the device's operation.

The workstation can also be used for simulation of an avionics device connected to DCS but not belonging to DCS itself.

It is practical to equip the mobile workstation with test suites for different DCS devices, each suite packaged for automatic execution (similar to the testbench for acceptance testing of series-produced DCS).

Testing logs and data exchange traces collected by the mobile workstation can be transferred by communication links to the main DCS development workgroup for analysis. To enable such analysis, the testing and monitoring tools used on the mobile workstation and on the stationary testbenches must be unified.

Unification of architecture and software tools on all the testbenches listed above, first of all the testing and monitoring tools, enables use of the same testing technology on different stages of DCS development at Sukhoi Design Bureau, as well as transfer of test suites between testbenches and DCS development stages with minimal adaptation.

DCS software, the requirements to this software, and test suites for verification of these requirements are subject to configuration management. A professional configuration management tool is included in the DCS software development toolchain applied by Sukhoi Design Bureau. Version of the test suite is an integral part of the DCS software version. Headers of test components comprising the test suite contain references to the requirements that apply to the software version.

7 Testing Dependability Features of Avionics Systems

This section describes techniques for testing of AS dependability features on a testbench based on ATT toolset.

To test a dependability feature of AS or its subsystem, the testbench must:

1. Generate a stimulus to activate the dependability feature;
2. Verify that the feature operates correctly.

The stimulus typically belongs to one of following classes:

1. Explicit command for dependability feature activation, delivered through onboard channel;
2. Simulation of a fault (also known as fault injection):
   - Software: delivery of corrupted data;
   - Hardware: simulation of noise on channels, power failures, device shutdown.

Generation of stimuli from classes listed above require control of "external" environment of the tested avionics devices, namely of data exchange channels and power supply systems. Other stimuli, like simulation of data corruption or hardware failures *inside* the devices, require instrumentation of the devices and are not considered here.

Special channel adapters installed on instrumental computers are used for injection of "bit inversion" or "noise on channel" faults into onboard channels connected to DCS devices. To generate a power failure for a specific device, its power cable is equipped with software-controlled switch which can temporarily "disrupt" the cable by a signal from SSH system belonging to the testbench. Both types of hardware fault injection are controlled by tests.

To test AS support for reconfigurability in case of faults as early as possible, simulation models of some AS devices can be used. If reconfiguration procedure involves coordinated actions of several devices, only one of which is available in hardware on the testbench, other devices can be substituted by simulation models. This enables testing reconfiguration-related functions of the hardware device in advance, to prepare it for testing on the "main" testbench for integration and acceptance testing. Avionics device simulation models are created using the hardware-in-the-loop simulation technology described in [3].

Table 1 lists several techniques for testing of AS dependability features, including basic support for reconfigurability. These features are usually implemented in software. Actual set of dependability features and expected AS responses depends on specifics of the target AS. Responses from AS are available to tests in form of data received from AS or obtained by monitoring of internal channels connecting AS devices.

Table 1 Testing avionics system dependability features

Dependability feature	Test stimulus	Expected response from AS
Resilience to sporadic noise on a channel	Injection of "bit inversion" error into the channel	Request for re-sending the corrupted message
Resilience to receiving a single incorrect set of data values	Delivery of message with incorrect (e.g. out of range) data values through an onboard channel	Correct output data values from AS, e.g. produced using last correct input data values
Switching to the secondary MIL STD-1553B bus in case of primary bus failure	Simultaneous shutdown of all primary bus terminals controlled by the testbench, or continuous injection of "noise" into the primary bus	Activation of data exchange on the secondary bus
Switching between primary and hot-spare devices in a redundant device pair (the pair of devices is tested)	Testbench acts as the bus controller and sends the command to start reconfiguration	Reconfiguration-related data exchange starts on the channel connecting the device pair. Former hot-spare device marks itself as primary one (and vice versa) in subsequent data exchange with other devices.
Initiation of switching between primary and hot-spare devices in a redundant device pair (central board computer is tested; device pair is simulated by the testbench)	Testbench acts on behalf of the device pair and simulates a malfunction of the primary device (shutdown or continuous generation of incorrect data)	CBC sends the command (or command sequence) to initiate the reconfiguration. After (simulated) reconfiguration is completed, CBC produces correct output data, matching the input data from the correctly operating (former hot-spare) device.
Switching to emergency mode	Simulation of multiple devices malfunction (or other condition that initiates the emergency mode according to AS specification)	CBC sends appropriate commands to all devices and switches to the emergency mode data exchange schedule
Recovery after power failure of a non-redundant device	Controlled instant disruption of a device's power supply	After specified time the device resumes operation (as a result of successful reboot), i.e. starts sending correct data through channels. The rest of AS resumes using the data produced by the device (this is observable on outputs of other devices).

8 Conclusion

In this paper we presented a toolset for target-based functional testing of avionics software. Architecture of testbenches based on this toolset was described.

Application of the testbenches to testing of avionics systems (AS) dependability features was illustrated by several examples.

The presented toolset is aimed at testing of software running on non-instrumented avionics devices. This capability is essential for acceptance and field testing of avionics devices. Brief analysis of two popular real-time testing toolsets performed in this work shows that they support this capability in a very limited way.

A family of testbenches based on the presented testing toolset was described. These testbenches are used in Sukhoi Design Bureau in the processes of aircraft data control system (DCS) software development and support.

Directions for future development of the testing toolset and testbench technology include support for:

- Declarative description of tests, as an alternative to procedural description by means of Test description language (TDL). An example of declarative test description is a table of test cases (a case contains input values for AS, processing delay, and condition to be checked on AS outputs) in format allowing automatic generation of TDL description.
- Use of a single description of onboard channel messages structure in all test projects for a specific version of AS software.
- Automatic binding of test components' interfaces to testbench hardware (channel adapter ports on instrumental computers) in order to minimize adaptation of test projects on transfer between testbenches. Descriptions of correspondence between testbench interfaces and AS interfaces must be created and automatically processed to implement this capability.
- Automatic analysis of exchange sequences recorded on channels for conformance to reference schedules from Interface control document database.
- Automatic testing of indication formats on DCS display devices. These devices create and send archive copies of displayed video frames. To support automatic testing of indication formats, tools must be created for automatic comparison of these archive frames with reference images.
- Open interfaces for integration of the testbench with external systems for AS workout. An example of such interface is the one specified by HLA standard [6].

References

[1] IBM, Embedded software test automation framework – IBM Rational Test Real Time (2011), http://www-01.ibm.com/software/awdtools/test/realtime/ (accessed December 19, 2011)

[2] Vector Software, How to Improve Embedded Software Unit/Integration Testing with Automation (2011), http://www.vectorcast.com/testing-solutions/unit-integration-embedded-software-testing.php (accessed December 19, 2011)

[3] Balashov, V.V., Bakhmurov, A.G., Chistolinov, M.V., Smeliansky, R.L., Volkanov, D.Y., Youshchenko, N.V.: A hardware-in-the-loop simulation environment for real-time systems development and architecture evaluation. Int. J. Crit. Comput.-Based Syst. 1(1/2/3), 5–23 (2010)
[4] Balashov, V.V., Balakhanov, V.A., Bakhmurov, A.G., Chistolinov, M.V., Shestov, P.E., Smeliansky, R.L., Youshchenko, N.V.: Tools for monitoring of data exchange in real-time avionics systems. In: Proc. European Conference for Aero-Space Sciences, EUCASS (2011)
[5] Balashov, V.V., Balakhanov, V.A., Kostenko, V.A., Smeliansky, R.L., Kokarev, V.A., Shestov, P.E.: A technology for scheduling of data exchange over bus with centralized control in onboard avionics systems. Proc. IMech. E Part G: J. Aerosp. Eng. 224(9), 993–1004 (2010)
[6] IEEE, IEEE Standard for Modeling and Simulation (M&S) High Level Architecture (HLA) – Framework and Rules (2010), doi:10.1109/IEEESTD.2010.5553440

Specification Means Definition for the Common Criteria Compliant Development Process – An Ontological Approach

Andrzej Białas

Abstract. The chapter presents a new ontology-based approach to the definition of specification means used in the IT security development process compliant with the Common Criteria standard. Introducing the ontological approach makes, generally, the IT security development process easier and more effective. The chapter provides multiple-use specification means to create Security Targets (STs) for different kinds of IT products or systems. First, the review of works concerning the ontological approach within the information security domain was performed. Then the chapter discusses the ITSDO workout: domain and scope definition, identification of terms within the domain, identification of the hierarchy of classes and its properties, creation of a set of individuals, and the ontology testing and validation. This way a prototype of the specification means knowledge base was proposed, developed in the Protégé Ontology Editor and Knowledge Acquisition System.

1 Introduction

The chapter concerns the IT security development process compliant with the Common Criteria (CC) standard [1], and deals with the ontological representation of the specification means which describe IT security features and behaviour.

The CC methodology is of key importance as it provides dependable IT solutions to the market, particularly those that are to be used by large businesses, critical infrastructures, and emerging e-services. All these critical IT applications need solutions providing the right assurance for their stakeholders and users. According to [1], the assurance is understood as the confidence that an entity, i.e. IT product or system, called TOE (Target of Evaluation), meets the security objectives specified for it. IT consumers want their IT products or systems work just as they were designed. The assurance foundation is created in a rigorous IT security development process whose results are independently verified according to the standard.

Andrzej Białas
Institute of Innovative Technologies EMAG, 40-189 Katowice, Leopolda 31, Poland
e-mail: a.bialas@emag.pl

W. Zamojski et al. (Eds.): Complex Systems and Dependability, AISC 170, pp. 37–53.
springerlink.com

The general recommendation is that IT products or systems should be developed in a rigorous manner. Rigorous means more precise, more coherent, mathematically based, supported by the rationale and verification processes, etc. Assurance is measurable in the EALn scale (Evaluation Assurance Level) in the range from EAL1 (min) to EAL7 (max).

The ontology represents explicit formal specifications of terms in the given domain and relations between them. Generally, the creation of ontologies allows to analyze, share and reuse knowledge in an explicit and mutually agreed manner, as well as to separate the domain knowledge from the operational one and make domain assumptions explicit [2]. A few methodologies were developed with a view to create ontologies and to perform reasoning. Ontologies were elaborated in many disciplines, such as web-based applications, medicine, public administration and biology, and, recently, in the information security domain.

The general motivation of the works is to improve the IT security development process using the advantages and new possibilities offered by the ontological approach. For this reason the ontology-based models were worked out. The chapter concerns the security ontologies discipline and presents the IT Security Development Ontology (ITSDO), with the focus on its part about specification means.

The chapter features some basic concepts and approaches of knowledge engineering used to improve the IT security development process. The most relevant works concern:

- the CC security functional requirements (Part 2 of [1]) modelling and their mapping to the specified security objectives with the use of the developed CC ontology tool (GenOM) [3]; the security requirements and the security objectives are represented by an ontology; as the paper focuses on the security objectives specification, it neither discusses the security problem definition (threats, policies, assumptions), whose analysis allows to specify these objectives, nor the security functions elaborated on the security requirements basis;
- the ontological representation of the CC assurance requirements catalogue (Part 3 of [1]) used by the developed CC ontology tool [4] to support evaluators during the certification process; the authors of the paper show the usefulness of this methodology especially in such activities as: planning of an evaluation process, review of relevant documents or creating reports; generally, this tool reduces the time and costs of the certification process; please note that the methodology focuses on the evaluation of assurance requirements only, not on the whole IT security development process.

Reviews of researches on security ontologies were provided in [5]. These ontologies concern: risk management issues, advanced security trade-off methods, information security management systems, especially those compliant with ISO/IEC 27001, or business continuity management systems compliant with BS 25999, common security issues, like: threats, attacks, policies, security of services, security agents, information objects, security algorithms, assurance and credentials, etc. The number of elaborated security ontologies is growing and they cover all security-relevant issues.

The most relevant papers [3] and [4] present how to build ontological models of the Common Criteria components only. Additionally, the work [3] shows how to map security requirements to security objectives, however, it does not discuss how to specify these objectives while a security problem is solved. None of these works concerns specification means for all stages of the CC-compliant IT security development process. This chapter concerns specification means for all development stages and security issues mapping between these stages.

The chapter is based on the author's earlier works summarized in [5] and in the monograph [6] presenting the CC-compliant IT security development framework (ITSDF). The UML/OCL approach was used for ITSDF to express all development stages. The models of specification means were elaborated. The means include not only CC components but also the introduced semiformal generics, called "enhanced generics", as they have features comparable to the CC components, allowing parameterization, derivation, iteration, etc. The paper [5] considers the ontological representation of the enhanced generics as a kind of design patterns used to specify security issues of intelligent sensors and sensor systems.

2 IT Security Development Process

The IT security development process is related to the Security Target (ST) or Protection Profile (PP) elaboration (or their simplified, low-assurance versions) [1]. The chapter concerns the specification means for the ST elaboration process, which is a more complicated case. The above mentioned ITSDF framework provides UML/OCL models of the specification means: enhanced generics – proposed by the author, and components – defined by the standard. They are implemented as library items. These are the main stages of the Security Target (ST) elaboration process and the used specification means:

- the preliminary ST introduction stage, for which an informal textual description is used;
- security problem definition (SPD), specifying the threats, OSPs (Organizational Security Policies) and assumptions; for this stage the ITSDF framework provides enhanced generics;
- setting, on this basis, the security objectives (SO) – for the TOE and its environment (development and operational); enhanced security objectives generics are used, covering the SPD items;
- working out the sets of functional (SFR) and assurance (SAR) requirements for the TOE and its environment, based on the CC components catalogues, and analyzing the above objectives;
- preparing the TOE summary specification (TSS) based on requirements, containing security functions (SF) generics that should be implemented in the IT product or system (the TOE development process).

The next section presents the creation of the ontology focusing on specification means for all IT security development stages.

3 Elaboration of the IT Security Development Ontology

The elaboration of the IT Security Development Ontology (ITSDO), compliant with the CC v. 3.1, was based on the author's experiences related to the ITSDF framework implementation and the use of the Protégé environment developed at Stanford University [7]. The ontology development process was performed with respect to the basic knowledge engineering rules [2]. It will be exemplified by some issues encountered during the ITSDO ontology validation (Section 3.7) using the MyFirewall project presented in [6].

3.1 The Domain and Scope of the Ontology

First, the domain and scope of the elaborated ontology have to be defined. For the ITSDO ontology they are related to the Common Criteria compliant IT security development process. The given ontology is able to answer specific questions called competency questions. All these answers define the ontology scope.

While elaborating this ontology, two goals should be achieved to improve the IT security development process:

1. ITSDO provides common taxonomy of the specification means. It allows to better understand terms and relationships between them.
2. Creating on this basis a knowledge database, containing specification means to be retrieved while the security specifications (ST/PP) are elaborated.

3.2 Possible Reuse of Existing Ontologies

As regards the reusability of third-party ontologies, the key issues are the range of compatibility, integration ability, quality, satisfied needs of the ontology users and, first and most – the availability of the given ontology.

3.3 Identifying Important Terms in the Ontology

The ontology development stage called "identification of important terms" was provided mostly during the ITSDF framework [6] elaboration by analyzing the IT security development process, functional and assurance components [1], evaluated products and systems, performed case studies, etc.

3.4 The Classes (Concepts) and the Class Hierarchy

Generally, a top-down ontology development approach was applied, though the entire process was iterative and some bottom-up activities were undertaken. To assure proper knowledge representation, the analyses of the identified terms and relationships were provided. During ITSDO development different analyses of terms and relations between terms should be performed, e.g.: class-individual, class-subclass, class-superclass, abstract classes, and classes having instances. It is important to decide what is to be expressed by a class and what by a property. The possibility of the future evolution of the class hierarchy, transitivity of class relations, avoiding common errors, naming convention, etc. should be considered too.

All concepts of ITSDO are grouped in the following superclasses:

- *CCSecComponent*, representing security assurance – SAR (*SARComponent*) and security functional requirements – SFR (*SFRComponent*);
- *EnhancedGeneric*, expressing enhanced generics used as the specification means for development stages other than the security requirements elaboration, defined previously for the ITSDF framework;
- *SecSpecification*, expressing structures of the ST/PP; this chapter is focused on specification means used to fill in these structures;
- *EvidenceDoc*, representing evidences which should be worked out to meet assurance requirements of the declared EAL (not discussed in this chapter);
- *AuxiliaryConcept*, representing a set of auxiliary terms used in the projects.

The *SARComponent* class represents all CC-defined SAR requirements which are grouped by CC assurance classes, e.g. *ACOClass*, *ADVClass*, etc. Each CC assurance class has its CC-families, e.g. *ADVClass* has *ADV_ARC*, *ADV_FSP*, *ADV_IMP*, etc. Please note that the CC-defined classes (functional, assurance) and their families are both represented by ontology classes. Assurance components are grouped in sets (packages) represented by EAL levels defined in the third part of the standard [1]. In the ITSDO ontology the SAR components are considered as individuals (instances) of the given family.

Example 1. Specification of the ADV_FSP_4 assurance component during ontology validation using the MyFirewall project (Figure 1).

The "ADV_FSP.4 Complete functional specification" CC assurance component is represented by *ADV_FSP_4*, i.e. by the individual of the *ADV_FSP* ontology class. It is exemplified with the use of the Protégé environment.

The left side of this figure shows a part of the ITSDO ontology class hierarchy. The main superclasses and ADV families, with numbers of their individuals in brackets, are visible. For the highlighted ADV_FSP CC family (i.e. ontology class) its individuals are presented (see "Instance Browser" – in the middle): from *ADV_FSP_1* to *ADV_FSP_6*. The right part of the window shows the *ADV_FSP_4* details. For each individual (SAR component and others) a set of properties is defined – according to the third part of the standard [1]. This issue will be discussed later in the second part of the example. □

Due to the restrictions concerning the naming convention, the dot notation commonly used for CC components, e.g. ADV_FSP.1 (and their parts called "elements", e.g. ADV_FSP.1.1) cannot be used in Protégé. The dot "." character is replaced by the "_" underscore character. A similar notation was assumed for the security functional requirements. The *SFRComponent* class represents all CC-defined SFR requirements which are grouped by CC functional classes, e.g. *FAUClass*, *FCOClass*, etc. Each CC functional class has its CC functional families, e.g. *FAUClass* has *FAU_ARP*, *FAU_GEN*, *FAU_SAA*, etc. For the ITSDO ontology it was assumed that particular CC components are individuals of the given family, e.g. *FAU_ARP_1* component is an individual of the *FAU_ARP* class. For each individual a set of properties is defined – according to the second part of the standard [1].

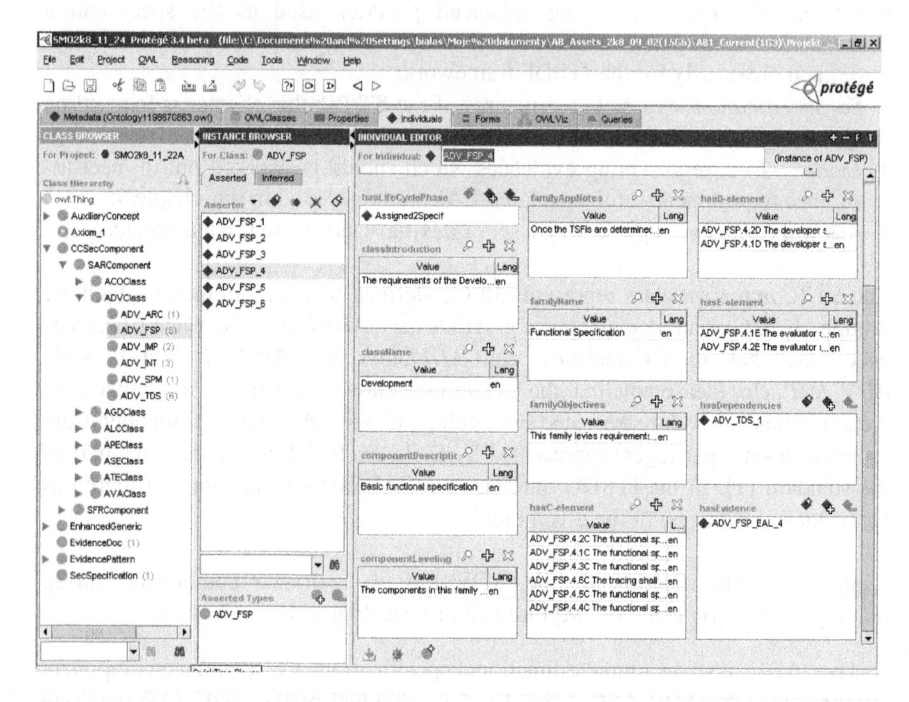

Fig. 1 Examples of ITSDO ontology classes and their instances within the Protégé environment (security assurance requirements)

The *EnhancedGeneric* class represents specification means other than CC components, called enhanced generics and introduced in the ITSDF framework. In comparison with the "generics" usually used by developers, they have some extra features, so they are called "enhanced". In comparison with the ITSDF enhanced generics, here presented enhanced generics have a little changed semantics, their set was optimized (reduced) and matched to the CC ver. 3.1. For this reason they need some explanation concerning the semantics elaborated for them.

Different kinds of assets are represented by the *DAgrGeneric* class which has three subclasses with individuals (Table 1). The part of the name after "_", i.e. the string "*Generic*", represents real names of individuals expressing different kinds of assets, called "Mnemonic" in the ITSDF framework. Subjects and other active entities are expressed by *SgrGeneric* subclasses (Table 2):

Table 1 Generics representing assets (passive entities): *DAgrGeneric* class

DAgrGeneric	Semantics
DTO_Generic	Data objects, services and other assets related to the TOE
DIT_Generic	Data objects, services and other assets placed in the TOE IT environment
DAP_Generic	Other assets placed in the physical environment of the TOE

Table 2 Generics representing subjects and other active entities: *SgrGeneric* class

SgrGeneric	Semantics
SAU_Generic	An authorized subject or other active entity; may be internal or external to the TOE; usually expresses legal users, administrators or processes
SNA_Generic	An unauthorized subject or other active entity; may be internal or external to the TOE; usually expresses threat agents
SNH_Generic	A non-human malicious entity within the system being the cause of unexpected events, different disturbances or technical failures, force majeures, etc.

The security problem definition needs enhanced generics for the specification of threats, organizational security policy rules and assumptions. Assumptions (*AgrGeneric* subclasses) are shown in Table 3. Threats concerning the TOE and its environment are expressed by three *TgrGeneric* subclasses (Table 4). Organizational security policy rules (OSPs) (*PgrGeneric* class) used for the security problem definition and security objectives for the TOE and its environment (*OgrGeneric* class) have similar subclasses, shown in Table 5.

Table 3 Generics representing assumptions: *AgrGeneric* class

AgrGeneric	Semantics
ACN_Generic	Connectivity aspects of the operational environment (IT aspects)
APH_Generic	Physical/organizational aspects of the operational environment
APR_Generic	Personnel aspects of the operational environment

Table 4 Generics representing threats: *TgrGeneric* class

TgrGeneric	Semantics
TDA_Generic	Direct attacks on the TOE
TIT_Generic	Threats to the IT environment of the TOE
TPH_Generic	Threats to the physical environment of the TOE, including procedural and organizational breaches

Table 5 Generics for OSPs (*PgrGeneric* class) and security objectives (*OgrGeneric* class)

PgrGeneric/OgrGeneric	Semantics
PACC_Generic /OACC_Generic	Access control and information flow control aspects
PIDA_Generic /OIDA_Generic	Identification and authentication issues
PADT_Generic /OADT_Generic	Accountability and security audit
PINT_Generic /OINT_Generic	Integrity
PAVB_Generic /OAVB_Generic	Availability
PPRV_Generic /OPRV_Generic	Privacy
PDEX_Generic /ODEX_Generic	General aspects of secure data exchange
PCON_Generic /OCON_Generic	Confidentiality
PEIT_Generic /OEIT_Generic	Right use of software and hardware within the TOE environment
PEPH_Generic /OEPH_Generic	Technical infrastructure and physical security of the TOE environment
PSMN_Generic /OSMN_Generic	Security maintenance, technical solutions and legislation, obligatorily used within the organization (other non-IT aspects)

Security functions (Table 6) representing the TOE security functionality (TSF) are expressed by *FgrGeneric* subclasses related to particular security functional requirements (part 2 of [1]).

Table 6 Generics representing security functions: *FgrGeneric* class

FgrGeneric	Semantics
SFAU_Generic	Security functions dealing with audit
SFCO_Generic	Security functions dealing with communication
SFCS_Generic	Security functions dealing with cryptographic support
SFDP_Generic	Security functions dealing with user data protection
SFIA_Generic	Security functions dealing with identification and authentication
SFMT_Generic	Security functions dealing with security management
SFPR_Generic	Security functions dealing with privacy issues
SFPT_Generic	Security functions dealing with the protection of the TSF
SFRU_Generic	Security functions dealing with the resource utilisation
SFTA_Generic	Security functions dealing with the TOE access
SFTP_Generic	Security functions dealing with trusted path/channels

The presented taxonomy is more concise and effective than ITSDF. The defined security functions subclasses facilitate their mapping to the principal SFRs.

The last main class is *AuxiliaryConcept* which contains EALs definitions (*EAL* subclass), security attributes (*SecurityAttribute* subclass), etc. The ITSDO ontology has also terms representing the ST/PP specifications and evidences.

The classes on the same generality level, called siblings, are usually disjointed, e.g. particular kinds of specification items. For example, a generic representing threats (*TgrGeneric*) cannot represent security policies (*PgrGeneric*) or assumptions (*AgrGeneric*). Please note that disjointed classes cannot have common instances. The well-defined classes hierarchy helps to avoid "class cycles".

It was assumed that only the classes of the lowest hierarchy level, e.g.: *ACN_Generic*, *DTO_Generic*, *TDA_Generic*, *ADV_FSP*, *ALC_DEL*, *SFCS_Generic*, can have individuals used as specification means, i.e. library elements to create ST/PP specifications. These specifications means have object-type properties to join other individuals, e.g. an asset generic being a parameter of a threat generic, a security objective generic covering a given threat generic. Besides, these individuals have many details assigned. Such details are expressed by data-type properties, like strings, numbers and dates.

3.5 The Class Properties and Their Restrictions

The above discussed hierarchy of classes defines the general taxonomy of the used concepts covering all CC-specification means issues. The next step is to define class properties and their restrictions, starting with the basic ones. The restrictions describe or limit a set of possible values for the given property. The ontology development is an iterative and incremental process. More complicated properties and restrictions can be added later, during the ontology refinement, taking into consideration the validation results. The modelled, more sophisticated relationships offer more and more benefits to the ontology users.

Two kinds of standard properties are used:

- object (instance-type) properties, expressing relationships between an individual member of the given class (the object) and other individuals; they are used to show parts of the structured concepts as well; they represent "complex properties", i.e. they contain or point to other objects;
- data-type properties, expressing simple intrinsic or extrinsic properties of the individuals of the most elementary classes; those are commonly used data types, like: integer, byte, float, time, date, enumeration, string, etc.; they represent "simple properties" or "attributes".

Moreover, annotation properties (to document different issues) which explain the meaning of the given concept can be used.

Please note that all subclasses of a given class inherit the properties of that class. If a class has multiple superclasses, it simply inherits properties from all of them. The classes to which an instance-type property is attached are called a domain, while the classes indicated by this property are called a range.

Example 2. Object properties used to express the relations between class individuals (mapping the issues).

The simple property *isCounteredBy*, having the *TgrGeneric* domain and the *OgrGeneric* range, allows to specify all security objectives (individuals) covering the given threat (an individual). This property has the *countersThreat* inverse property which allows to retrieve knowledge on all threats (individuals) that can be countered by the considered security objective (an individual). Similar properties are defined for many other relations, i.e. between security policy rules and security objectives, security objectives and security functional requirements, security requirements and functions. □

Example 3. Object properties of the enumerative class range (knowledge organization and retrieving).

The ontology allows to define enumerative classes [2] including a strictly defined type and number of individuals which have abstract meaning. Such individuals help to organize and retrieve knowledge. ITSDO has defined, e.g.:

- the *EAL* class, encompassing individuals: from *EAL1*, *EAL2*,... to *EAL7*,
- the *SecurityAttribute* class including: *Availability*, *Confidentiality*, *Integrity* individuals,
- the *LifeCyclePhase* of the specification means with three allowed individuals: *UnderDevelopment*, *DefinedInLibrary*, *Assigned2Specif*.

The *hasLifeCyclePhase* property, having the domain of two classes: *EnhancedGeneric* and *CCSecComponent*, has a range of the enumerative class *LifeCyclePhase*. This property is used to manipulate the specification means, e.g. moving items to or removing them from the specification. □

The second kind of properties, i.e. data-type properties, can be used to assign numeric, enumerative values and strings to the individuals of the given class.

Example 4. Data-type properties used to express numerical, textual, logical values.

1. The *hasE-element* property (the *SARComponent* domain) has a range of a data-type *string* and allows to assign an textual description expressing the E-type element (of the component) to the component class.
2. The *hasRisk* property of *TgrGeneric* has a range of an *integer*, representing the risk value inherent to the threat item.
3. The *EnhancedGeneric* class has the *hasDescription* property, representing the verbal description of the generic (the data-type string). □

The next example presents how a property can be used to express details of the assurance components.

Example 1 (continuation). Specification of the *ADV_FSP_4* assurance component during ontology validation using the MyFirewall project (Figure 1).

Please note some details concerning properties of *ADV_FSP_4*. The *hasLifeCyclePhase* property value indicates that this component is assigned to the elaborated security specification. The *hasDependencies* property points to the *ADV_TDS_1* (ADV_TDS.1 Basic Design) component (dependencies), while the *hasEvidence* property – to *ADV_FSP_EAL_4*, an individual of the *EvidencePattern* class. The *EvidencePattern* class represents evidences sampled for this component dealing with EAL4 and influenced by the D, C and E elements (see string-type properties expressing these elements of the component) with respect to the given TOE (i.e. validated MyFirewall system). □

3.6 Creating Individuals (Instances)

During the ontology defining process it is very important to identify concepts that have individuals (instances). Please note that an instance of a subclass is an instance of a superclass. The ITSDO ontology is focused on specification means, i.e. functional components, assurance components and enhanced generics classes. As it was mentioned above, they belong to the lowest levels of the class hierarchy and have individuals which are elementary issues of the created knowledge database.

Example 5. Specification of the *TDA_IllegAcc* enhanced generic expressing a threat against the MyFirewall system (Figure 2).

Enhanced generics will be exemplified by threats within the Protégé environment. The left side of Figure 2 shows a part of the ITSDO ontology class hierarchy, presenting different enhanced generics classes (i.e. groups of generics [6]) and their subclasses (i.e. families of generics [6]). Please note that only generics families (and earlier mentioned component families) can have instances. The numbers of individuals of particular families are in brackets. For the highlighted *TDA_Generic* family (i.e. ontology class, seen at the bottom of the left window), 62 individuals were defined up until now in the library (i.e. in the ITSDO related knowledge base), which represent direct attacks against different kinds of TOEs. The highlighted *TDA_IllegAcc* individual (as well as the *TDA_FwlAdminImpers*, *TDA.NewAttMeth* individuals) were used to specify threats for the MyFirewall firewall during the ontology validation (see "Instance browser" in the middle). The right window ("Protégé Individual Editor") shows details on the highlighted *TDA_IllegAcc* individual of the *TDA_Generic* ontology class.

The *TDA_IllegAcc* meaning is expressed by the string-type property *hasDescription*, having the following value partially visible there (defined on the library level):

"*An attacker [SNA] on the hostile network may exploit flaws in service implementations (e.g. using a well-known port number for a protocol other than the one defined to use that port) to gain access to hosts or services [D*]*".

The meaning of the discussed threat is supplemented by the refinement, expressed by the string-type property *hasRefinement*, having the following value also partially visible:

"*The MyFWL firewall protects external assets, located in the protected network*".

Please note that parameters in square brackets can be substituted by proper subjects/assets generics. The first [SNA] parameter is expressed by the *hasThreatAgent* property and has the *SNA_HighPotenIntrud_D1* value, while the second, [D*] (* means any kind of data assets) parameter is expressed by the *threatenedAsset* property of the *DIT_ProtectedNet* value.

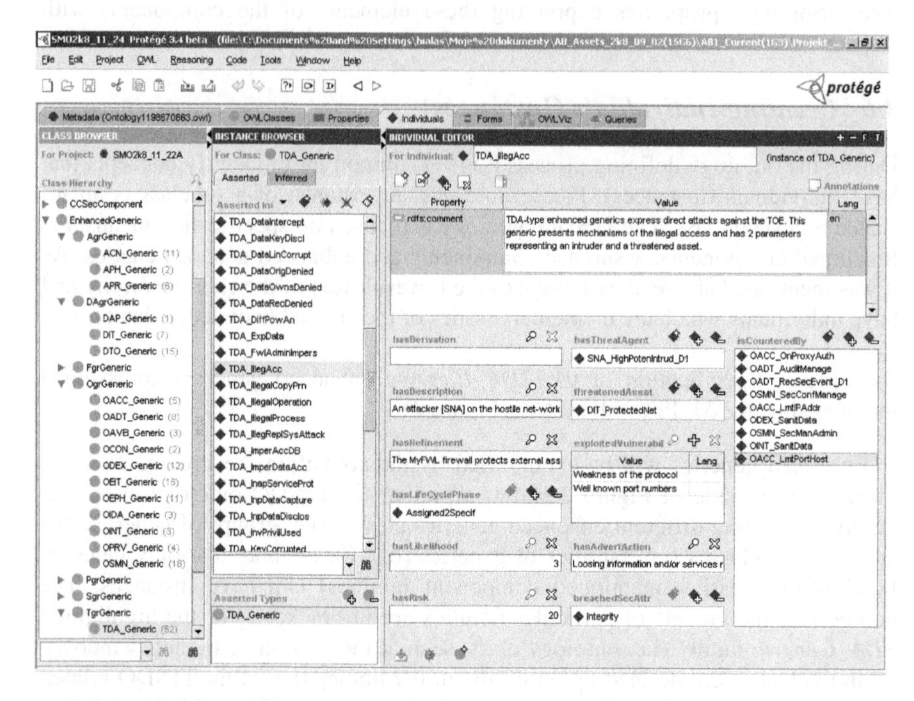

Fig. 2 Examples of ITSDO ontology classes and their instances within the Protégé environment (enhanced generic expressing threat)

The *SNA_HighPotenIntrud_D1* is derived from the *SNA_HighPotenIntrud* generic and means: "*Attacker having high level skills, enough resources and deep motivation to perform a deliberate attack*", expressed by its property *hasDescription* (not shown in Figure 2). The *DIT_ProtectedNet* generic has the *hasDescription* property value: "*Hosts, workstations, their data and services on the private network protected by the firewall*" (not shown either). Please note that it is a "DIT" family generic, concerning the TOE IT environment, because these protected assets are external with respect to the firewall (TOE).

Going back to the *TDA_IllegAcc* enhanced generic, please look at its other properties which contain much information concerning this threat. First, note that this generic is native, i.e. it was not derived from another, because the *hasDerivation* property is empty. Please note that the *breachedSecAttr* property has the I*ntegrity* value, being one of three possible individuals of the enumerative class *SecurityAttribute* (see Example 3).

The *exploitedVulnerability* and *hasAdvertAction* properties present details on the threat, while the *hasLikelihood* and *hasRisk* properties – information about risk inherent in this threat. The ITSDO ontology has a simple risk analysis facility built-in. The risk value displayed within the *hasRisk* property can be calculated in the following way: multiplying the value introduced into the *hasLikelihood* property of the threat by the value introduced into the *hasValue* property of the generic representing assets, e.g. *DIT_ProtectedNet*, and used as a threat parameter, i.e.:

$$hasRisk := hasLikelihood * hasValue.$$

All numbers related to the risk calculation are expressed in predefined measures (e.g. 1, 2, 3). In the future this risk facility will allow to order threats by risk value that can be helpful during the security objectives workout.

A very important property is *isCounteredBy* which specifies security objectives covering this threat. This is an inverse property to the *countersThreat* property placed for any security objective family. Both are applied in bidirectional "navigation" through the ontological model. Figure 2 shows the security objectives selected as a solution to the elementary security issue expressed by the considered *TDA_IllegAcc* threat. Each of these individuals is easily accessed by clicking it and using the "Protégé Individual Editor". Moreover (not shown) it is possible to access functional components, which cover them, from any security objective individual, and, finally, from these components to the security functions individuals – the navigation through the entire model is very easy.

Please note the annotation-type *rdfs:comment* property in the upper part of the "Protégé Individual Editor" window, used to add extra information to any issue of the ontology. □

The developed ITSDO ontology allows to build a knowledge base encompassing all CC-defined functional/assurance components and authors-defined generics designed to express the security problem definition, security objectives for the TOE and its environment, security requirements and security functions. Together they cover all issues needed to specify the security targets (or protection profiles) for different IT products or systems. Moreover, the knowledge database contains ST/PP patterns and evidence patterns for EAL1-EAL7. The knowledge database can be used as a library of specification means allowing to retrieve solutions for elementary security issues.

3.7 Test and Validation of the Developed Ontology

During the ITSDO ontology development the Protégé environment was useful to perform the following types of basic tests:

1. Manual ontology inspections – to avoid commonly known errors [2], such as: cycles in the class hierarchy, violation of property constraints, interval restrictions issuing empty intervals, e.g. min val > max val, terms not properly defined, classes with a single subclass.
2. Ontology tests offered by the Protégé menu functions, like "Checking consistency", "Run ontology tests", were used to remove any inconsistency on early development stages on the basis of the test report.
3. Usability tests were performed by creating individuals of the given class and checking if proper structures of the enhanced generics, functional and assurance requirements, as well as evidence documents are composed.

4. Knowledge retrieving tests, based on the Protégé built-in query mechanism, are designed to check if proper data are issued by different queries applied.

Generally, ontology designing has subjective meaning but the ontology should be objectively correct [2]. The best way to check this correctness is the application for which the ontology was designed with respect to the ontology users' needs and expectations. For this reason, the ontology validation was performed using the typical, nearly realistic MyFirewall project, based on the example described in [6]. This kind of checking was focused on two issues:

- the selection of proper specification means for the ST or PP, from the huge number of available generics and components included in the knowledge base; the specification means for: the security problem definition, security objectives, requirements and functions are checked, along with evidences;
- the creation of the right relationships dealing with the generics parameterization and mapping (covering) of the specification means; the following issues are checked: covering the security problem definition issues by proper security objectives, security objectives by functional components and, finally, functional components by security functions.

Example 6. Building security specification (i.e. ST parts) for the MyFirewall system using the ITSDO ontology (Figure 3).

The "Protégé Queries" facility was used to retrieve knowledge from the elaborated MyFirewall security target model. For the given class and property, a query can be defined to find individuals meeting the condition specified for them. Figure 3 shows one of them, i.e. the query called "CoverageOfTDA_IllegAcc" allowing to get an answer to the question: "Which security objectives cover the elementary security problem expressed by the *TDA_IllegAcc* threat in the considered firewall ST?" The results, i.e. 7 security objectives covering this threat, are shown in the right part of the main window. Clicking the selected individual, e.g. *OACC_OnProxyAuth*, the developer can see its details in the "Protégé Individual Editor" (see extra window on the right side of Figure 3). Please note that this security objective is addressed by two components: *FIA_UAU_2* and *FIA_UID_2*. Clicking any of them, the new "Protégé Individual Editor" window is displayed (not show there) with a security function implementing these functional requirements, i.e.: *SFDP_FwlOnProxyAuth*, responsible for the network users authentication on the firewall proxy. Besides, for any search results item (the upper right corner) it is possible to display all its references (see bottom left part of the figure) as well as to export the property (older name: "slot") value to the spreadsheet. □

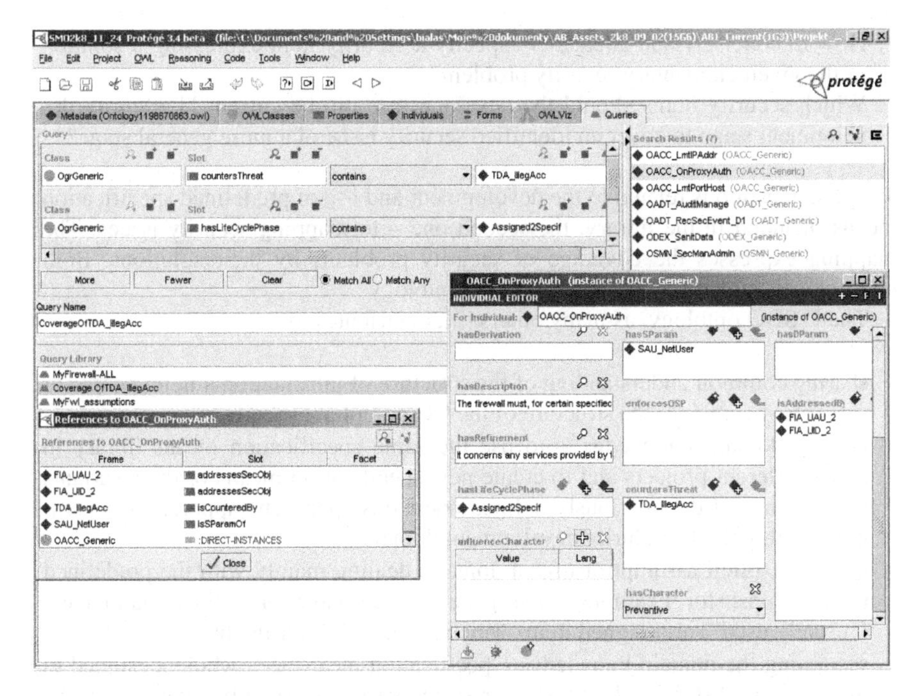

Fig. 3 Using the Protégé query facility to retrieve specification items for the MyFirewall system

A few other queries are defined, e.g. to retrieve specification means of the given IT security development stage or all used in the MyFirewall project. Please note that queries may concern all defined ontology properties.

4 Conclusions

The chapter concerns an object-oriented ontology related to the IT security development process compliant with the Common Criteria standard. The elaboration of the IT Security Development Ontology (ITSDO) was presented, exemplified by the selected issues dealing with the ontology validation on the firewall project. The knowledge base, created on the ITSDO basis, can be considered a specification means library used to create security specifications of different IT products or systems (TOEs).

The enhanced generics and functional and assurance security components provide the basic set of specification means for the entire IT security development process for any IT product or system. The ITSDO users can add their own items into this specification means library.

ITSDO considers many details of the CC-compliant IT security development process, which can be generalized and grouped around two competency questions:

1. How to concisely and adequately specify a security item expressing a solution of the given elementary security problem?
2. Which security items should the developer use on the current IT security development stage to cover an identified security issue of a more general stage?

The first question is related to the development and use of predefined specification means included in the library, the second one – to mapping security issues. The mapping expresses the coverage of security problems by their solutions. Both questions are supported by the introduced ontology.

The ITSDO ontology, as many others [2], contributes to:

- sharing common understanding of the structure of information among people or software, i.e. mainly the structure of the CC-compliant specification means;
- reusing domain knowledge – to use the same specification means library in many different projects and to create new specification means or their variants with the use of the previously defined ones; this approach supports interoperability within the IT security development domain;
- making explicit assumptions for a domain, dealing mainly with the predefined relationships; for example some generics can have predefined parameters ("usually used") or assigned items which "usually" cover them;
- separating the domain knowledge (specification means as a whole, designed to use in many different IT products or systems security specifications) from the operational knowledge (how to use this domain knowledge to compose new specification or to create new specification means);
- analyzing the domain knowledge, e.g. variants, semantics, relationships of the developed specification means.

The development of an ontology in the considered domain allows different correct solutions, more or less convenient for the assumed application and related competency questions. As it was mentioned earlier, the best test is the application for which the ontology is designed. It means further validation on different use cases. The worked out ontology and related knowledge bases are prototypes. They need more investigation, especially from the users' point of view.

Apart from validations and knowledge base optimization on real projects, the ontology integration and enhancement should be continued to allow more sophisticated competency questions. Besides, other issues should be solved, e.g.: ontology integration, refining an ontological model of evidences, risk value calculation and interoperability with the selected third party ontologies.

To improve knowledge acquisition it is necessary to use automated knowledge acquisition techniques, like knowledge-acquisition templates for experts specifying only part of the knowledge required. Moreover, tools generating ontologies from structured documents (e.g. XML documents) are needed. This generation is possible because the CC standard is provided as an XML document as well.

The ontology development is usually just the beginning of further implementation works concerning knowledge bases and different kinds of applications, including those which support decisions.

ITSDO is used in the CCMODE project [8], especially to elaborate models of the knowledge base and data structures for the supporting tool. Additionally, these results are used in other IT security related projects to develop tools for business continuity and information security management in an organization [9].

Acknowledgments. This work was conducted using the Protégé resource, which is supported by the grant LM007885 from the United States National Library of Medicine.

References

[1] ISO/IEC 15408 Common Criteria for IT security evaluation, v.3.1. Part 1-3 (2007)

[2] Noy, N.F., McGuiness, D.L.: Ontology Development 101: A Guide to Creating Your First Ontology. Knowledge Systems Laboratory. Stanford University, Stanford (2001), http://www-ksl.stanford.edu/people/dlm/papers/ontology-tutorial-noy-mcguinness-abstract.html (accessed January 19, 2012)

[3] Yavagal, D.S., Lee, S.W., Ahn, G.J., Gandhi, R.A.: Common Criteria Requirements Modeling and its Uses for Quality of Information Assurance (QoIA). In: Proc. of the 43rd Annual ACM Southeast Conference (ACMSE 2005), vol. 2, pp. 130–135. Kennesaw State University Kennesaw, Georgia (2005)

[4] Ekelhart, A., Fenz, S., Goluch, G., Weippl, E.: Ontological Mapping of Common Criteria's Security Assurance Requirements. In: Venter, H., Eloff, M., Labuschagne, L., Eloff, J., von Solms, R. (eds.) New Approaches for Security, Privacy and Trust in Complex Environments, pp. 85–95. Springer, Boston (2007)

[5] Bialas, A.: Common Criteria Related Security Design Patterns for Intelligent Sensors—Knowledge Engineering-Based Implementation. Sensors 11, 8085–8114 (2011), http://www.mdpi.com/1424-8220/11/8/8085/ (accessed January 19, 2012)

[6] Bialas, A.: Semiformal Common Criteria Compliant IT Security Development Framework. Studia Informatica, vol. 29(2B(77)). Silesian University of Technology Press, Gliwice (2008), http://www.znsi.aei.polsl.pl/ (accessed January 19, 2012)

[7] Protégé Ontology Editor and Knowledge Acquisition System, Stanford University, http://protege.stanford.edu/ (accessed January 19, 2012)

[8] CCMODE (Common Criteria compliant, Modular, Open IT security Development Environment) Project, http://www.commoncriteria.pl/ (accessed January 19, 2012)

[9] OSCAD Project (the computer supported business continuity and information security management system), http://www.oscad.eu/ (accessed January 19, 2012)

Real-Time Gastrointestinal Tract Video Analysis on a Cluster Supercomputer

Adam Blokus, Adam Brzeski, Jan Cychnerski, Tomasz Dziubich, and Mateusz Jędrzejewski

Abstract. The article presents a novel approach to medical video data analysis and recognition. Emphasis has been put on adapting existing algorithms detecting lesions and bleedings for real time usage in a medical doctor's office during an endoscopic examination. A system for diagnosis recommendation and disease detection has been designed taking into account the limited mobility of the endoscope and the doctor's requirements. The main goal of the performed research was to establish the possibility of assuring the necessary performance to introduce the solution into real-life diagnostics.

The structure of an exemplary algorithm has been analyzed to distinguish and discuss parallelization options. After introducing them to the algorithm, the usage of a supercomputer multimedia processing platform allowed to acquire the throughput and latency values required for real-time usage. Four different configurations of the algorithm have been tested and all their measured parameters have been provided and discussed.

1 Introduction

Today, medical image processing tasks are not only getting larger but they are also getting more complicated as we strive to achieve a more accurate representation of reality.

Constantly developed inventions provide the doctors with new means of examination and allow diagnosing parts of the body that were unreachable so far. One of the interesting technological advances is the Wireless Capsule Endoscopy (WCE) [1]. It is a new method of diagnosis that involves the patient swallowing a medium-sized (11mm x 26mm) capsule equipped with a light, camera, and wireless transmitter that traverses the patient's gastrointestinal (GI) tract capturing

Adam Blokus · Adam Brzeski · Jan Cychnerski · Tomasz Dziubich · Mateusz Jędrzejewski
Department of Computer Architecture, Gdańsk University of Technology, Poland, Gdańsk, G. Narutowicza street 11/12
e-mail: ablokus@eti.pg.gda.pl, brzeski@eti.pg.gda.pl, jan.cychnerski@eti.pg.gda.pl, dziubich@eti.pg.gda.pl, matjedrz@eti.pg.gda.pl

W. Zamojski et al. (Eds.): Complex Systems and Dependability, AISC 170, pp. 55–68.
springerlink.com

images and sending them to a recording device outside of the patient's body. It allows the doctor to see the whole GI tract and proves to be superior to methods used so far to diagnose conditions of the small intestine.

As the method becomes more popular, scientist work on reducing the workload put on the diagnostician who has to spend 45 to 120 min analyzing the 8h long recording [2]. The current research has two main directions. The first one is the elimination of uninformative frames and therefore the reduction of the total amount of images to analyze by removing series of similar frames and keeping only their most typical representatives. An image mining approach firstly proposed in [3] has been successfully improved by the introduction of distributed computing [4] and primary division of the video into separately processed segments. The summarized final recordings, having ca 30% of their initial length, turn out to cover all the findings marked by professional doctors. Other solutions are proposed in [5], which utilize the Gabor and wavelet filters to detect the frames which are blurry or contain only bubble-shaped intestinal liquids.

The second direction in research involves the automatic recognition of images by specialized classifiers and means of artificial intelligence. The main idea behind many algorithms is the extraction of a set of features from the image and training the system to recognize the ones classified as pictures of lesions or bleedings. After the algorithm analyzes the video, the detected frames are presented to the medical doctor, who is responsible for their final classification. There are various works that relate to the problem of automated disease detection [2]. Most implemented algorithms perform remarkably well, achieving rates of sensitivity of over 90%.

Our team has put work into incorporating these methods into the traditional endoscopy, by adapting the recognizing algorithms to be useful not only in offline processing but also during endoscopic examinations. While performing his duties, the doctor could be supported by a computer system giving hints and marking abnormal regions in the video image. Such system could vastly improve the diagnosis and disease detection rate but only if it was highly efficient and would not slow down the examination procedure.

Our aim was to create a system capable of performing a real-time analysis of the video data coming directly from the endoscope. To achieve this we have established the values of the throughput and the latency that the system must comply with. In the considered case, the throughput should be equal to at least 30 fps. An interview of medical specialists cooperating with our project indicated that for an online analysis to be helpful during the examinations, the latency parameter should fall within the range of 2 seconds. Such amount of time typically corresponds to the movement of the endoscope by approximately 5cm in the gastrointestinal tract. If we include the inevitable time of transferring data back and forth, about 0.2 seconds are left for the video processing and recognition algorithms. Of course, from the medical specialist's point of view the overall latency of the system should be kept as low as possible.

Although systems that divide the data among computational nodes, like the one presented in [4], or analyze only a subset of video frames can increase the throughput, the latency value is still going to be too high if the algorithm analyzing a single frame proves be too slow. Also, in a continuous processing of a video stream analyzing only selected frames would not be desirable. Therefore, the key is to parallelize the processing algorithm itself. This could be done by dividing it in a workflow fashion into logically consistent blocks and executing them on different computational nodes.

Most of the workflow-building solutions found in the papers, notably ones such as Pegasus [6], Triana [7], Taverna [8] and Kepler [9], operate in a grid environment. However, computations in a heterogeneous environment can often prove to be non-deterministic and not always reliable. Furthermore, grid nodes are usually connected by a slow, high-latency network that can hinder the whole system's effectiveness. Hence, we propose the use of a uniform and reliable environment of a dedicated cluster computer.

The research which shows the capability of real-time endoscopic video stream processing has been conducted on the multimedia stream processing platform called KASKADA [10], which is described in section 3. The platform is deployed on a cluster supercomputer environment which consists of multiple computational nodes connected by a fast low-latency network. It is capable of processing incoming data with the use of services created by the user and provides methods for an easy parallelization of the algorithms.

During the initial tests, the aforementioned typical video analysis and disease detection algorithms proved to be too slow to satisfy the parameters imposed by real-time processing. Therefore we have chosen a representative algorithm which was subjected to parallelization with the use of methods provided by the KASKADA platform. Then the possibility for the real-time processing was tested by measuring latency and throughput of the endoscopic video analysis. The tests have shown that the prerequisites for the real-time processing can be met using the chosen methods.

The rest of the paper is organized as follows. In section 2 the purely sequential version of the chosen video analysis algorithm is briefly described. Section 3 contains the characterization of the execution environment of the KASKADA platform and its capabilities. Next, in section 4, a parallel pipeline version of the algorithm is proposed. The testing procedure along with the definition of the chosen measures is described in section 5 whereas the results and the discussion are presented in section 6. Finally, the section 7 encompasses the paper conclusions and suggestions for future research.

2 The Sequential Algorithm

There are several algorithms addressing the problem of abnormality detection on videos acquired from gastrointestinal tract examinations such as endoscopy, colonoscopy or wireless endoscopy, that can be found in scientific literature [2].

Most of them consist of two main steps. The first one is finding a set of features in an image acquired from the examination video. The second one is the classification step in which a set of features calculated in the previous step is assigned to one of the predefined classes that represent diseases as well as healthy tissue.

For conducting the research, an algorithm developed by Baopu Li and Max Q.-H. Meng [11] was chosen as it proved to be highly effective both in the original paper and in our tests. Furthermore, it shows similarities to other solutions, so a successful parallelization attempt would open path for accelerating other algorithms.

In this section, the selected algorithm will be explained in detail to give a perspective on how speed improvements can be made. Later, in section 4, several algorithm acceleration enhancements proposals will be presented.

The feature set calculation of the selected algorithm is composed of three major steps. The first one is preprocessing which involves only a lossless recalculation of image pixels to a desired color space like RGB or HSV. Then each subsequent channel is subjected to a DWT (Discrete Wavelet Transform) [12] algorithm which highlights and amplifies texture features found in the processed image. DWT decomposes the image into 4 smaller images using a combination of low- and highpass filters based on a wavelet transform. An image obtained by a combination of two highpass filters (called LL) is a smaller and less detailed input image equivalent, whereas images created with other filter combinations (named LH, HL, HH) contain valuable texture information and are taken for the later analysis.

At the last step, texture features from DWT images (HL, LH and HH) created from all of the channels are calculated separately using a uniform LBP (Local Binary Pattern) [13]. Basically, LBP gathers statistical information about relevant points (pixels) found in an image, such as corners, edges, holes, etc. Additionally, it is invariant to monotonic grayscale transformations and the uniform version of the algorithm can gather information about same classes of points regardless of their rotation and transposition.

The LBP used in the article and implemented by us for further analysis, gathers information about 10 different classes of points found in an image. Thus, a combination of features (statistics of pixel data) found in all DWT images of all color channels gives a vector containing 90 features (3 color channels * 3 DWT images * 10 classes of points).

The next step of the algorithm is image classification. It is done with methods chosen directly from the field of artificial intelligence. In the analyzed algorithm an SVM classifier was used, which proved to have a high accuracy potential. From the designed system and real-time analysis point of view, this classifier has other benefits as well, such as being able to perform classification in a time efficient fashion. We are still in the phase of developing a more elaborate decision system, so for the current research we have chosen a single SVM classifier which was tuned for recognition between healthy large intestine tissue and cancer tissue located in the same organ.

3 Primary Execution Environment

All experiments have been carried out in the environment provided by the multimedia processing platform KASKADA (polish acronym for: *Contextual Analysis of Data Streams from Video Cameras for Alarm Defining Applications*), which was developed as a part of the MAYDAY EURO 2012 project at the Gdańsk University of Technology.

The KASKADA platform has been primarily designed as an execution environment for algorithms which process multimedia streams and is currently working on the cluster supercomputer Galera [14] (672 nodes, each with two Intel Xeon Quad Core 2,33 GHz processors, at least 8GB RAM, InfiniBand network). The main execution units on the KASKADA platform are the processing services. Each user can create simple services which incorporate chosen algorithms. They can realize a particular task such as image conversion, or perform more complicated ones like a complete face detection algorithm.

Simple services can be organized into more sophisticated complex services forming a well-known parallel workflow system. The platform is responsible for managing the life cycle of all the services, the connections between them and the input streams from various data sources. The incoming streams are decoded and given as input for the services. Each simple service can create new data streams, which can become the inputs for other simple services or the outputs of the whole complex service. Computations in the simple services are performed in parallel, as each service can be bound to different nodes of the supercomputer. This feature of the KASKADA framework encourages the design of complex algorithms in a fashion that allows to parallelize their execution by introducing logically separate blocks and pipeline processing of the incoming data.

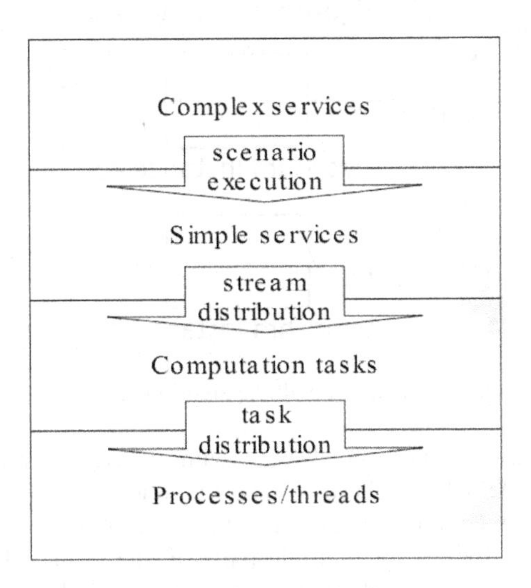

Fig. 1 The execution model of the KASKADA platform [10]

The services are also capable of signaling detected contextual events (a dangerous situation, a disease, etc.). Those events are sent to a queue server specified by the user and can be read by his application.

The structure of the platform makes it reasonable to perceive it as a possible component of complex diagnostical systems that can aid medical doctors with real-time analysis of video data from endoscopic examinations. Thus, all the efforts to improve the algorithm are focused on its parallelization with the tools provided by the KASKADA platform.

4 Parallelization Options

Initial speed tests of the sequential algorithm (presented in section 6) have shown, that the method cannot be used straightforwardly in real-time video analysis, which is why we present the following proposals for accelerating computations. Our aim in all the tests was to use parallelization and speed-up mechanisms provided by the aforementioned KASKADA platform and to assess their efficiency in the context of the analyzed problem.

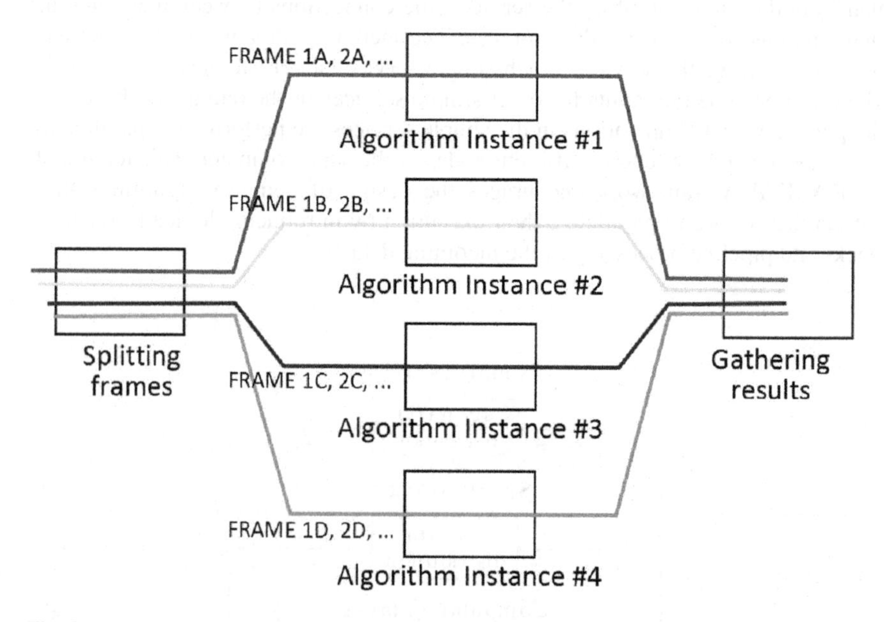

Fig. 2 Distributing a frame sequence to multiple algorithm instances

First experiments were performed solely with the use of parallelization methods provided by the KASKADA platform. The first and most straightforward way of decomposing the recognition algorithm into parallelizable modules involves processing multiple frames (or multiple parts of one frame - like different color channels) concurrently by distributing them among multiple instances of one algorithm. As it is shown in Fig. 2, the incoming video stream is evenly distributed by sending every portion of consecutive frames to different instances of the same processing algorithm. After all computations are done, the resulting output is gathered by another node and sent to the output or for further computations.

For the purposes of our research, the scheme presented above has been applied twice, to achieve a finer division of the original stream. The algorithm chosen in section 2 was decomposed into logically consistent blocks which were then implemented and executed as separate simple services forming one complex service on the KASKADA platform. Service startup, data flow, event message passing and, most notably, computation nodes allocation were all maintained by the execution environment of the platform. The platform also ensures that all processing blocks are given the predefined computation resources. Those mechanisms guaranteed that each computing service block used during the experiments had an exclusive access to at least one processing core.

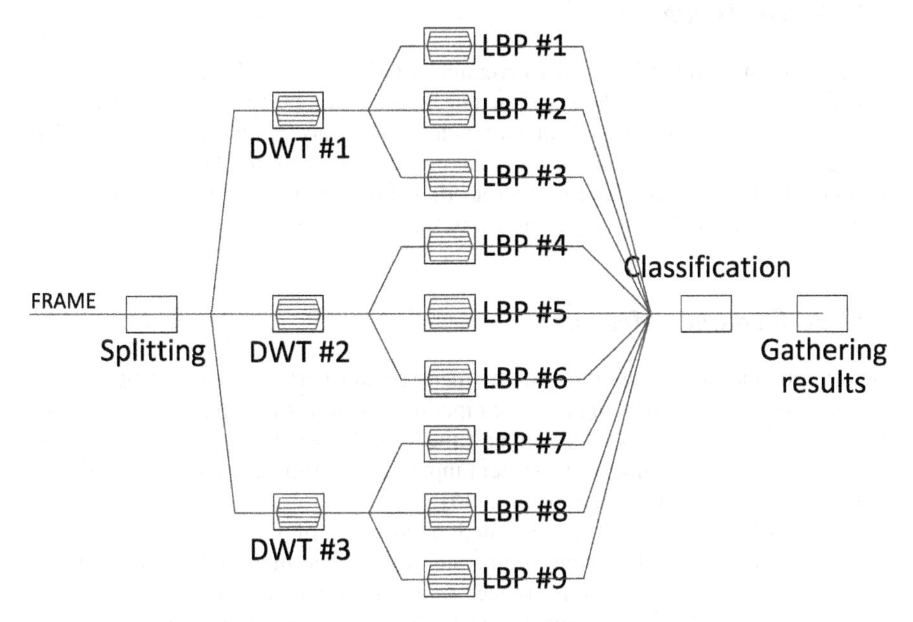

Fig. 3 Proposed parallel version of the chosen algorithm

The data flow of the parallel algorithm is presented in Fig. 3. In this version of the algorithm, calculations were modeled as a pipeline system with elements of data decomposition concerning separate color channels and DWT images. Such processing model should lead to significant enhancements in terms of both latency and throughput.

To make further acceleration improvements we decided to parallelize calculations in the services. Each simple service deployed on the KASKADA platform is assigned to a multiprocessor cluster node of the Galera supercomputer at run-time. In consequence, classic process parallelization techniques such as dividing computations into multiple threads can be efficiently employed. To achieve this, we chose an OpenMP library which allows smooth parallelization of the most computationally intensive parts of code, such as loops.

The DWT and LBP services were parallelized in this fashion. During the experiments, two numbers of OpenMP threads utilized by these services have been tested.

5 The Testing Procedure

The testing procedure consisted of three steps: a) collecting a set of endoscopic videos that can be processed by the system, b) processing these videos multiple times with prepared scenarios, and c) calculating performance statistics. Some important aspects of these operations are described below.

5.1 Videos Database

In cooperation with the Medical University of Gdańsk we collected a large set of endoscopic videos (ca. a thousand of check-ups). For performing the tests, we have chosen a set of 5 representative video fragments, each approximately 20 seconds long. The test videos have been saved in the original DVD format (i.e. MPEG2, PAL 720x576, 25 fps interlaced, 9400 Kbps). Afterwards, a medical doctor has analyzed each video and has marked frames containing images of diseased tissues.

5.2 Performance Measures

Throughput and latency are the two key measures for performance evaluation of a pipeline system. The *throughput* of the pipeline system is the maximum amount of data that system can process in a given time, and the *latency* is the time between the arrival of a video frame at the system input and the time at which the detection report is available at the system output. [15]

It is worth noting that the evaluated parameters of the chosen algorithm's depend only on the resolution and framerate of the video input, not on the content of the frames. Therefore, there is no need to involve a larger number of test cases, as any influence of external factors can be limited by repeating the test several times for the same video stream.

5.3 Stream Processing

In the KASKADA platform, stream processing can be simplistically illustrated as shown in Fig. 4. Each of the consecutive frames arrives at the input of the system is being processed (possibly, in parallel with other frames) for the time l_i, after which the output corresponding to that frame is acquired at the time T_i, d_i after the previous frame's output. The overall throughput is determined by one of the internal blocks, that has the worst time of processing a single incoming frame. The latency – by the longest path in the graph on Fig. 3 (in the case of the analyzed algorithm, all paths are actually of the same length).

As every processing block has an internal queue of awaiting frames, it is impossible to clearly determine the time of a frame being input to the service, because all frames arrive almost at once at the input, and continue waiting in queues and being processed by consecutive blocks. Therefore, all measurements have been performed on the time values acquired at the output of the service, relative to the starting time of the whole processing.

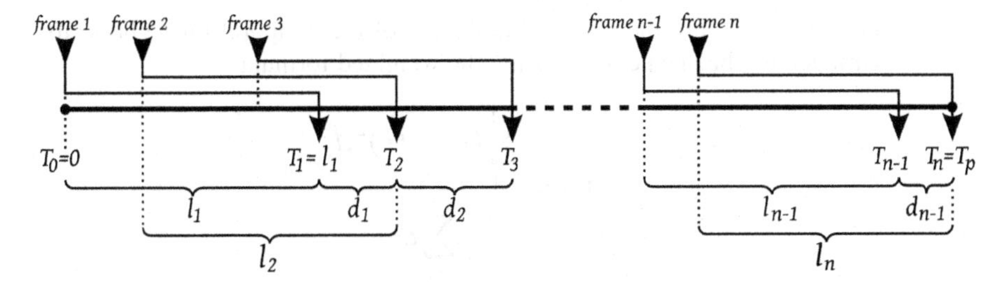

Fig. 4 Stream processing in the KASKADA platform
n - number of frames in the whole stream
T_0 - time at which the first frame arrives to the system, fixed to 0
T_i - time at i-th frame's report is available
T_p - time of the whole video processing, $T_p = T_n - T_0$
l_i - processing time of i-th frame
d_i - time interval between two succeeding reports, $d_i = T_{i+1} - T_i$

5.4 *The Measurements*

Each configuration scenario has been tested in the environment of the KASKADA platform. The execution environment of the platform ensured that each separate thread in every service performing computations had an exclusive use of one processing core.

Each test consisted of two steps:

1. **throughput** measurement - videos were processed separately, with maximal input frame rate (>100 fps). Every video was processed 5 times. Finally, throughput H and σ_H (standard deviation of h) were calculated. While processing the video, temporal throughput values h_i have been calculated using the relationship:

$$h_i = \frac{1}{d_i}$$

After processing the whole video, the average throughput value H has been computed as the mean value of the temporal values weighted by their time:

$$H = \frac{\sum_{i=1}^{n-1} h_i d_i}{\sum_{i=1}^{n-1} d_i} = \frac{\sum_{i=1}^{n-1} \frac{1}{d_i} d_i}{\sum_{i=1}^{n-1} d_i} = \frac{n-1}{T_n - T_1}$$

which gives the ratio of the number of frames to the amount of time needed to process them - the well-known FPS (Frames Per Second) measure.

Finally, to acquire the standard deviation of the temporal throughput, its variance has been calculated using the weighted formula:

$$\sigma_H^2 = \frac{\sum_{i=1}^{n-1} (h_i - H)^2 d_i}{\sum_{i=1}^{n-1} d_i}$$

2. **latency** measurement - videos were processed separately, with their original frame rate (25 fps) for all the versions of the algorithm which proved to provide a sufficient throughput. Other configurations have been skipped as being unsuitable for real life use. Every video was processed 5 times. Finally, the average (L) and the standard deviation (σ_L) of the latency were calculated using the simple formulas:

$$L = \frac{\sum_{i=1}^{n} l_i}{n}$$

and

$$\sigma_L^2 = \frac{\sum_{i=1}^{n} (l_i - L)^2}{n}$$

6 Test Results

All the experiments have been carried out according to the procedure described in the previous section. The results of all the experiments have been summarized in Table 1, which shows the average values of the chosen measures from all performed tests. More detailed results of the throughput measurement have been presented in Fig. 5 which shows the outcome of each test together with estimated error ranging three standard deviations of the measurement.

Table 1 Sample data

Algorithm version	H[fps]	σ_H	L[ms]	σ_L
Sequential	0.9	0.01	-	-
Pipeline	8.94	0.39	-	-
Pipeline+OpenMP (4 threads)	32.51	1.58	56.11	3.9
Pipeline+OpenMP (8 threads)	58.55	4.58	38.96	3.09

The first tests involved the sequential version of the algorithm and confirmed its poor performance for live video processing. Therefore, the first steps have been taken to parallelize the algorithm. Pipeline processing and data decomposition have been introduced by dividing the algorithm into logically separate blocks as presented in section 4. The throughput has been re-measured, acquiring a result which was better by an order of magnitude, but still not suitable for real-time applications. The last experiment involved additional parallelization through the use of OpenMP. The resulting throughput turned out to be adequate. Therefore, further experiments have been carried out to determine the average latency of the system.

Fig. 6 presents the latency measurement results of the two tested OpenMP thread configurations. Average latencies of particular tests have been shown together with corresponding intervals of values within a one standard deviation. As we can observe, the latency is kept far below 100ms.

As shown in Table 1 and Fig. 5 and 6, the parameters of the final algorithm fit into the established boundaries. This holds for both tested numbers of threads. Still, the version with more threads available for OpenMP's parallelization of loops has proven to be more time-efficient.

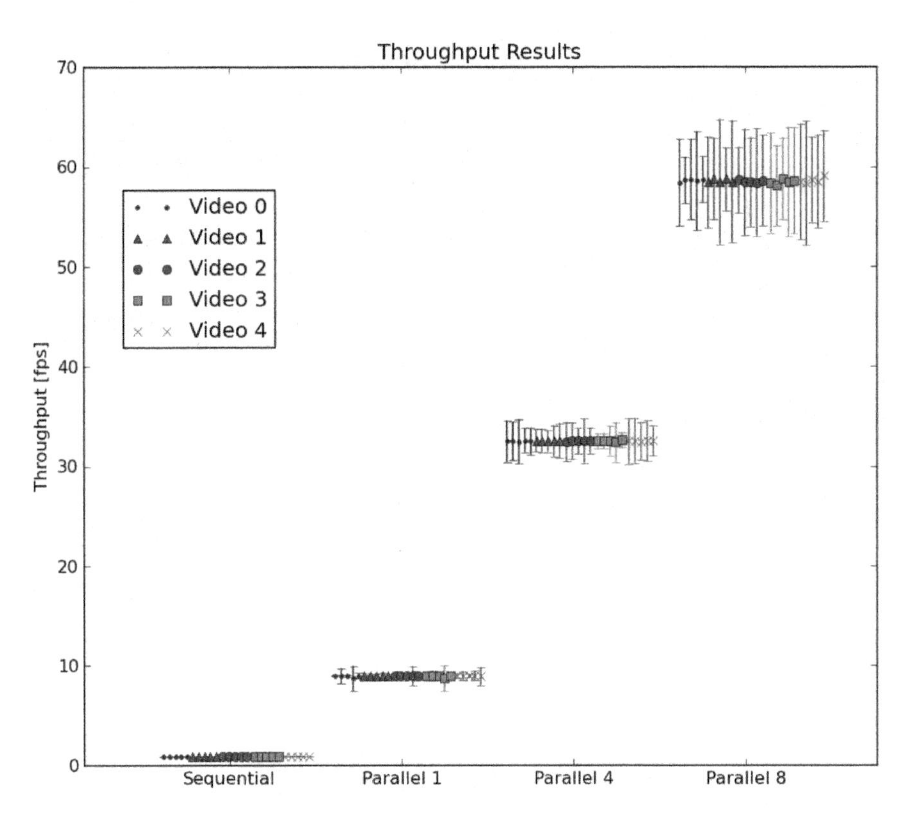

Fig. 5 Diagram of the throughput(H) values acquired in the experiments, with their intervals of values within a standard deviation

7 Conclusions

In the course of the performed experiments, it has been proven that real-time processing of video from endoscopic examinations is possible with a proper choice of hardware, a parallel multimedia processing platform and a parallelizable algorithm. Although these prerequisites are quite extensive and demanding, the acquired results are valuable as the first ones in this direction of research, as most other experiments focus purely on the accuracy of the off-line processing algorithms.

The determined value of the system's processing latency is relatively low in comparison with the boundary value. Therefore, the Internet connection is allowed to introduce a latency of not more than 1.9 s - a value not difficult to achieve.

All of the results were made possible by the use of the supercomputer multimedia processing platform KASKADA. It allowed an easy and straightforward parallelization of the chosen algorithm and provided the runtime environment for it. Other algorithms (e.g. [16][17][18]) , similarly designed, can be parallelized in an analogous manner. Further research in this direction is planned to involve a real life installation of the system at a Medical University of Gdańsk examination room to evaluate its usability for the medical specialist.

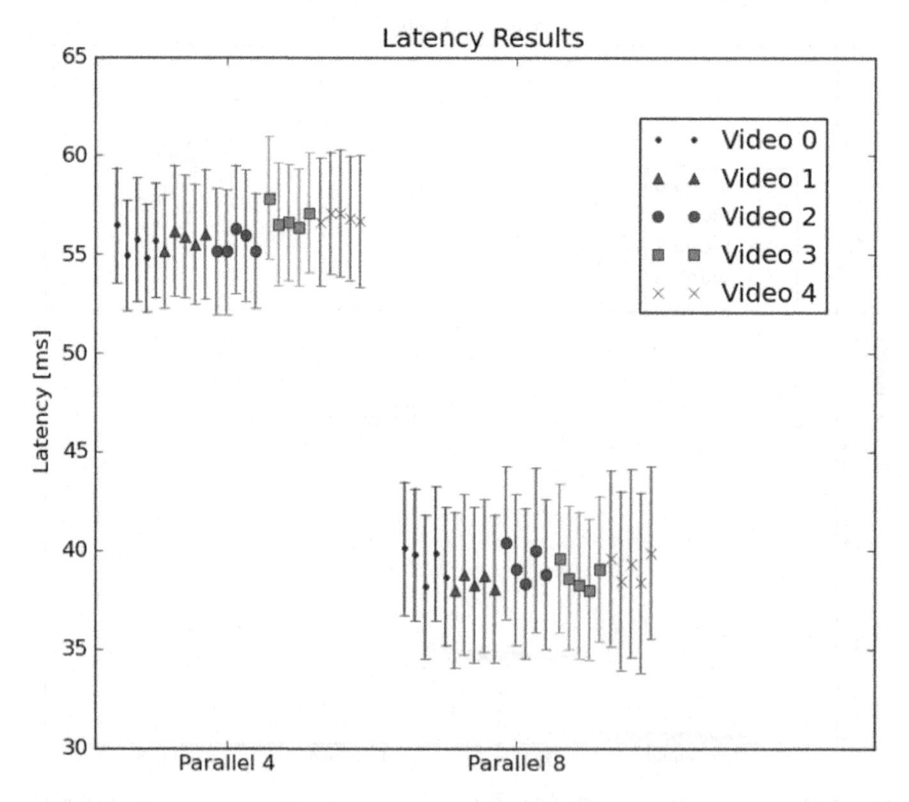

Fig. 6 Diagram of the latency(L) values acquired in the experiments, with their intervals of values within a standard deviation

Acknowledgments. Research funded within the project No. POIG.02.03.03-00-008/08, entitled "MAYDAY EURO 2012- the supercomputer platform of context-depended analysis of multimedia data streams for identifying specified objects or safety threads". The project is subsidized by the European regional development fund and by the Polish State budget".

References

[1] Iddan, G., Meron, G., Glukhovsky, A., Swain, P.: Wireless capsule endoscopy. Nature 405(6785), 405–417 (2000)

[2] Karargyris, A., Bourbakis, N.: Wireless capsule endoscopy and endoscopic imaging: A survey on various methodologies presented. IEEE Engineering in Medicine and Biology Magazine 29(1), 72–83 (2010)

[3] Iakovidisa, D., Tsevasa, S., Polydorouc, A.: Reduction of capsule endoscopy reading times by unsupervised image mining. Computerized Medical Imaging and Graphics 34, 471–476 (2010)

[4] Ioannis, K., Tsevas, S., Maglogiannis, I., Iakovidis, D.: Enabling distributed summarization of wireless capsule endoscopy video. In: 2010 IEEE International Conference on Imaging Systems and Techniques (IST), pp. 17–21 (July 2010)

[5] Bashar, M.K., Kitasaka, T., Suenaga, Y., Mekada, Y., Mori, K.: Automatic detection of informative frames from wireless capsule endoscopy images. Medical Image Analysis (January 2010)

[6] Deelman, E., Blythe, J., Gil, A., Kesselman, C., Mehta, G., Patil, S., Hui Su, M., Vahi, K., Livny, M.: Pegasus: Mapping scientific workflows onto the grid, pp. 11–20 (2004)

[7] Majithia, S., Shields, M., Taylor, I., Wang, I.: Triana: A graphical web service composition and execution toolkit, Web Services. IEEE International Conference on Web Services, 514 (2004)

[8] Oinn, T., Addis, M., Ferris, J., Marvin, D., Senger, M., Greenwood, M., Carver, T., Glover, K., Pocock, M.R., Wipat, A., Li, P.: Taverna: A tool for the composition and enactment of bioinformatics workflows. Bioinformatics 20(17), 3045–3054 (2004)

[9] Ludäscher, B., Altintas, I., Berkley, C., Higgins, D., Jaeger, E., Jones, M., Lee, E.A., Tao, J., Zhao, Y.: Scientific workflow management and the kepler system: Research articles. Concurr. Comput.: Pract. Exper. 18, 1039–1065 (2006)

[10] Krawczyk, H., Proficz, J.: Kaskada - multimedia processing platform architecture. In: SIGMAP (2010)

[11] Li, B., Meng, M.Q.-H.: Capsule endoscopy images classification by color texture and support vector machine. In: IEEE International Conference on Automation and Logistics (2010)

[12] Mallat, S.G.: A theory for multiresolution signal decomposition: the wavelet representation. IEEE Transactions on Pattern Analysis and Machine Intelligence 11, 674–693 (1989)

[13] Ojala, T., Pietikainen, M., Harwood, D.: A comparative study of texture measures with classification based on feature distributions. Pattern Recognition 29, 51–59 (1996)

[14] TOP500 supercomputer list (2012) top500.org

[15] Liao, W.-K., Choudhary, A., Weiner, D., Varshney, P.: Performance evaluation of a parallel pipeline computational model for space-time adaptive processing. The Journal of Supercomputing 31(2), 145 (2005)
[16] Kodogiannis, V.S., Boulougoura, M.: An adaptive neurofuzzy approach for the diagnosis in wireless capsule endoscopy imaging. Int. J. Inf. Technol. 13(1) (2007)
[17] Magoulas, G.: Neuronal networks and textural descriptors for automated tissue classification in endoscopy. Oncol. Rep. 15, 997–1000 (2006)
[18] Magoulas, G.D., Plagianakos, V.P., Vrahatis, M.N.: Neural network based colonoscopic diagnosis using on-line learning and differential evolution. Appl. Soft Comput. 4(4), 369–379 (2011)

Detection of Anomalies in a SOA System by Learning Algorithms

Ilona Bluemke and Marcin Tarka

Abstract. The objective of this chapter is to present the detection of anomalies in SOA system by learning algorithms. As it was not possible to inject errors into the "real" SOA system and to measure them, a special model of SOA system was designed and implemented. In this systems several anomalies were introduced and the effectiveness of algorithms in detecting them were measured. The results of experiments may be used to select efficient algorithm for anomaly detection. Two algorithms: K-Means clustering and emerging patterns were used to detect anomalies in the frequency of service call. The results of this experiment are discussed.

1 Introduction

With the growth of computer networking, electronic commerce, and web services, security of networking systems has become very important. Many companies now rely on web services as a major source of revenue. Computer hacking poses significant problems to these companies, as distributed attacks can make their systems or services inoperable for some period of time. This happens often, so an entire area of research, called Intrusion Detection, is devoted to detect this activity.

Nowadays many system are based on SOA idea. A system based on a SOA provides functionality as a suite of interoperable services that can be used within multiple, separate systems from several business domains. SOA also generally provides a way for consumers of services, such as web-based applications, to be aware of available SOA-based services. Service-orientation requires loose coupling of services with operating systems, and other technologies that underlie applications. SOA separates functions into distinct units, or services, which developers make accessible over a network in order to allow users to combine and reuse them in the production of applications.

There are several definitions of service oriented architecture (SOA). OASIS [1] defines SOA as the following: *A paradigm for organizing and utilizing distributed capabilities that may be under the control of different ownership domains.*

Ilona Bluemke · Marcin Tarka
Institute of Computer Science, Warsaw University of Technology, Nowowiejska 15/19, 00-665 Warsaw, Poland
e-mail: I.Bluemke@ii.pw.edu.pl

W. Zamojski et al. (Eds.): Complex Systems and Dependability, AISC 170, pp. 69–85.
springerlink.com

It provides a uniform means to offer, discover, interact with and use capabilities to produce desired effects consistent with measurable preconditions and expectations. Definition of SOA can be also found in SOA Manifesto [2].

The objective of this chapter is to present the detection of anomalies in SOA system by learning algorithms. Related work is presented in section 2 and in section 3, a special model of SOA system which was used in experiments, is presented. In this systems several anomalies were introduced. Four algorithms: Chi-Square statistics, k-means clustering, emerging patterns and Kohonen networks were used to detect anomalies. Detection of anomalies by k-means and emergent patterns is presented in section 4 and some conclusions are given in section 5.

2 Related Work

There are many anomaly detection algorithms proposed in the literature that differ according to the information used for analysis and according to techniques that are employed to detect deviations from normal behavior. Lim and Jones in [3] proposed two types of anomaly detection techniques based on employed techniques:

- learning model method
- specification model.

The *learning* approach is based on the application of machine learning techniques, to automatically obtain a representation of normal behaviours from the analysis of system activity. The *specification-based* approach requires that someone manually provides specifications of correct behaviour. Approaches that concern the model construction are presented in Fig. 1.

The *specification* approach depends more on human observation and expertise than on mathematical models. It was first proposed by C. Ko et. al. [4] and uses a logic based description of expected behaviour to construct a base model. This specification-based anomaly detector monitors multiple system elements, ranging from application to network traffic.

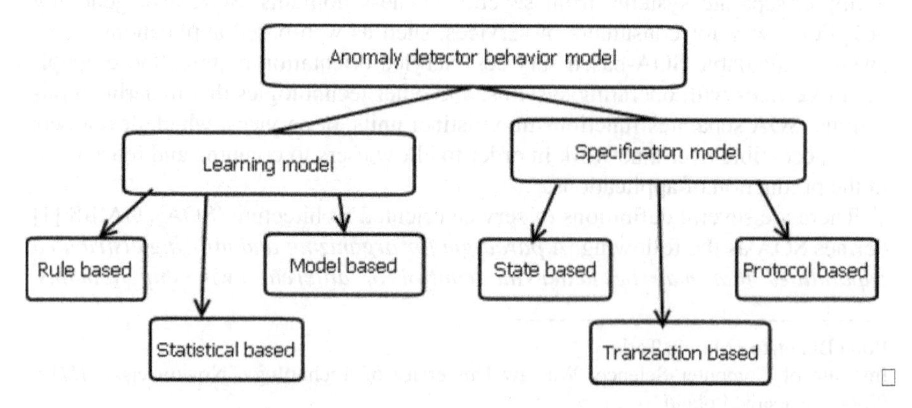

Fig. 1 Taxonomy of anomaly detection behavioral model (based on [3])

In the *protocol based* approach [5] a normal use model is built from the protocol specification e.g. TCP/IP. Lemonnier [5] proposed a protocol anomaly filter able to specifically analyse a protocol and model the normal usage of a specific protocol. This technique can be seen as a filter looking for protocol misuse. Protocol could be interpreted as any official set of rules describing the interaction between the elements of a computer system. Protocols always have theoretical rules governing their usage which can either refer to their official description or the practical usage of this protocol. Hence any use of this protocol outside the defined area can be considered as a protocol anomaly. Protocol anomaly filters are able to detect all attacks by deviations from the theirs normal usage.

In protocols, in networks often certain events must take place at certain times so many protocol anomaly detectors are built as *state* machines [6]. Each state corresponds to a part of the connection, such as a server waiting for a response from client. The transitions between the states describe the legal and expected changes between states. Besides, Z. Shan et. al. [7] uses network state based model approach to describe intrusion attacks. This model uses finite automata theory and can detect unknown attacks, the attacks and intrusions are described by the states and state transitions of network protocols and operating systems.

In the *transaction* based approach the "positive" behaviour is formally described. The desired actions and sequence of actions are specified by the definition of transactions. Such explicit definition makes the transaction an integral part of security policy. Transactions are a well known concept originating from the field of database management systems and now widely applied in other environments. In the research proposed by R. Buschkes et. al. [8] the detection of anomalies is based on the definition of correct transactional behaviour. This definition of correct and desired behaviour defines the system's multi-level security policy, which is monitored during runtime. The system is monitored only for potential conflicts.

The *learning* model must be trained on the specific network. In the training phase, the behaviour of the system is observed and logged and machine learning techniques are used to create a profile of normal behaviours. In the process of creating an effective anomaly detection model, *rulebased*, *model-based*, and *statistical-based* approaches have been adopted to create the baseline profile.

Rule-based systems used in anomaly detection describe the normal behaviour of users, networks and/or computer systems by a set of rules. These predefined rules typically look for the high-level state change patterns observed in the audit data. The *expert* system is an extension of rule-based systems. The system state is represented in a *knowledge base* consisting of a *fact base* and a *rule base*. A fact base is a collection of assertions that can be made on accumulated data from the audit records or directly from system activity monitoring. The rule base contains the rules that describe known intrusion scenario or generic techniques. When a pattern of a rule's antecedent matches the asserted fact, a rule-fact binding is created. After this binding is made, if all the patterns of the rule have been matched, then a binding analysis is performed to make sure that all the associated variables with the rule are consistent with the binding.

SRI International's Next-generation Intrusion Detection Expert System (NIDES) [9] is an innovative statistical algorithm for anomaly detection, and an expert system that encodes known intrusion scenarios. The features considered by NIDES are related to user activity, for instance CPU usage and file usage. The usage rate or intensity features derived are used to match with the long term profile and deviations for the system are learned and then summarized in a chi-squared statistic. A variant of incorporating expert system in anomaly detection is presented in [10]. Owens et. al. presented an adaptive expert system for intrusion detection based on fuzzy sets. The Fuzzy set theory permits the gradual assessment of the membership of elements in a set and it is described with the aid of a membership function valued in the real interval 0 and 1. Therefore, this system has the ability to adapt to the type and degree of threat, and it is relatively simple to implement it in anomaly detection system which has a high degree of uncertainty and ambiguity.

Model based anomaly detector models intrusions at a higher level of abstraction than audit records in rule based approach. It restricts execution to a pre-computed model of expected behaviour. In this approach more data can be processed, because the technique allows to focus on some data. More intuitive explanations of intrusion attempts are possible and system can predict the intruder's next action. A challenging task is to construct a model well balancing needs of detection ability and efficiency. Many researchers used different types of models to characterize the normal behaviour of the monitored system like *data mining*, *neural networks*, *pattern matching, etc.* to build predictive models.

In *data mining* approach models are automatically extracted from a large amount of data. Current intrusion detection system requires frequent adaptation to resist to new attacks so data mining techniques that can adaptively build new detection models are very useful. The basics idea is to extract an extensive set of features that describes each network, connection, or session, and apply a data mining program to learn rules that accurately capture the behaviour of intrusions and normal activities. An example of data mining system was proposed by Lee et. al. and is presented in [11]. The key idea is to mine network audit data, then use the patterns to compute inductively learned classifiers that can recognize anomalies and known intrusions.

Neural network is trained on a sequence of information, which may be more abstract than an audit record. Once the neural net is trained on a set of representative command sequences of a user, the net constitutes the profile of the user, and the fraction of incorrectly predicted events then measures, the variance of the user behaviour from his profile. Kohonen networks used to detect anomalies in information system is described in [12]. There are several libraries to built neural networks e.g. [13, 14, 15]. Other neural networks based systems for intrusion detection are described in [16,17,18,19].

In *pattern matching* approach, learning is used to build a traffic profile for a given network. Traffic profiles are built using s features such as packet loss, link utilization, number of collisions. Normal behaviour is captured as templates and tolerance limits are set, based on different levels of standard deviation. These profiles are then categorized by time of day, day of week and so on. When newly acquired data fails to fit within some confidence interval of the developed profiles then an anomaly is pointed out. The efficiency of the pattern matching approach depends on the accuracy of the profile generated. For a new network, significantly time may be necessary to build traffic profiles and this method might not scale well with the evolving network topologies and traffic conditions. The usage of emerging patterns in anomaly detection is described in [20].

First *statistical based* anomaly detection was proposed by Denning and Neumann [21] in 1985. The anomaly detector observes subjects and generates profiles for them that represent their behaviour. These profiles should use little memory to store their internal state, and be efficient in updating because every profile may potentially be updated for every audit record. System periodically generates a quantitative measure of the normal profile. The well known techniques in statistics can often be applied; e.g. data points that lie beyond a multiple of the standard deviation on either side of the mean might be considered anomalous. There are many statistical technique like *Bayesian statistics*, *covariance matrices and Chi-square statistics* [22] for the profiling of anomaly detection. Example of *Chi-square statistic* in anomaly detection is described in [23]. Statistical approaches disadvantage is the insensitivity to the order of occurrence of events. Sequential interrelationships among events should be considered for more accurate detection. It is also difficult to determine a threshold above which an anomaly should be considered.

Commercial products applying rule based, statistical based, model based and neural networks approach to detected anomalies (known till 2008) are briefly described in [3].

3 Research Model and Environment

It was not possible to detect anomalies on real SOA system so a special system was implemented. The business idea of this system VTV (Virtual TV) is presented in Fig. 2. The exemplary company is a virtual TV provider . Company does not have its own technical infrastructure and is associating TV digital provider with

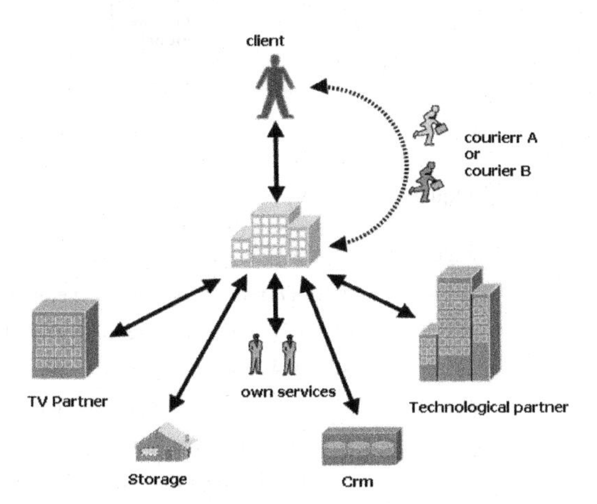

Fig. 2 The business relations of an exemplary company

the telecommunication operator. The receiving equipment is delivered to the client by one of courier companies. The company is also using two applications: CRM containing all clients data and a storage management system.

The VTV system simulates a real SOA system and enables to inject typical anomalies into its regular operation. Configurable are frequencies of :

- single services calls,
- group of services calls,
- processes calls,
- services in a context.

3.1 Architecture of VTV System

The architecture of VTV system is presented in Fig. 3. The *request generator* simulates activities of clients. It generates different types of request (e.g. create, modify, deactivate) for available services (e.g. TV or hardware). Depending of the value of request some requests can be identified as VIP. The generated address of a client determines the currier group. The configurable parameters of the request generator shown in Table 1, enable to simulate real operation and to inject some anomalies.

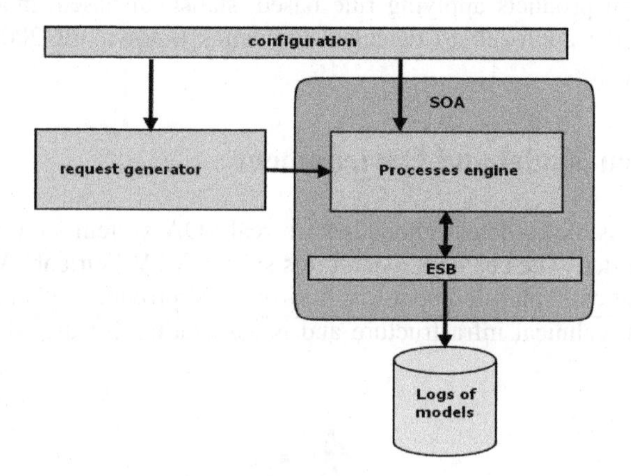

Fig. 3 Architecture of VTV system

Table 1 Configuration of the request generator

parameter	range	description
Create_share	<0:1>	CREATE requests share in generated requests
Modify_share	<0:1>	MODIFY requests share in generated request
Deactivate_share	<0:1>	DEACTIVATE requests share in generated request
Tv_share	<0:1>	Request for TV service
Hardware_share	<0:1>	Request for hardware service
courrerABprobability	<0:1>	Probability of the usage of A courier
sleeptime	<0:100000>	Sleep time between request generation

The generated request is transferred to the *business processes engine* (Fig.3) which composes processes from services available on ESB bus. Outputs of processes are logs. Logs of model contain information similar to logs from monitoring real SOA systems.

3.2 Main Process

In Fig. 4 the main process of integration layer is presented in BPMN notation [24]. This process exists in three versions: service creation, modification and deactivation. The graphs of these processes are similar but the probabilities of the execution of sub-process can differ. The probabilities of the execution of three configurable processes are given in configuration file:

1. *CourierPartnerConfig* – process is responsible for the delivery, by a courier company, to the client, necessary hardware. The rule decides to call the process if in a request HARDWARE service is present.
2. *TechPartnerConfig* – configuration of the technological partner, shown in Fig. 5. Process communicates with technological partner. The rule decides to call the process if in a request HARDWARE service is present.
3. *TvPartnerConfig* - configuration of TV partner. Process communicates with TV partner. The rule decides to call the process if in a request TV service is present.

The remaining five sub_processes are not configurable directly (Fig. 4) :

1. *FeasibilityStudy* – process is called for each request, shown in Fig.6.
2. *CrmConfig* – the configuration of application managing clients data. Process is called if the request is feasible.
3. *Activation* – service activation. Process is included for successful request elaboration.
4. *ClientCompensation* – process is called if the request is infeasible, parameter *feasibilityRate* defines the number of requests accepted in *FeasibilityStudy*.
5. *ErrorHandling* – process is called if an error appeared during the elaboration of a request. Information about the error can be generated by each service in the model. The probability of an error in a service is set in configuration.

All the above listed processes are described in details in [25].

3.3 Environment of Experiment

All experiments were conducted on PC with Intel Core 2 Duo T7200, 512 Mb of memory under Fedora Core operating system . The examined algorithms are using text input files prepared from logs of the VTV system (Fig.3). In this log file information like number of request, name of process, name of service called, execution time are written. These logs files are then processed by scripts, implemented in R [26] environment into transactions (example in Fig.7.) or summarized reports (example in Fig. 8.) and written into files used by detection algorithms. The flow of data from logs of the model is shown in Fig. 9.

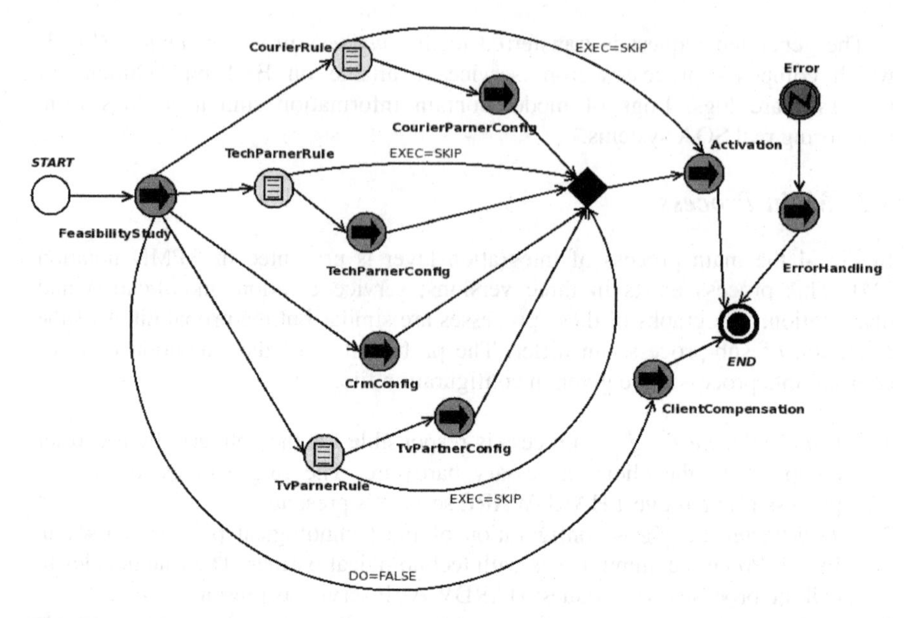

Fig. 4 Main process

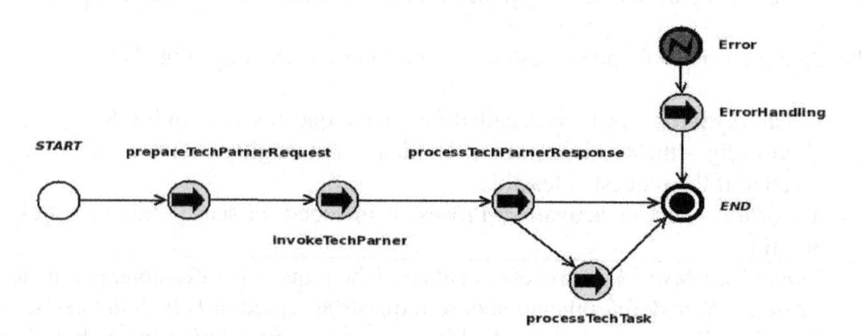

Fig. 5 Process *TechPartnerConfig*

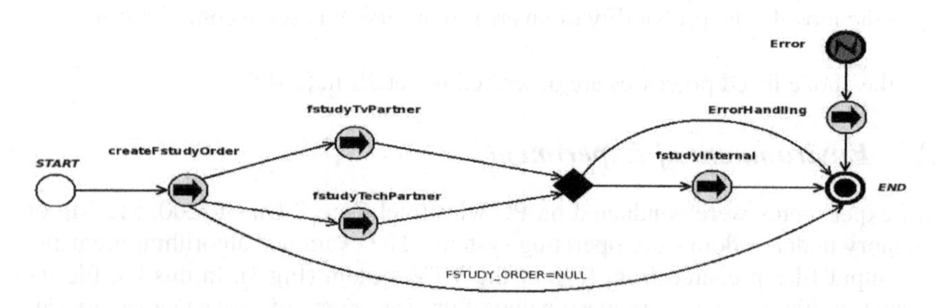

Fig. 6 Process *FeasibilityStudy*

```
1       " createFStudyOrder " , " fstudyTvPartner " , " fstudyTechParner " ,
        " prepareTvPartnerRequest " , " invokeTvPartner " ,
        " processTvPartnerResponse " , " storeClientData " ,
        " getStandardOffer " , " sendStandardOffer " ,
5       "magazineOrder " , " courierASystemOrder " ,
        " courierASystemResponse " , " commitTvPartner " ,
        "commitCrm" , " sendNotification " ;
        " createFStudyOrder " , " fstudyTvPartner " , " fstudyTechParner " ,
10      " prepareTvPartnerRequest " , " invokeTvPartner " ,
        " processTvPartnerResponse " , " storeClientData " ,
        " getStandardOffer " , " sendStandardOffer " , "magazineOrder " ,
        " courierASystemOrder " , " courierASystemResponse " ,
        " commitTvPartner " , "commitCrm" , " sendNotification " ;
```

Fig. 7 Logs in transactional format

```
1       " createFStudyOrder ":0.089
        " fstudyTvPartner ":0.065
        " fstudyTechParner ":0.056
        " prepareTvPartnerRequest ":0.064
```

Fig. 8 Logs processed into a summary report.

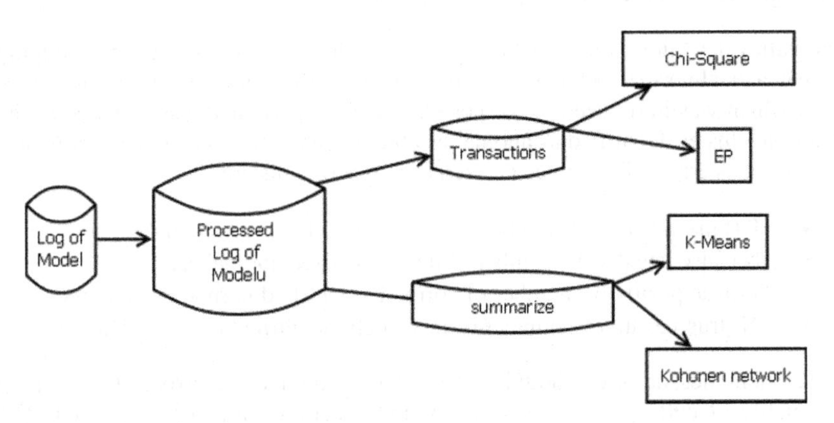

Fig. 9 Flow of data from logs

The implementation of algorithms for anomaly detection *k-means*, *Chi-Square* and *Kohonen* networks algorithms was made in R environment while the *emerging patterns* algorithm, which is more complicated, was implemented in C++.

k-means algorithm was prepared based on [27], for *emerging patterns* algorithm information from [20, 28] were used, *Chi-Square* detector was taken from [23].

4 Experiment

The research system presented in section 3 was used to explore four cases typical for SOA systems i.e.:

1. change in the frequency of service call,
2. change in the frequency of a group of services,
3. change of the context of services calls,
4. not used functionality.

For each of the above listed cases it was expected, that anomalies detector provides information useful for the maintenance team. Four learning algorithms (section 2) were used in the anomalies detector:

- *Chi-square statistic*
- *Kohonen network*
- *Emerging patterns*
- *k-means clustering*

Each of the above algorithms represents different approach to anomalies detection. The goal of experiment was to examine advantages and disadvantages of each of these algorithms.

4.1 Quality Evaluation of Algorithms

Anomalies detection is a kind of clustering with two types of clusters grouping normal and abnormal behaviours. Correctly identified anomaly is an abnormal event which was introduced by purpose by case' scenario and was assigned to the abnormal cluster. Identified as anomalies other events or not recognized anomalies are treated as errors. The values measured in experiments are:

- FP (false positive) – number of incorrectly identified anomalies,
- FN (false negative) - number of not recognized anomalies,
- TP (true positive) - number of correctly identified normal behaviours,
- TN (true negative) - number of incorrectly identified normal behaviours.

Good anomalies detector should generate small number of errors. To compare the quality of detectors often sensitivity and specificity measures are used. The sensitivity and specificity are defined accordingly as:

$$sensitivity = \frac{TP}{TP + FN} \tag{1}$$

and

$$specificity = \frac{TN}{TN + FP} \tag{2}$$

The relation between specificity and sensitivity – ROC (Receiver Operating Characterisctics) curve [29] can be used to compare the quality of models. ROC as a technique to analyze data was introduced during the second world war to identify if signal seen on a radar comes from an enemy, an alliance or it is noise. Currently ROC curves are used to analyze different types of data e.g. radiological data.

The construction of ROC curves during experiments was as follows:

1. create entities with algorithm' parameters
2. for each entity :
   - perform experiment
   - calculate specificity and sensitivity
   - mark the point on the diagram
3. draw the line connecting points.

For each examined algorithm also the learning time was calculated.

4.2 Plan for Experiment

The examination of anomaly detection algorithms was based on four test cases: changes in frequency of service and groups of services calls, change of the context of services calls, and lacking functionality. Each of these cases simulates one type of anomaly typical for SOA.

The experiment was conducted in following steps:

1. Create data for regular behaviour
2. For each of test case' scenario :
   - Create data for abnormal behaviour in this scenario
   - For each algorithm execute:
     - Using regular data perform the learning phase
     - Perform detection phase on data for abnormal behaviour
     - Evaluate the quality of detection
3. Compare algorithms.

Below the results for the first examined scenario "*change of the frequencies of service calls*" are presented. In some SOA systems e.g. telecommunications or transactional in in some time periods, the frequencies of services is almost stable. Changes in frequencies in such systems may indicate an anomaly, important for the maintenance team. Such anomaly can be caused by e.g. are errors in the configuration or inefficient allocation of resources. In our research model the changes were applied to two independent services. The frequency of calls of one process *fstudyInternal* in *FeasibilityStudy* process (Fig. 6) was increased (the probability of call of this service was raised from 0.2 into 0.8.). The frequency of calls of one service *processTechTask* in process *TechPartnerConfig* (Fig.5) was decreased by decreasing its probability call from 0.5 into 0.2.

4.3 Results of Experiments

The goal of examination was to find high level of detection with minimal number of false alarms. If the ideal detection was not possible preferred were the results with no false alarms. In practice, if system is generating many false alarms the user will neglect any alarm.

k-Means Algorithm

k-means clustering [27] is a clustering analysis algorithm that groups objects based on their feature values into *k*disjoint clusters. Objects that are classified into the same cluster have similar feature values. *k*is a positive integer number specifying the number of clusters, and has to be given in advance. All object are assigned to their closest cluster according to the ordinary Euclidean distance metric.

At the beginning the summarized reports used by algorithm were created for all logs in the system. The results are given in Table 2. High sensitivity (1) values were obtained for many false alarms. The Euclidean distance metric is used to calculate differences between summarized reports, elements close to each other are assigned to the same clusters. Such approach does not take into account the independence of events. E.g. two summarization with the same number of calls of service *fstudyInternal* may be assigned to different clusters because may have different characteristics of VIP accounts. Number of independent functionalities is usually significant so tuning the algorithm for real system may be impractical.

When the summarized reports were prepared separately for processes the results improved. The ROC curve for k=4 and 150 training logs is shown in Fig.10. Point A represents configuration detecting all anomalies but with false alarms constituting 25% of all anomalies. In point B number of false alarms is zero but the sensitivity of detection is only 0.6. Point C is optimal, with the sensitivity= 0.8 and specificity= 0.87, the costs of false qualification of anomaly and false qualification of normal behaviour are equal.

Emerging Patterns Algorithm

Emerging patterns (EP) are defined as item-sets whose supports increase significantly from one dataset to another. EP can capture emerging trends over time or useful contrasts in data sets. EP are item-sets whose growth rates the ratios of the two supports are larger than a given threshold. EP algorithm is working on transactions (e.g. in Fig 8) obtained from the monitored system and containing names of services. The first attempts to use EP approach, based on [20,28], showed that many pattern were discovered, some of them with low value of score given in formula (3).

Table 2 Results for clustering algorithm

k	logSize	maxL	TP	TN	FP	FN	Sensitivity	Specificity
6	50	3	44	1	43	0	1	0.02
6	50	3.5	32	13	31	12	0.7	0.3
7	50	3	44	4	40	4	0.92	0.09
7	50	3.5	30	18	26	14	0.68	0.41
7	50	4	18	22	22	26	0.41	0.5
7	80	3.5	26	1	26	1	0.96	0.04
7	80	5	21	14	13	6	0.78	0.52
6	250	11	7	6	3	2	0.78	0.67
6	250	15	3	6	2	7	0.3	0.75

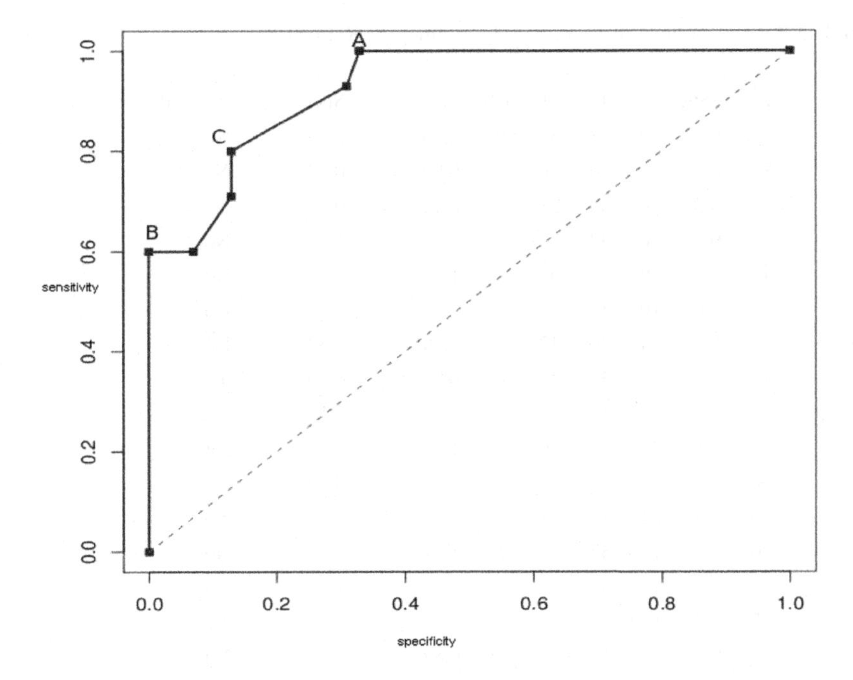

Fig. 10 ROC curve for k-means algorithm, summarization for each process

$$score = \frac{\frac{\sup p_c(A)}{\sup p_{\sim c}(A)}}{1+\frac{\sup p_c(A)}{\sup p_{\sim c}(A)}} \sup p_c(A) \tag{3}$$

Where: $\sup p_c(A)$ - support of pattern A in class *c*,

$\sup p_{\sim c}(A)$ - support of pattern A in negation of class *c*.

Majority of these pattern were characteristic for new data, potential anomalies. They would generate many false alarms. Filters in EP algorithms were added:

- Pattern with score values lower than threshold are not used in classification,
- Transactions with none class assigned to them are treated as normal.

After the above modification EP algorithm was able to give good anomalies' detection. The results are presented in Table 3. In all examined settings the parameter *growRatio* equals 1.85. The value of this parameter can be estimated if the characteristics of normal and abnormal behaviours are known. In training data support for calls of *internalFstudy* is increased twice. Support for call service *procesTechTask* is decreased in similar ratio. Hence the value of *growRatio*, minimal necessary increase of support, is 1.85.

Table 3 Results for emerging patterns algorithm, growRatio=1.85

lT	cSup	TP	TN	FP	FN	Sensitivity	Specificity	Anomaly 2
7	0.2	70	18	36	19	0.79	0.33	No
7	0.5	80	30	26	7	0.92	0.54	No
7	0.55	78	30	28	7	0.92	0.52	No
7	0.575	106	37	0	0	1	1	No
7	0.6	106	37	0	0	1	1	No
7	0.65	106	37	0	0	1	1	No
7	0.7	64	20	42	17	0.79	0.32	No
10	0.2	63	36	43	1	0.98	0.46	Yes
10	0.4	77	37	29	0	1	0.56	Yes
10	0.42	80	37	26	0	1	0.59	Yes
10	0.43	80	37	26	0	1	0.59	Yes
10	0.44	82	37	24	0	1	0.61	No
10	0.45	106	37	0	0	1	1	No
10	0.5	106	37	0	0	1	1	No
10	0.7	106	37	0	0	1	1	No
10	0.8	94	37	0	12	0.89	1	No
10	0.85	0	37	0	106	0	1	No
10	0.9	0	37	0	106	0	1	No
13	0.2	98	27	8	10	0.91	0.77	Yes
13	0.22	100	18	6	19	0.84	0.75	Yes
13	0.25	106	37	0	0	1	1	No
13	0.5	106	37	0	0	1	1	No

Emerging pattern algorithm was able to provide satisfactory detection of anomalies introduced in process *FeasibilityStudy* (service *internalFstudy).* Several setting were ideal. Unfortunately it was not possible to make the algorithm capable of finding both anomalies. In Fig 11 the ROC curve is shown. This curve was obtained for *lT* parameter equal to 10, this is appropriate to parameter *LogSize* = 150 in *k-means* algorithm.

Point A represents the ideal assignments of anomalies in process *FeasibilityStudy* and it was possible to obtain it for several settings. Unfortunately the anomalies in the second process were not detected for these settings. Point B was obtained for high value of *cSup*, equal to 0.8. The increase of the value of *cSup* results in zero values for sensitivity and specificity (point C). Pattern characterizing anomalies in our experiment have support less than 0.8.

Point D was created for low value of *cSup* (0.44), which caused the generation of many emerging patterns, then decreased the specificity to 0.61. Further decrease of *cSup* created point E and others. The decrease of minimal patterns' support was aimed to obtain the detection of the second anomaly (service *procesTechTask*). For *cSup* equal or less than 0.43 the second anomaly was detected.

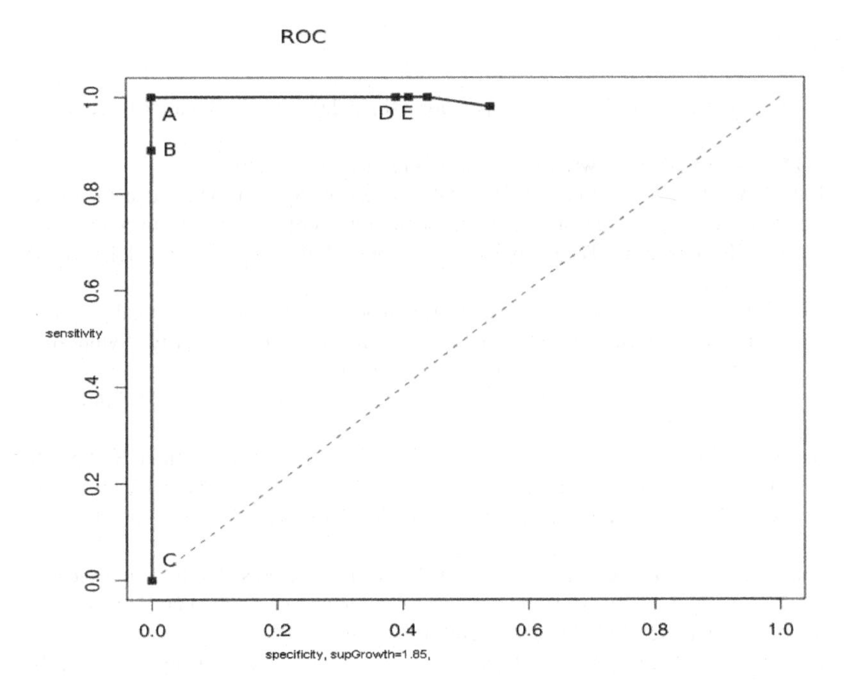

Fig. 11 ROC curve for EP algorithm

5 Conclusions

In this chapters some results of the experiment – detection of anomalies in SOA system are presented. The experiment was conducted in special environment built to introduce several types of anomalies, detect them and measure. The environment was described in section 3. In the detection of anomalies four learning algorithms, from different types (section 2): *emerging patterns*, *k-means* clustering, *Kohonen* networks and statistical *Chi-Square* were used. In this chapter the results of only two: the worst *(k-means* clustering) and the best one are shown, for one type of anomaly – the change in frequencies of services calls. The *emerging pattern* algorithm was able to produce the best quality of detection even for several settings. The results for two other algorithms (*Kohonen* networks and *Chi_Sqare*) in detecting the same anomaly are discussed in [25]. Statistical algorithm *Chi_Sqare* appeared to be inappropriate for the tested anomaly, Kohonen produced quite good results. The exemplary SOA system (section 2) enables to conduct other experiments examining the suitability of learning algorithms in the detection of other anomalies. The results of these experiments will be available very soon.

References

[1] BPEL Standard, http://docs.oasis-open.org/wsbpel/2.0/wsbpel-v2.0.html (access July 2011)

[2] SOA manifesto, http://www.soa-manifesto.org (access July 2011)

[3] Lim, S.Y., Jones, A.: Network Anomaly Detection System: The State of Art of Network Behaviour Analysis. In: Proc. of the Int. Conference on Convergence and Hybrid Information Technology 2008, pp. 459–465 (2008), doi:10.1109/ICHIT2008.249

[4] Ko, C., Ruschitzka, M., Levitt, K.: Execution monitoring of security-critical programs in distributed systems: a specification-based approach. In: Proc. of IEEE Symposium on Security and Privacy, Oakland, CA, USA (1997)

[5] Lemonnier, E.: Protocol Anomaly Detection in Network-based IDSs. Defcom white paper (2001)

[6] Sekar, R., Gupta, A., Frullo, J., Shanbag, T., Tiwari, A., Yang, H., Zhou, S.: Specification-based anomaly detection: A New Approach for Detecting Network Intrusions. In: ACM Computer and Communication Security Conference, Washington, DC, USA (2002)

[7] Shan, Z., Chen, P., Xu, Y., Xu, K.: A Network State Based Intrusion Detection Model. In: Proc. of the 2001 International Conference on Computer Networks and Mobile Computing, ICCNMC 2001 (2001)

[8] Buschkes, R., Borning, M., Kesdogan, D.: Transaction-based Anomaly Detection. In: Proc. of the Workshop on Intrusion Detection and Network Monitoring, Santa Clara, California, USA (1999)

[9] Anderson, D., Frivold, T., Valdes: A Next-generation Intrusion Detection Expert System, NIDES (2005)

[10] Owens, S., Levary, R.: An adaptive expert system approach for intrusion detection. International Journal of Security and Networks 1(3-4) (2006)

[11] Lee, W., Stolfo, S.J.: Data mining approaches for intrusion detection. In: Proc. of the 7th USENIX Security Symposium (1998)

[12] Bivens, A., Palagrini, C., Smith, R., Szymański, B., Embrechts, M.: Network-based intrusion detection using neural networks. In: Proc. Intelligent Eng. Systems through Neural Networks, ANNIE 2002, St. Louis, MO, vol. 12, pp. 579–584. ASME Press, NY (2002)

[13] C Neural network library, http://franck.fleurey.free.fr/NeuralNetwork/

[14] NeuroBox, http://www.cdrnet.net/projects/neuro/

[15] Fast Artificial Neural Network Library, http://sourceforge.net/projects/fann/

[16] Ryan, J., Lin, M., Miikkulainen, M.: Intrusion Detection with Neural Networks. In: Advances in Neural Information Processing Systems, vol. 10 (1998)

[17] Ghosh, A.K., Schwartzbard, A.: A Study in Using Neural Networks for Anomaly and Misuse Detection. In: Proc. of the 8th USENIX Security Symposium, Washington, D.C., USA (1999)

[18] Han, S.-J., Cho, S.-B.: Evolutionary Neural Networks for Anomaly Detection Based on the Behaviour of a Program. IEEE Transactions on Systems, Man and Cybernetics (2006)

[19] Bivens, A., et al.: Network-based intrusion detection using neural networks. In: Proc. of Intelligent Engineering Systems through Artificial Neural Networks, ANNIE 2002, St.Luis, MO, vol. 12, pp. 579–584. ASME press, New York (2002)

[20] Ceci, M., Appice, A., Caruso, C., Malerba, D.: Discovering Emerging Patterns for Anomaly Detection in Network Connection Data. In: An, A., Matwin, S., Raś, Z.W., Ślęzak, D. (eds.) Foundations of Intelligent Systems. LNCS (LNAI), vol. 4994, pp. 179–188. Springer, Heidelberg (2008)
[21] Denning, D., Neumann, P.: Requirements and Model for IDES-A Real-Time Intrusion-Detection Expert System. SRI Project 6169, SRI International, Menlo Park, CA (1985)
[22] Masum, S., Ye, E.M., Chen, Q., Noh, K.: Chi-square statistical profiling for anomaly detection. In: Proceedings of the 2000 IEEE Workshop on Information Assurance and Security (2000)
[23] Ye, N., Chen, Q.: An anomaly detection technique based on a chi-square statistic for detecting intrusions into information systems. Qual. Reliab. Engng. Int. 17, 105–112 (2001)
[24] http://www.bpmn.org/
[25] Tarka, M.: Anomaly detection in SOA systems. Msc Thesis, Institute of Computer Science, Warsaw University of Technology (2011)
[26] The R Project for Statistical Computing, http://gcc.gnu.org/ (access September 2011)
[27] Munz, G., Li, S., Carle, G.: Traffic Anomaly Detection Using K-Means Clustering, Wilhelm Schickard Institute for Computer Science, University of Tuebingen (2007)
[28] Guozhu, D., Jinyan, L.: Efficient Mining of Emerging Patterns: Discovering Trends and Differences. Wright State University, The University of Melbourne (2007)
[29] Hanley, J.A.: Receiver operating characteristic (ROC) methodology: the state of the art. Crit. Rev. Diagn. Imaging (1989)

Service Availability Model to Support Reconfiguration

Dariusz Caban and Tomasz Walkowiak

Abstract. Web based information systems are exposed to various dependability issues during their lifetime. Reconfiguration is then used by the administration to ensure continuity of service. Such forced reconfigurations can cause unforeseen side-effects, such as server overloading. To prevent this, it is proposed to use simulation techniques to analyze the reconfigurations and construct a safe reconfiguration strategy. Extensions to the available network simulation tools are proposed to support this. The authors present the results of multiple experiments with web-based systems, which were conducted to develop a model of client-server interactions that would adequately describe the relationship between the server response time and resource utilization. This model was implemented in the simulation tools and its accuracy verified against a testbed system configuration.

1 Introduction

Whenever a web based information system experiences some dependability issue, caused by a hardware failure, a software error or by a deliberate vulnerability attack, the administrator is faced with the difficult problem, how to maintain the continuity of critical business services. Isolation of the affected hardware and software is usually the first reaction (to prevent propagation of the problem to yet unaffected parts of the system). It then follows that the most important services have to be moved from the isolated hosts/servers to those that are still available. This is achieved by system reconfiguration [1, 2].

Redeployment of service components onto the available hosts changes the workload of the various servers. In consequence some of them are over-utilized and cannot handle all the incoming requests, or handle them with an unacceptable response delay. It is very difficult to predict these side-effects. One of the feasible approaches is to use simulation techniques [3]: to study what are the possible

Dariusz Caban · Tomasz Walkowiak
Wrocław University of Technology, Wybrzeże Wyspiańskiego 27, 50-320 Wrocław, Poland
e-mail: dariusz.caban@pwr.wroc.pl, tomasz.walkowiak@pwr.wroc.pl

W. Zamojski et al. (Eds.): Complex Systems and Dependability, AISC 170, pp. 87–101.
springerlink.com

effects of a change of system configuration. Available network simulators are usually capable of analyzing the impact of reconfiguration on the proper functioning of the services, the settings of the network devices and on security.

The network simulators as a rule can predict transmission delays and traffic congestions – that is natural, since it is their primary field of application. They have a very limited capability to simulate tasks processing by the host computers. This can be modified by developing some simulator extensions – models that provide processing delays dependent on the number of concurrently serviced requests [8, 9].

Accurate prediction of the response times in a simulator is in general quite unlikely: there are too many factors that can affect it. Moreover, a lot of these factors are unpredictable, being specific to some algorithm or unique software feature. This can be overcome in case of predictions made for the purpose of reconfiguration. A lot of system information can be collected on the running system prior to reconfiguration (by analyzing its performance). This information can be used to fine tune the simulation models.

2 System Model

The paper considers a class of information systems that is based on web interactions, both at the system – human user (client) interface and between the various distributed systems components. This is fully compliant with the service oriented architecture, though it does not imply the use of protocols associated with SOA systems. The system is described at 3 levels [1]. On the high level, it is represented by the interacting service components. At the physical layer, it is described by the hosts, on which the services are deployed, and by the visibility and communication throughput between them (provided by the networking resources). The third element of the system description is the mapping between the first two layers.

2.1 Service Model

The system is composed of a number of service components. Interaction between the components is based on the client-server paradigm, i.e. one component requests a service from some other components and uses their responses to produce its own results, either output to the end-user or used to respond to yet another request. The client (user) requests are serviced by some components, while others may be used solely in the background.

A service component is a piece of software that is entirely deployed on a single web node (host) and all of its communication is done by exchange of messages. The over-all description of the interaction between the service components is determined by its choreography, i.e. the scenarios of interactions that produce all the possible usages of the system [9].

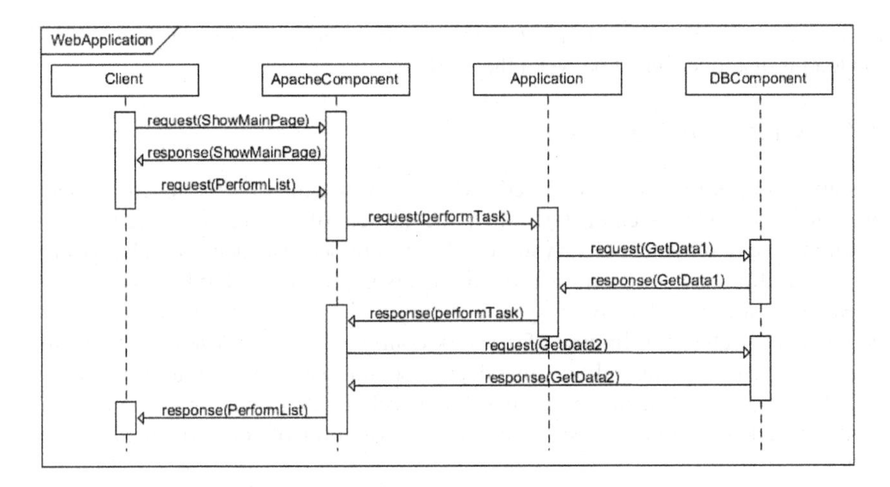

Fig. 1 An UML diagram representing a simple service choreography

A very simple choreography description is given in Fig.1 for illustration purposes. It represents a very common dynamic web service architecture based on an Apache server with Tomcat servlet container and a back-end SQL database. The system serves static HTML pages (e.g. MainPage) and information requiring computation and database access (e.g. PerformList).

The service components interact in accordance with the choreography. As the result, they generate demand on the networking resources and on the computational power of the hosts running the components.

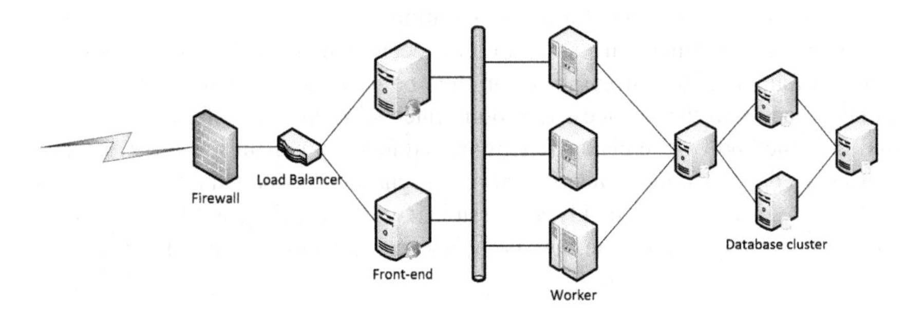

Fig. 2 A simple system infrastructure based on multiple levels of load balancing

2.2 Network Infrastructure

The service components are deployed on a network of computers. This underlying communication and computing hardware is abstracted as a collection of interconnected computing hosts (or server software deployed on the hosts to run the service components). Fig. 2 presents a possible network that may be used to provide the services described in Fig. 1. The configuration encompasses a load balancer, used to distribute requests between the front-end hosts. At the back-end the load is also distributed between the workers.

2.3 System Configuration

System configuration is determined by the deployment of service components onto the hosts. This is characterized by the subsets of services deployed at each location. The deployment clearly affects the system performance, as it changes the communication and computational requirements imposed on the infrastructure.

Reconfiguration (change of system configuration) takes place when service deployment is changed. In case of network configurations that use load balancing to improve service availability, some degree of automatic reconfiguration is introduced by the infrastructure. Load balancing techniques usually implement some fault fallback mechanisms, which prevent distribution of load to failed worker nodes.

More sophisticated reconfiguration can be achieved by redistribution of tasks to be performed by each worker host. This is fairly easily achieved by reconfiguring the front-end servers, responsible for workload distribution among the worker nodes.

3 System Reconfiguration

Mostly, reconfiguration occurs as a routine procedure of system maintenance. Services are redeployed when the administrators observe that some hosts are either overloaded or under-utilized. This is usually done in a leisurely manner, planned and tested well in advance. As such, it does not require any tools or support specifically designed for reconfiguration.

A much more demanding situation can occur when the system experiences some malfunction. Reconfiguration can then be used to overcome its effects and maintain access to the services. It is quite interesting that reconfiguration can be used as a method of exercising functional redundancy existing in the system, to improve its dependability. This type of reconfiguration is done in a hurry, to bring up the system service as quickly as possible. Consequently, it is likely that some side-effects of reconfiguration may be overseen, especially if these are connected with the system performance.

3.1 System Faults

System reconfiguration may be triggered by a wide class of adverse events. We consider various sources of system faults [1, 2]: transient and persistent hardware faults, software bugs, human mistakes and exploitation of software vulnerabilities. There are also attacks on services, based on draining their limited resources, e. g. DOS attacks.

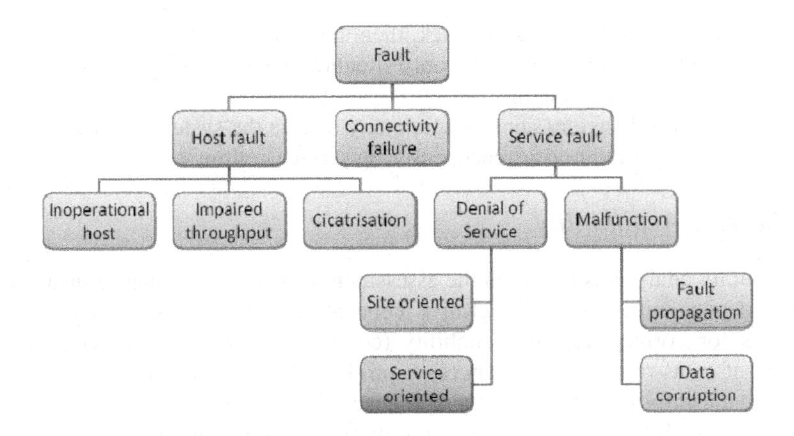

Fig. 3 A classification of system faults reflecting their impact on reconfiguration

In the considered approach to complex information systems modeling, the hosts are the basic components of the system infrastructure. Thus, all the faults are attributed to them (and not to hardware or software components). This is the basis for the classification of faults presented in Fig. 3.

Inoperational host fault – the host cannot process services that are located on it, these in turn do not produce any responses to queries from the services located on other nodes. Reconfiguration is required, all the service components have to be moved to unaffected hosts.

Impaired performance fault – the host can operate, but it cannot provide the full computational resources, causing some services to fail or increasing their response time above the acceptable limits. Some service components may tolerate the fault, especially if other ones are moved elsewhere.

Connectivity fault – the host cannot communicate with other hosts with the required throughput. In effect, the node may become unreachable, requiring service redeployment.

Service malfunction fault – the service components located on the node can produce incorrect or inconsistent responses due to accumulated software errors, effects of transient malfunctions and possible exploitation of vulnerabilities. The operation of services located at such nodes becomes unpredictable and potentially dangerous to services at other ones (that communicate with them). Thus, the fault may propagate to other connected hosts and services. It is advisable to isolate the affected service to prevent problem escalation.

DOS fault – a special case of a service fault, where the service component loses its ability to respond to requests. It is usually caused by some exploitation of security vulnerability, often a proliferation of bogus service requests that lock up all the server resources. This is a common consequence of insufficient security measures (firewall rules, antivirus software, etc.). A very important aspect of this class of faults is that the attack may be either IP site locked or service locked. Reconfiguration is effective only in the first case: moving the affected services to other network addresses can prevent further damage. On the other hand, if a service is moved in case of a service locked attack, then the fault will also be propagated to the new location. In effect, this is the only situation when reconfiguration may increase the damage to the system.

Effectively, the change of configuration may restore the system functionality in almost every type of fault occurrence, ensuring service continuity.

3.2 Service Availability

Dependability analysis is based on the assessment of some performance measures. In fact, dependability is an integrative concept that encompasses: availability (readiness for correct service), reliability (continuity of correct service), safety (absence of catastrophic consequences), confidentiality (absence of unauthorized disclosure of information), integrity (absence of improper system state alterations), maintainability (ability to undergo repairs and modifications).

Any of the considered faults will cause the system to fail if/when they propagate to the system output (affecting its ability to generate correct responses to the client requests). This is best characterized by the availability function $A(t)$, defined as the probability that the system is operational (provides correct responses) at a specific time t. In stationary conditions, most interesting from the practical point of view, the function is time invariant, characterized by a constant coefficient, denoted as A.

The measure has a direct application both from the business perspective and from the administrator viewpoint. The asymptotic property of the steady-state availability A:

$$A = \lim_{t \to \infty} \frac{t_{up}}{t}, \tag{1}$$

gives a prediction of the total uptime t_{up}. This is a very useful business level measure. From the administrators' perspective, the asymptotic property may be further transformed, assuming a uniform rate of service requests [1]:

$$A = \lim_{t \to \infty} \frac{n_{ok}}{n} \tag{2}$$

This yields a common understanding of availability as the number of properly handled requests n_{ok} expressed as a percentage of all the requests n. Equations (1) and (2) are equivalent only if the operational system handles all the requests correctly.

Availability does not reflect the comfort of using the service by the end-users. This has to be analyzed using a different measure of the quality of service. The most natural is to use the average response time, i.e. the time elapsed from the moment of sending a request until the response is completely delivered to the client [8]. The mean value is calculated only on the basis of correctly handled response times. The error response times are excluded from the assessment (or assessed as a separate average).

3.3 System Reconfiguration Strategy

System reconfiguration is realized by changing the system from one configuration to another, in which all the services can still be delivered. There are various situations when a reconfiguration may be desirable, we concentrate on the dependability oriented reconfiguration. This implies that the reconfiguration is forced by the occurrence of a dependability issue, i.e. a fault occurrence which causes some services to fail in the current configuration. Reconfiguration is achieved by isolating the faulty hosts and servers, and then moving the affected services to other hosts.

The reconfiguration strategy should ensure that the target configuration improves the availability and response times of the services, as compared to the state, at which the system ends after a fault occurrence. The target configuration, if it exists, should ensure the following:

- It should be able to handle all the end client requests, i.e. it should not limit the system functionality.
- It should maintain the quality of service at the highest possible level, given the degraded condition of the infrastructure.

The first requirement is met if all the service components are deployed on unaffected hosts, they do not lead to compatibility issues with other components, and the communication resources ensure their reachability. Thus, it is a combinatorial problem of eliminating all inherently conflicting configurations. The set of permissible configurations can then be determined. Within these configurations, some are affected by a specific dependability issue. All the others are potential candidates for reconfiguration. If the resulting set is empty, then the considered faults cannot be tolerated and the system fails (reconfiguration cannot bring it up).

The reconfiguration strategy is constructed by choosing just one configuration from the set corresponding to the various dependability issues. Usually, there are numerous different reconfiguration strategies that can be constructed in this way. Any one of them will ensure the continuity of service. Optimal strategy is obtained by choosing the configuration that ensures the best quality of service, i.e. with the shortest average response times. This can be achieved if there is an efficient tool for predicting the service availability and response time. One of the feasible approaches is to use network simulation.

4 Network Simulation Techniques

There are a large number of network simulators available on the market, both open-source (ns3, Omnet+, SSFNet) and commercial. Most of them are based on the package transport model – simulation of transport algorithms and package queues [3]. These simulators can fairly well predict the network traffic, even in case of load balancing [6]. What they lack is a comprehensive understanding of the computational demands placed on the service hosts, and how it impacts the system performance. In effect they are useful to predict if a system configuration provides access by all the end-users to all the system functionality, i.e. if a configuration is permissible.

However, these network simulators cannot be directly used to develop or test a reconfiguration strategy, since they cannot predict the quality of service (availability and response times) of the target configurations. This is the consequence of the lack of models for predicting tasks processing time, based on resource consumption. The simulators need to be extended, by writing special purpose models to accommodate this functionality [8, 9].

Alternative approach is to test the target configurations on a testbed or in a virtual environment. In this case the software processing times need not be predicted: they result from running the production software on the testbed/virtual hosts. This approach has drawbacks: the time overhead of testing the target configuration may be inacceptable, considering that the time to react to a dependability incident is very limited. Furthermore, it is hardly feasible to perform the emulation on hardware, which provides similar level of computational power to the production system – thus, the results have to be scaled, which is a large problem.

Response time prediction in network simulators is based on the proper models of the end-user clients, service components, processing hosts (servers), network resources. The client models generate the traffic, which is transmitted by the network models to the various service components. The components react to the requests by doing some processing locally, and by querying other components for the necessary data (this is determined by the system choreography, which parameterizes both the client models and the service component models). The request processing time at the service components is not fixed, though. It depends on the number of other requests being handled concurrently and on the loading of other components deployed on the same hosts.

The network simulator has a number of parameters that have to be set to get realistic results. These parameters are attributed to the various models, mentioned above. In the proposed approach we assume that it is possible to formulate such (fairly simple) models describing the clients and service components, which will not be unduly affected by reconfiguration. Then, we can identify the values of the parameters on the production system. Simulating the target configuration with these parameters should provide reliable predictions of the effects of reconfiguration.

5 Modeling Client – Server Interaction

The basis of operation of all the web oriented systems is the interaction between a client and a server. This is in the form of a sequence of requests and responses: the client sends a request for some data to the server and, after some delay, the server responds with the required data. The most important characteristic of this interaction is the time needed by the server to respond (deliver the data to the client) – this is the response time that can be determined experimentally or estimated using the simulation techniques.

The response time depends on a number of different factors: the processing to be done at the server site, response time of other services that need to be queried to determine the response, etc. Even in a very simple situation, where the response is generated locally by the server, it usually has an unpredictable component (random factor). The understanding of these simple client-server interactions is paramount to building a simulation model that will be capable of analyzing more complex situations.

Actually, the server response time is strongly related to the client behaviour, as determined by the request-response interaction. Such factors as connection persistence, session tracking, client concurrency or client patience/think times have a documented impact on the reaction. For example, it has been shown in [5] that if user will not receive answer for the service in less than 10 seconds he or she will probably resign from active interaction with the service and will be distracted by other ones.

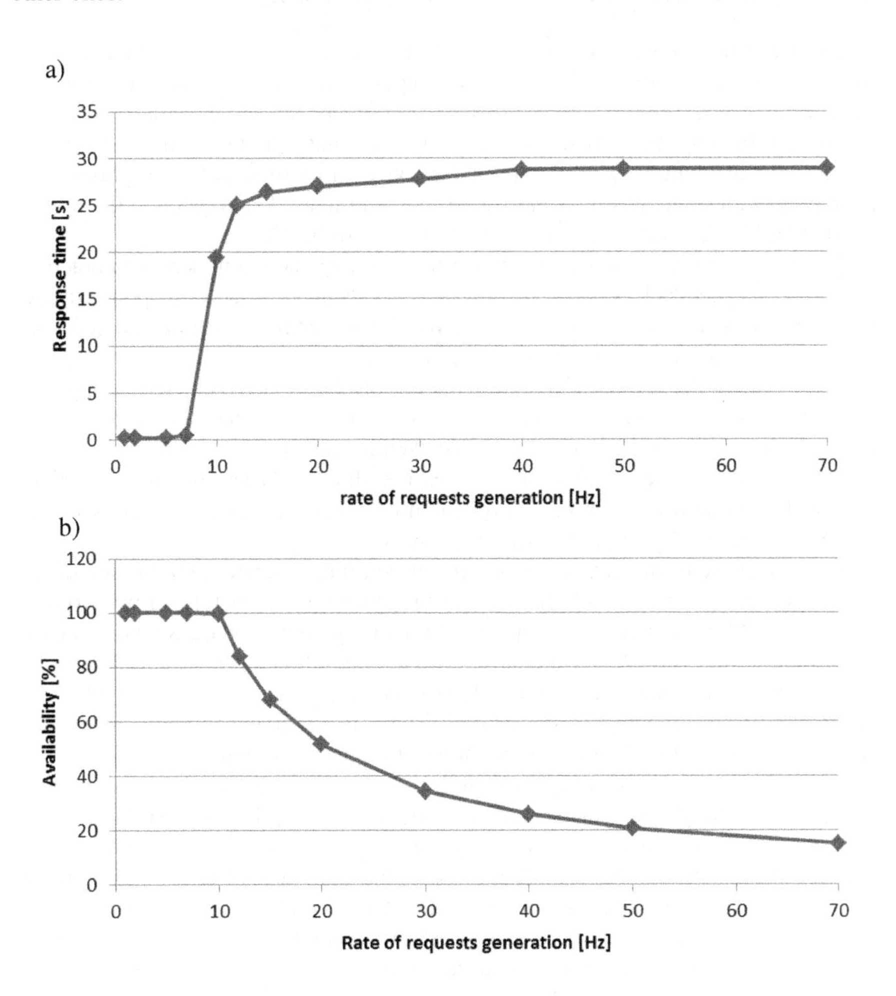

Fig. 4 The performance of an off-the-shelf web service under varying rates of incoming client requests: a) the upper graph shows the response time, b) the lower – service availability

Let's consider the models used in these simple interactions in more detail. For this purpose, we have set up a simple testbed, consisting of a virtual machine running an Apache server. The server hosts a PHP script application, on which we can accurately regulate the processing time needed to produce a result. This application is exposed to a stream of requests, generated by a choice of client applications (a Python script written by the authors, open source traffic generators such as Funkload and jMeter). Full control is maintained of the available processor resources (via the virtualization hypervisor). This ensures that the client software is not limited by insufficient processing capabilities, while the server resources are regulated to determine their impact.

5.1 Client Model Used in Server Benchmarking

The simplest model is adopted by the software used for server/service benchmarking, i.e. to determine the performance of computers used to run some web application. In this case, it is a common practice to bombard the server with a stream of requests, reflecting the statistics of the software usage (the proportion of the different types of requests, periods of burst activity, think times, etc.). Sophisticated examples of these models of client-server interaction are documented in the industry standard benchmarks, such as the retired SPECweb2009 [7].

The important factor in this approach to modeling the client-server interaction is lack of any feedback between the rate of requests and the server response times. In other words, the client does not wait for the server response, but proceeds to send further requests even if the response is delayed.

Fig. 4 shows the results of experiments performed on a typical server application exposed to this type of traffic. It should be noted that the results were obtained in the testbed, discussed above. While they reflect the normal server behaviour in such stress tests, the processing thresholds are much lower than expected in modern web servers. This should be expected, since the virtual server, being used, has a very limited processing power.

Fig. 4 a) presents the changes in the response time, depending on the rate of requests generation. It should be noted that the system is characterized by two distinct thresholds in the requests rate. Up to approximately 6 requests per second, the response time very slowly increases with the rate of requests. This is the range, where the server processing is not fully utilized: the processor is mainly idle and handles requests immediately on arrival. There is a gradual increase in the response time due to the increased probability of requests overlapping.

When the requests rate is higher than the underutilization threshold, the processor is fully utilized, the requests are queued and processed concurrently. The increase in the response time is caused by time sharing: it is proportional to the number of concurrently handled requests and the time needed to process a single one. This holds true, until the server reaches the second threshold – overutilization. This corresponds roughly to 12 requests per second in the presented Figure.

Above the overutilization threshold the server is no longer capable of handling the full stream of requests. In consequence, some requests are timed-out or rejected. Further increase in the request rate does not increase the number of concurrently handled ones. Thus, the response time remains almost constant. On the other hand, the percentage of requests handled incorrectly increases proportionately to the request rate. This is illustrated in Fig. 4 b).

In fact, there are also some further thresholds within the overutilization range. This is caused by the fact that there can be different mechanisms of failing to handle a request. Initially, connection time-out is the dominating factor in the studied servers (Apache, MySQL, simpleHTTPD). As the requests rate increases, rejects and exceptions become more common. This is omitted from the presented results, as it is assumed that the web based system should never be allowed in this range of request rates. Thus, there is no point in accurate modeling of these phenomena for the purposes of simulation. Rather, the simulator should flag the situations when the overutilization occurs.

In the underutilization range, another phenomenon can be observed. There is a very high dispersion of the response times for small request rates. This is caused by the phenomenon of server "warm-up". Requests are initially handled much more slowly. It is probably a side-effect of compiling scripts on the fly and server side caching. This impacts the performance in the underutilization range.

5.2 Client Models Reflecting Human Reactions

The real behaviour of clients differs significantly from the model discussed so far. In fact, the client sends a burst of related requests to the server, then it waits for the server to respond and, after some "think" time for disseminating the response, sends a new request. This implies that the request rate depends on the response time.

This type of model is implemented in a number of traffic generators available both commercially and open-sourced (Apache JMeter, Funkload). The workload is characterized by the number of concurrent clients, sending requests to the server. The actual requests rate depends on the response time and the think time.

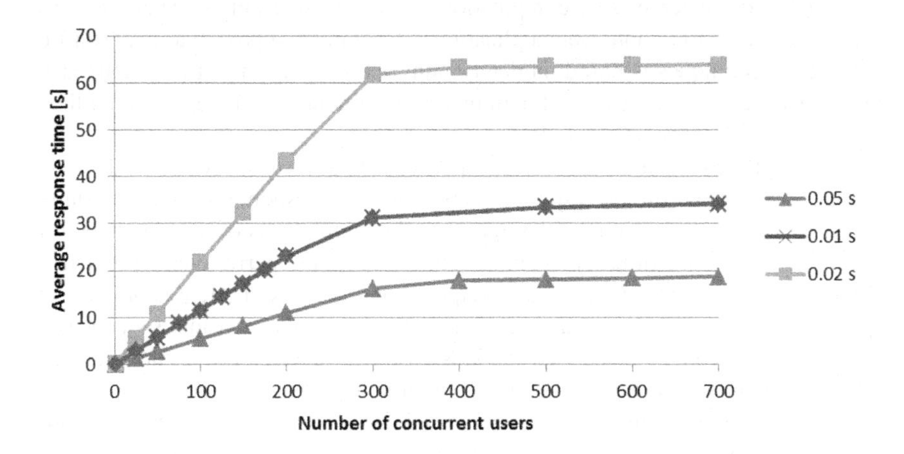

Fig. 5 Average service response when interacting with various number of concurrent clients, waiting for service response before issuing another request

Fig. 5 shows how the response time typically depends on the number of concurrent clients. In this case we have set the "think" time to 0, i.e. a new request is generated by the client directly on receiving the response to a previous one. Quite interestingly, the server operates practically only in the normal utilization range, until it reaches the maximum number of clients that it can handle correctly (roughly 300 clients in the considered testbed). Thereafter, increasing the number of clients (concurrent requests) leads to a commensurate increase in the number of request rejects (represented by the error responses).

For the purpose of correctly simulating this behavior, it is not enough to know the thresholds of under- and overutilization. It is also necessary to model the time of error responses. As commented in 5.1, in general this is very difficult since there are different mechanisms coming into play (time-outs, rejects triggered by hard-coded limits or by computing exceptions). A heavily over utilized server sends an unpredictable mix of error responses, some of them practically with no delay, others after a fixed delay time. In some cases the server becomes unstable and does not respond at all to some requests.

Performed experiments show that this behaviour occurs only in situations of heavy server overutilization. The dominating phenomenon, observed when the server load only slightly exceeds the overutilization threshold, is based on queuing the requests for a fixed time-period and error-responding thereafter. This behaviour, enhanced by flagging the state of server overutilization, is the basis of the proposed client-server interaction model in case of error responding. It is characterized by one parameter – the error response time.

5.3 Client Models Derived from Choreography Description

The client-server interaction model has to consider the various tasks initiated by the client. In a typical web application, these tasks can exercise the server resources in a wildly varied manner: some will require serving of static web pages, some will require server-side computation, yet others will initiate database transactions or access to remote web applications. A common approach to load (traffic) generation techniques is based on determining the proportion of the various tasks in a typical server workload, and then mixing the client models representing these tasks in the same proportion [4, 7].

This approach assumes that the proportion of tasks in a workload does not change significantly due to response delays and error-responding. It also assumes that it is possible to accurately classify the tasks on the basis of the observed traffic, a daunting problem that can significantly impact the performance prediction. Direct traffic analysis can distinguish requests on the basis of client addresses, response times, size of requests and responses, etc. It can also consider sequences of requests identified by connections and sessions. Traffic analysis does not yield any information on the semantics of client-server interactions, which should be the basis for determining the client models used for load generation. In effect, this produces a mix of tasks, in no way connected to the aims of the clients. It can be improved using the service choreography description.

It is assumed that the analyzed web service is described by its choreography description, using one of the formal languages developed for this purpose (we consider WS-CDL and BPEL descriptions). This description determines all the sequences of requests and responses performed in the various tasks, described in the choreography. This is further called the set of business tasks, as opposed to the tasks obtained from the classification of traffic. Traffic analysis can then be employed to classify the observed request-response sequences to the business tasks identified in the choreography description. This procedure determines the typical proportion of the various business tasks in the workload that is much less affected by the service response times or proportions of error responses.

An even better description of client behaviour can be achieved if we have a semantic model of client impatience, i.e. how the client reacts to waiting for a server response. Currently, in case of end-clients (human users of the service) this is modeled very simplistically by setting a threshold delay, after which the client stops waiting for the server response and starts over the requests sequence needed to perform a business task. A more sophisticated approach would have to identify the changing client perspective caused by the problems in accessing a service, e.g. a client may reduce the number of queries on products, before deciding to make a business commitment, or on the other hand, he may abandon the commitment. These decisions could significantly influence the workload proportions.

The same problem occurs during interactions between the web service components. In this case one component becomes the client of another. The same phenomena can be observed. The client component usually has a built-in response time-out period which corresponds to the impatience time. The significant difference is that, in this case, the choreography description defines the reaction of the client component. Thus, the client impatience model is fully determined, derived from this description.

5.4 Resource Consumption Model – Server Response Prediction

The client-server interaction is paramount to the proper simulation of a complex web service. The analysis of the behaviour of typical servers led to the formulation of a simplified model that is used in our analysis:

- The server response time is described by 3 ranges: a constant response time below the underutilization threshold, a linearly increasing response time in the normal operation range and a constant limit response time when the server is over utilized.
- If the model is to be used for determining the load limits, the response delay in the range below the underutilization threshold does not affect the results.
- If the model is to be used for determining the response time in the underutilization range, the warmup time has to be added. The model may be anyway inadequate, since this is not the application area that we are targeting.
- The model responds with error messages to some requests when the server is over utilized. The error response is always delayed by a random error delay time fixed to a constant average.

- Client is described by an impatience time delay, after which it assumes the server is not responding and continues as if it received an error message.

The deployment of multiple services on the same host leads to a time-sharing of processor time between them. This does not affect noticeable the thresholds for under- and overutilization of the services. Mainly, it changes the level at which the response time of the service stabilizes after the load exceeds the overutilization threshold. Further work is needed to observe the possible impact of service prioritization.

6 Dependability Analysis

The model proposed in 5.4 can be used to simulate all the interactions between the service components. This is the basis of the extended SSFNet simulation tool, used by us to predict the results of reconfiguration. The performance of this simulator is currently under study and the results are very promising. It is still too early to conclude, though, whether these models are sufficiently accurate in general.

a)

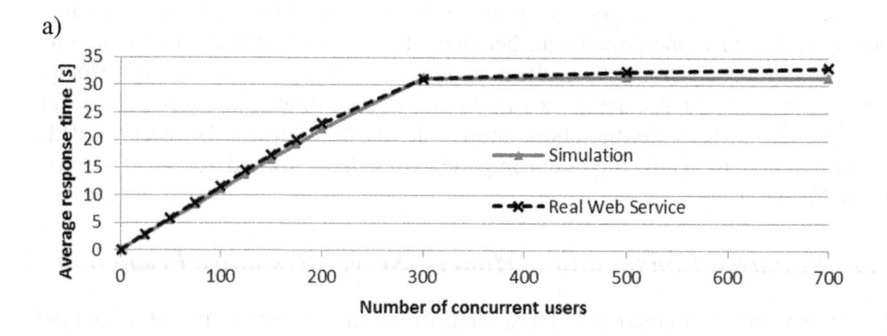

b)

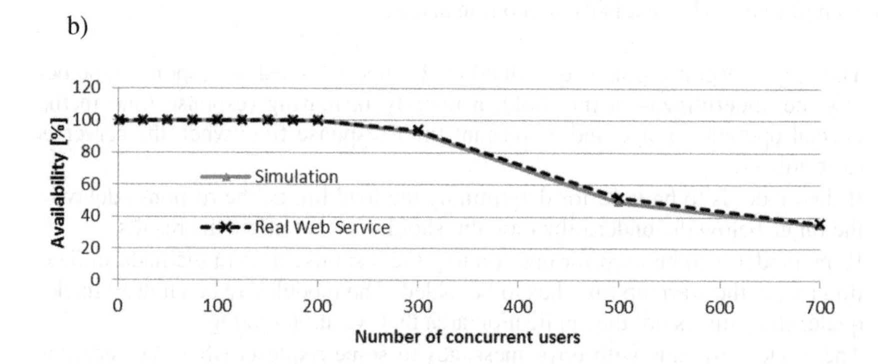

Fig. 6 The performance a real web service (dashed line) and simulated one (solid line): a) the upper graph shows the response time, b) the lower – service availability

As an illustration, let's consider the results of simulating the client – server interactions discussed in 5.2. The interaction model is based on the thresholds identified in 5.1. So, how do the simulation results bear out the response times observed in reality?

This is shown in Fig. 6. The results are very accurate considering that we are approximating the complex behaviour of a software component with just a few parameters. More to the point, the observed accuracy is fully satisfactory for the purpose of reconfiguration analysis.

The presented work was funded by the Polish National Science Centre under grant no. N N516 475940.

References

[1] Caban, D.: Enhanced service reconfiguration to improve SOA systems depedability. In: Problems of Dependability and Modelling, pp. 27–39. Oficyna Wydawnicza Politechniki Wrocławskiej, Wroclaw (2011)

[2] Caban, D., Walkowiak, T.: Dependability oriented reconfiguration of SOA systems. In: Grzech, A. (ed.) Information Systems Architecture and Technology: Networks and Networks' Services, pp. 15–25. Oficyna Wydawnicza Politechniki Wrocławskiej, Wroclaw (2010)

[3] Lavenberg, S.S.: A perspective on queueing models of computer performance. Performance Evaluation 10(1), 53–76 (1989)

[4] Lutteroth, C., Weber, G.: Modeling a Realistic Workload for Performance Testing. In: 12th International IEEE Enterprise Distributed Object Computing Conference (2008)

[5] Nielsen, J.: Usability Engineering. Morgan Kaufmann, San Francisco (1994)

[6] Rahmawan, H., Gondokaryono, Y.S.: The simulation of static load balancing algorithms. In: International Conference on Electrical Engineering and Informatics, pp. 640–645 (2009)

[7] SPEC, SPECweb2009 Release 1.20 Benchmark Design Document vers. 1.20 (2010), http://www.spec.org/web2009/docs/design/SPECweb2009_Design.html (accessed February 10, 2012)

[8] Walkowiak, T.: Information systems performance analysis using task-level simulator. In: DepCoS – RELCOMEX, pp. 218–225. IEEE Computer Society Press (2009)

[9] Walkowiak, T., Michalska, K.: Functional based reliability analysis of Web based information systems. In: Dependable Computer Systems, pp. 257–269. Springer, Heidelberg (2011)

On Some Statistical Aspects of Software Testing and Reliability

Frank P.A. Coolen

Abstract. This article discusses the author's views on possible contributions statistics can make to software testing and reliability. Several difficulties are highlighted and several research challenges are discussed. Overall, the message is that statistical methods cannot provide proper support for software testing or provide useful inferences on software reliability if the statistical methods are considered to be an 'add-on'; careful treatment of the uncertainties and the adequate use of statistical methods have to be right at the center of the software development and test processes to ensure better tested and more reliable software. In line with this requirement, the development of suitable statistical methods requires collaboration of software developers and testers with statisticians on real-world problems.

1 Introduction

Ten years ago, a paper entitled 'Bayesian graphical models for software testing', written by my colleagues David Wooff, Michael Goldstein and me, was published in IEEE Transactions on Software Engineering [16]. This presented the pinnacle of a substantial multi-year research project in collaboration with an industrial partner, in which we explored the possibilities to use statistical methods to support software testers. As testing software is effectively all about uncertainty and information, it seemed obvious that statistics, which may be regarded as the art of dealing with uncertainty and information, could help testers with their very complex tasks. This project was successful, particularly in setting a direction for future collaboration between statisticians and software testers; an overview of the project was presented in the paper 'Using Bayesian statistics to support testing of software systems' (by the same authors), published in the Journal of Risk and Reliability [8].

Frank P.A. Coolen
Dept. of Mathematical Sciences, Durham University, Durham, United Kingdom
frank.coolen@durham.ac.uk

W. Zamojski et al. (Eds.): Complex Systems and Dependability, AISC 170, pp. 103–113.
springerlink.com

After initial difficulties due to very different jargons and cultures between the academic statisticians and the real-world software testers, very useful meetings followed during which the software testers explained more and more which aspects involving uncertainty were really difficult, what their current practice was with regard to all aspects of software testing, and which specific aims and restrictions there were for the software testing. It should be emphasized that the 'current practice' was considered to be good, certainly in line with the state-of-the-art at the time of the project (late 1990's), and indeed that there were quite many and different aims, particularly when managers with different responsibilities in the company got involved and expressed the hope that the methodology we were developing in collaboration with the software testers would also be useful to assist them in their specific duties related to the software testing process. For example, one such a duty involved setting the budget for a specific software testing project, well in advance of the software actually becoming available. I refer the interested reader to the above cited papers for more details on the specific project, and particularly on the Bayesian graphical modelling approach we developed. A substantial monograph that addresses many more aspects of that project, including for example methodology to predict the time required for software testing including considerations of re-testing after failures, and indeed how to do such re-testing efficiently, is in preparation [17].

In this article, I reflect a bit further on aspects from the mentioned long-term project, and on research questions that arose from it. Beyond this, I reflect on several aspects of the interaction between statistics and software testing and reliability, raising some important challenges for research and application that, to the best of my knowledge, are still open topics. This article does not answer many questions, but I hope that it provides some possible directions towards answers, or at least some issues to reflect upon and which might steer future developments and collaborations in the field.

2 Statistics, Software Testing and Reliability

The question how statistics can help software testing and be used to assess software reliability appears, at first, a simple question. However, it is not. To start, we have to consider what statistics is, and indeed what software testing and software reliability are. These are all generic terms that encompass many different activities, problems, methods, theories and so on. Generally speaking, however, statistics can be regarded as the theory of quantification of uncertainty, which includes the effect of information on uncertainty. As such, it is natural to use statistics to support all software testing and software reliability problems where dealing correctly with uncertainty and information is a major issue. It should be emphasized that not all problems in testing and software require, or can benefit from, the use of statistical methods. For example, if one is absolutely certain that checking code will reveal all possible problems, then one has no need to quantify uncertainty and therefore statistics has nothing to offer. Of course, the moment the 'absolute certainty' disappears then one should consider the use of statistical methods.

In statistics, there are two predominant foundational frameworks, which some might see as competing theories yet they both have their own important place in uncertainty quantification in general, hence also with regard to software testing and reliability. The frequentist framework of statistics generally provides methods that are calibrated in case of repeated use, under suitable assumptions. Results are traditionally in terms of e.g. confidence or significance levels, which are difficult to interpret for specific applications and indicate an approximate proportion of times that the specific method, if applied to many different problems, will give the correct answer. While this has been criticized by many, it is not a bad principle: if one has a tool available which is known to deliver a good outcome for a specific kind of task approximately 95 out of 100 times it is being applied, many will consider it a suitable tool. One problem with many of the classical frequentist approaches is that uncertainties often involve unobservable quantities such as model parameters, which add a further level of difficulty to understanding of the results in many cases.

The Bayesian framework of statistics has very different foundational starting points, with emphasis on subjectivity of probability to quantify uncertainty jointly for all random quantities in a specific application, where inferences follow from probabilistic conditioning on the information that becomes available. There is no frequency interpretation as a starting point, but in practice results from Bayesian statistics are often either fully or approximately in agreement with their frequentist counterparts. This is not surprising when there is a substantial amount of data and the wish not to let modelling assumptions or additional information in the form of expert opinions influence the inferences. However, in situations with few data the methods provide quite differing opportunities, where the explicit inclusion of expert opinions in the Bayesian framework can be a substantial advantage. It should be remarked that both the frequentist and Bayesian methods can be generalized by the use of sets of probabilities instead of a single one, which provides more robust inferences and models more adequately any lack of information. In particular some generalized methods for uncertainty quantification provide attractive opportunities to take unobserved or even unknown possible outcomes or risks into account [1, 2, 4, 11]. Such imprecise probabilistic methods are gaining popularity but have thus far not yet been implemented for software testing [9]. For general reliability problems, including a little attention to basic software reliability, such methods have been developed [10].

Software testing, and related aspects of software reliability, vary widely in different applications. The number of inputs that can be tested can vary from very few to effectively an unlimited number; some functionality might not be testable; it may be anything from trivial to impossible to know if an output is correct; the tests may take place on the actual software system or on a special test system which may resemble but not be identical to the actual system; the software may be fully open for inspection, it may be entirely black-box or anything in between, et cetera. Furthermore, the effects of errors in the output may be anything from neglectable to catastrophic, and may even be genuinely unknown. Statistical methods can be used to support decision making in all such scenarios, but it requires a high level of statistical expertise to ensure correct choice and application of methods. Importantly, it must be emphasized that statistics is not a miracle cure, it can only deal with the information that is available and its use should be aimed at supporting software testers.

The power of statistical methods lies mostly in careful learning from information and guidance with regard to further testing, which involve problems that even for basic test scenarios are soon too complex for the human mind to solve without the aid of sound statistical methods. Dealing correctly with uncertainties in applications is, in most cases, difficult and time consuming. The benefits can be substantial but of course will require justification with regard to the effort. Such considerations may be used to guide the level of detail of statistical modelling. This necessarily must be done related to a specific statistical approach; as an example of such guidance I refer to [7] where the considerations for deciding an appropriate level of detail in Bayesian graphical models for software testing are discussed.

In the following section, I will discuss a number of topics on the interface of statistics and software testing and reliability, which I believe require thought and attention, and indeed further research. It is crucial that such further research is linked to real-world applications with genuine collaboration between statisticians and software testers and engineers. These topics are discussed briefly without attempts to provide a full picture of the problems and state-of-the-art. It will be followed by a brief personal view on the way ahead in Section 4.

3 Some Topics That Require Attention

The first thing to consider in software testing is what actually can be tested. This is particularly important if testing leads to quantification of reliability of the software system in practical use. Usually, testing is limited in scope and executed during a relatively short period of time, possibly even without full linkage of a specific system with other real-world systems, databases et cetera. Statistical methods using only information from such limited tests cannot be expected to extrapolate to more practical issues. Software systems may also fail due to failure modes occurring at the interface between software and hardware, e.g. electricity provision might be interrupted during software processes, or due to background activities (e.g. memory space may slowly fill up); such aspects often cannot be discovered during practical testing, hence deducing a level of reliability from test results requires care.

A very basic question which, to our surprise, took quite some time to resolve in the aforementioned application of Bayesian graphical models to support software testing, is what to model? In some applications, particularly where very many tests can be done, one might simply be able to consider the input domain and model for each element of it (there may be 'infinitely' many, for example if inputs are real-valued quantities) whether or not the software provides the correct output. The software testing we were involved with, however, considered integration testing of new or upgraded functionality and code, particularly with regard to communication between databases. Inputs were typically related to some processes, and customers could possibly provide all kinds of inputs and in different orders, so it was difficult to define a theoretical input space. Testers would create scenarios to go through in their testing, and of course such test scenarios reflected their expertise in detail. We modelled the testers' knowledge and beliefs about the software system and its functioning, in particular how they distinguished between related but different inputs and where they believed possible errors in output would originate from. The software systems were complex and (almost entirely) black-box, with parts developed over quite many years without detailed knowledge about the development and core software remaining in the company. The statistical methodology was then used explicitly to assist the testers in their extremely complex tasks with regard to selecting efficient tests, ordering these, inferring what to do if a test outcome was wrong, and indicating when testing could be stopped[1]. As the statistical approach clearly modelled the testers' expertise and actions and enhanced their expertise, without taking over (and it was clear to the testers that the approach would not be taking over their roles in the future, due to the individual aspects of specific software systems and testing tasks), they understood the supporting role of the statistical methodology, helping them to do their job better and more efficiently. This is crucial for any collaboration, of course, but perhaps particularly so when experts are sceptical about novel methods that claim to be able to support them but which they may find difficult to understand. Actually, we found that the software testers quite quickly got to grips with the Bayesian graphical models that we built, and with the possibilities of support these gave them. They also understood well that the method could only provide useful support if they included useful and honest information on which to base the models. A lot more research is required on this topic, particularly considering a wide variety of software testing and reliability scenarios, and it is hard to say which aspects can and should be modelled without considering a specific scenario.

[1] In early papers in the software testing literature the view was sometimes expressed that testing was of no use if it did not reveal failures. This is quite a remarkable view, as no failures is of course the ultimate aim. Probably the more important question is when to stop testing in such cases, this links to the topic of high reliability demonstration where, of course, statistical methods can also provide guidance [5].

An important consideration in practical testing is what the consequences will be of a software error, and indeed if such errors can be recognized. In the project we were involved with it was not a problem to discover software errors. Generally, if it is not known with certainty whether or not software output is correct (think e.g. at complicated computational software, particularly cases were computation for one input can take a very long time), Bayesian statistical methods can provide some modelling opportunities, mostly by comparing results with prior expectations for them and trying to identify if results differ from expectations by a substantial amount. I have not yet seen detailed applied research about this, it will be fascinating and important.

Consequences are often, although not necessarily explicitly, grouped into e.g. cosmetic, minor, major or catastrophic. Typically, under severe time and costs constraints, tests are not aimed at discovering failures with cosmetic and perhaps even minor consequences, but any such errors are noted and dealt with at a possible future opportunity. Test suites designed by testers normally reflect the severity of consequences of failures, particularly so if such failures occurred in the (recent) past. Stastical methods can be set up in order to deal with different consequences of failures, and hence the importance of discovering such failures. The concept of 'utility' is a natural part of the Bayesian framework and is precisely used to take such consequences into account. However, in a company with different levels of management, with several different people responsible for different aspects of the output of the software system that is being tested, such utilities may not be discussed explicitly and they may differ based on the personal views and responsibilities of specific people. What to some people might appear just a failure with minor consequences, could by others be regarded as a major problem or worse. Such aspects must be discussed as it is crucial that there is a unified and clear goal for the software testing. One may be tempted to define 'delivering fault-free software' as such a goal, but this is mostly unrealistic for substantial software systems and would be extremely difficult to verify from test results, unless time and budget for testing are extremely generous. When using statistical methods to support software testing, the utilities related to the outcomes, and indeed to finding failures in order to prevent future problems with specific outputs, are important and must be determined at a relatively early stage in the process as they will guide the focus and level of detail of the testing and therefore of the statistical modelling process.

An intriguing situation occurs if one may not know (expected) possible consequences of failures, or may not know what kind of failures can occur. This is very natural, in particular when testing new black-box systems where discovering the functionality may be part of the testing aims. Statistical methods that can support such testing have not really been developed, and hence this provides very important and interesting research challenges. Imprecise probabilistic methods, including nonparametric predictive inference [4], have some specific features which make them promising for inference on the occurrence of unknown events. The theory of decision making with utilities has also been nearly entirely restricted to known outcomes and known utilities. Recently, a generalization to adaptive utilities has been presented, both within the Bayesian and nonparametric frequentist statistical frameworks [13, 14, 15]. In this work, utility is uncertain and one can learn about it, which of course is quite often a major reason for experimenting and testing (e.g. to learn about unknown side-effects of medication; there is a suitable parallel to software). We have ideas to link this work to software testing, but it would be from theoretical perspective at first; we hope to link it to real-world test situations in the future.

As mentioned, there may be several managers in an organisation who each have different responsibilities and hence may have different needs for statistical methods to support their activities related to software testing and reliability. For example, someone managing the actual test team will have quite different specific needs for support compared to someone managing the overall development and test budgets, and indeed having to make cases to set such budgets, with the latter typically needing to make decisions quite far in advance to the actual testing. Statistical methods, in particular the Bayesian framework, can support all such different decision processes, but again this may require problem specific modelling, will not be easy and will not be a miracle cure. As part of our industrial collaboration[2] we developed an approach to provide, approximately, an expected value and corresponding variance for the length of a future testing process, supported by the Bayesian graphical modelling approach. This approximation took into account re-testing after corrections of software failures discovered during testing. Of course, this requires substantial input as expected times must be provided for all possible events together with information on probabilities, not only for failures to occur but also with regard to success of correcting actions. Managers should not expect statistical methods to be able to provide answers to such difficult questions without such inputs, it underlines that careful modelling is difficult and time consuming, as is generally the case if one wants to deal well with uncertainty.

In addition to such quite general aspects there are a number of issues that, quite remarkably, still do not seem to have been solved satisfactorily, or at least about which there is substantial confusion amongst software testers and even some researchers who present methods with a suggestion of statistical expertise. For example, if one can do ample testing and has detailed knowledge of the input profile for the software system in daily practical use, it tends to be advocated that testing should be according to that input profile. However, this will rarely be optimal use of test time and budget (unless these are effectively limitless). Intuition on a most simple example probably suffices to explain this point: suppose that a software system will only have to provide outputs for two groups of inputs, say groups A and B. Suppose that the input profile from practical use of the system is such that 99% of inputs

[2] This will be reported in the monograph [17] that is being prepared.

are from group A and only 1% from group B. Suppose further that 100 inputs can be tested. In this case, testing in line with the input profile would lead to 99 inputs from group A and one input from group B to be tested. Intuitively it will be clear that the 99 tests of inputs from group A probably leave little remaining uncertainty about the success rate of the software on dealing correctly with inputs from this group. But the single test from group B is not so informative for other inputs from this group. Hence, if one wishes to have high quality software which rarely makes mistakes, then one would probably wish to do some more tests for group B and reduce the number of tests from group A accordingly. It is not difficult to prove this mathematically, but the crucial part is to recognize that the input profile is to be used related to the optimality criterion, that is proportions of applications in the future, after the testing, with related utilities for avoiding failures in the outputs. Then the optimisation variables should be the proportions of inputs from each group within the test, and this will rarely lead to testing in the same proportions as the input profile [3].

In this reasoning with regard to profiles there is an aspect that is at the heart of software testing and the statistical support for it, namely the required judgements on exchangeability of the different inputs [6, 12]. A key judgement that software testers have to make, and that must be reflected in any statistical model and method to support them, is how similar different inputs are, for a specific software system, with regard to the expected quality of the corresponding output, and with regard to what knowledge about the corresponding output quality reveals about such output quality for other inputs. One extreme situation would be that the software's output either will be incorrect for all inputs or will be correct for all inputs. Under such judgement, clearly a single test suffices if one can correctly classify the output. The other extreme situation would be that every tested input only reveals whether or not the software provides the correct output for that specific input value, while not revealing anything about other inputs. In the first case the software's performance is identical for all inputs, in the second case it is independent for all inputs. In practice the truth, and with it the assumptions testers are confident to make, tends to be somewhere between these two extremes. Statistically, the concept of exchangeability (and partial exchangeability [12]) provides the opportunity to model such judgements, and these can be taken into account in a variety of ways in statistical methods. However, it is not easy to do so, this is often overlooked [3]. In particular, one might naturally judge that the input space can be divided into a partition, with all inputs in one element of the partition being more similar than inputs in different elements, but with neither being identical or independent. Bayesian graphical models provide a modelling framework for this, but are not as flexible or easy to deal with in case of learning from many test results as some possible alternatives [6]. Further research is required on this crucial aspect which sits at the heart of uncertainty in software testing.

Practical statistical support for software testing highlights one major research challenge for statistics which, perhaps remarkably, appears not yet to have attracted much attention. Any decisions on design of test suite (and beyond software testing, general design of experiments) depend on modelling assumptions, usually based on expert judgements, experience and convenience of the corresponding statistical methods. Ideally, one would want to use the test suite also to confirm the underlying model assumptions, in particular where tests are performed sequentially this could lead to changes if it was discovered that the assumptions were not fully adequate. In such cases, one would still hope that the earlier test results would be of use while one could adapt further tests in the light of the new information, and so on. This idea to use test outcomes in case the underlying assumptions were not fully adequate is an issue of robustness of the test suite with regard to aspects of the model specification. The idea to take this into account is particularly relevant for sequencing of tests, where at the early stages of testing one may want to include tests which can indicate whether or not the modelling assumptions are adequate. It has, to the best of my knowledge, not been considered in the literature, on software testing and even on general design of experiments. A small step to this idea is the concept of adaptive utility [13, 14] in Bayesian methods, where in sequential decision processes it is shown that it can be optimal to first look for observations that enable one to learn the utilities better, to the possible benefit of later decisions. Due to the specific features of software testing, it would be exciting if this aspect of designing suitable test suites were investigated in direct connection to a real-world application.

4 The Way Ahead

I should make an important comment about the (mostly academic) research literature on software testing and reliability: In practice there is often scepticism about the methods presented, and I believe rightly so. Most of these methods have been developed by mathematicians and statisticians based on assumptions that are inadequate in many challenging software testing problems. For example, the many models based on the idea of fault counting and removal or on assumed reliability growth appear not to have a substantial impact on practical software testing. Our industrial collaborators, who were very experienced software testers and engineers, did not even think in terms of numbers of faults in the software, as it was largely black-box and defining a fault would be non-trivial.

I believe that the only way ahead is through genuine collaboration between software testers and statisticians throughout the process of testing, including all stages of test preparation. This is difficult and might be expensive, so might only be considered feasible for substantial applications, particularly with safety or security risks in case software failures occur. However, sound statistical support for the software testing process is likely to lead to both more efficient testing and more reliable software, as such it will probably lead to cost reduction, both for the actual test process and with regard to the consequences of software failures. It would be particularly beneficial to have long-term collaborations between the same teams of testers and statisticians, both for working on upgrades of some software systems and for working on a variety of systems. Testing upgrades normally benefits greatly from experiences on testing of earlier versions, while working on a variety of systems will also provide great challenges for the testers and statisticians, which is likely to provide useful further insights towards more generic approaches for statistically supported software testing.

I strongly hope that during the next decade(s) a variety of such long-term collaborations will be reported in the literature and will lead to important further methods and insights. Possibly some generic aspects might be automated, to facilitate easier implementation of sound statistical support for testing of software with some specific generic features. I am sceptical however about the possibility to fully automate such statistical support. I am sceptical about the possibility to fully automate software testing, and as dealing adequately with the uncertainties involved adds substantial complexity to the problem, full automation is highly unlikely. It is beyond doubt, however, that thorough long-term collaborations between software testers and statisticians can lead to very substantially improved software testing, hence leading to more reliable software which will justify the effort. Clearly, this is a great field to work in due to the many challenges and opportunities, both for research and practical applications which have to be developed together.

Acknowledgements. I am grateful to the members of the organising committee of the Seventh International Conference on Dependability and Complex Systems (DepCoS-RELCOMEX 2012), in particular Professor Wojciech Zamojski, for their kind invitation to share my experiences and views on the interaction between statistics and software testing and reliability with the conference participants, both through a presentation at the conference and this article. The contents of this contribution are fully from the author's perspective, but are based on several long-standing and ongoing research collaborations, as reflected by the bibliography of this article. I sincerely thank all collaborators for their substantial inputs in jointly authored research publications, and beyond that for the many discussions which have played a major part in shaping my research activities and views on interesting research challenges.

References

1. Coolen, F.P.A.: On probabilistic safety assessment in case of zero failures. Journal of Risk and Reliability 220, 105–114 (2006)
2. Coolen, F.P.A.: Nonparametric prediction of unobserved failure modes. Journal of Risk and Reliability 221, 207–216 (2007)

3. Coolen, F.P.A.: Discussion on: 'A discussion of statistical testing on a safety-related application' by Kuball and May. Journal of Risk and Reliability 222, 265–267 (2008)
4. Coolen, F.P.A.: Nonparametric predictive inference. In: Lovric (ed.) International Encyclopedia of Statistical Science, pp. 968–970. Springer, Berlin (2011)
5. Coolen, F.P.A., Coolen-Schrijner, P.: On zero-failure testing for Bayesian high reliability demonstration. Journal of Risk and Reliability 220, 35–44 (2006)
6. Coolen, F.P.A., Goldstein, M., Munro, M.: Generalized partition testing via Bayes linear methods. Information and Software Technology 43, 783–793 (2001)
7. Coolen, F.P.A., Goldstein, M., Wooff, D.A.: Project viability assessment for support of software testing via Bayesian graphical modelling. In: Bedford, van Gelder (eds.) Safety and Reliability, pp. 417–422. Swets & Zeitlinger, Lisse (2003)
8. Coolen, F.P.A., Goldstein, M., Wooff, D.A.: Using Bayesian statistics to support testing of software systems. Journal of Risk and Reliability 221, 85–93 (2007)
9. Coolen, F.P.A., Troffaes, M.C.M., Augustin, T.: Imprecise probability. In: Lovric (ed.) International Encyclopedia of Statistical Science, pp. 645–648. Springer, Berlin (2011)
10. Coolen, F.P.A., Utkin, L.V.: Imprecise reliability. In: Lovric (ed.) International Encyclopedia of Statistical Science, pp. 649–650. Springer, Berlin (2011)
11. Coolen-Maturi, T., Coolen, F.P.A.: Unobserved, re-defined, unknown or removed failure modes in competing risks. Journal of Risk and Reliability 225, 461–474 (2011)
12. De Finetti, B.: Theory of probability (2 volumes). Wiley, London (1974)
13. Houlding, B., Coolen, F.P.A.: Sequential adaptive utility decision making for system failure correction. Journal of Risk and Reliability 221, 285–295 (2007)
14. Houlding, B., Coolen, F.P.A.: Adaptive utility and trial aversion. Journal of Statistical Planning and Inference 141, 734–747 (2011)
15. Houlding, B., Coolen, F.P.A.: Nonparametric predictive utility inference. European Journal of Operational Research (to appear, 2012)
16. Wooff, D.A., Goldstein, M., Coolen, F.P.A.: Bayesian graphical models for software testing. IEEE Transactions on Software Engineering 28, 510–525 (2002)
17. Wooff, D.A., Goldstein, M., Coolen, F.P.A.: Bayesian graphical models for high-complexity software and system testing. Springer (in preparation)

Generalizing the Signature to Systems with Multiple Types of Components

Frank P.A. Coolen and Tahani Coolen-Maturi

Abstract. The concept of signature was introduced to simplify quantification of reliability for coherent systems and networks consisting of a single type of components, and for comparison of such systems' reliabilities. The signature describes the structure of the system and can be combined with order statistics of the component failure times to derive inferences on the reliability of a system and to compare multiple systems. However, the restriction to use for systems with a single type of component prevents its application to most practical systems. We discuss the difficulty of generalization of the signature to systems with multiple types of components. We present an alternative, called the survival signature, which has similar characteristics and is closely related to the signature. The survival signature provides a feasible generalization to systems with multiple types of components.

1 Introduction

Theory of signatures was introduced as an attractive tool for quantification of reliability of coherent systems and networks consisting of components with random failure times that are independent and identically distributed (*iid*), which can be regarded informally as components of 'a single type'. Samaniego [14] provides an excellent introduction and overview to the theory[1]. The main idea of the use of signatures is separation of aspects of the components' failure time distribution and the structure of the system, with the latter quantified through the signature. Let us first introduce some notation and concepts.

Frank P.A. Coolen · Tahani Coolen-Maturi
Department of Mathematical Sciences, Durham University, Durham, United Kingdom. Kent Business School, University of Kent, Canterbury, United Kingdom
e-mail: `frank.coolen@durham.ac.uk`, `T.Coolen-Maturi@kent.ac.uk`

[1] Samaniego [14] assumes *iid* failure times of components, which we follow in this paper; the theory of signatures applies also under the weaker assumption of exchangeability [9]

W. Zamojski et al. (Eds.): Complex Systems and Dependability, AISC 170, pp. 115–130.
springerlink.com

For a system with m components, let state vector $\underline{x} = (x_1, x_2, \ldots, x_m) \in \{0,1\}^m$, with $x_i = 1$ if the ith component functions and $x_i = 0$ if not. The labelling of the components is arbitrary but must be fixed to define $\underline{x}$. The structure function $\phi : \{0,1\}^m \to \{0,1\}$, defined for all possible $\underline{x}$, takes the value 1 if the system functions and 0 if the system does not function for state vector $\underline{x}$. We restrict attention to coherent systems, which means that $\phi(\underline{x})$ is not decreasing in any of the components of x, so system functioning cannot be improved by worse performance of one or more of its components[2]. We further assume that $\phi(\underline{0}) = 0$ and $\phi(\underline{1}) = 1$, so the system fails if all its components fail and it functions if all its components function[3].

Let $T_S > 0$ be the random failure time of the system and $T_{j:m}$ the j-th order statistic of the m random component failure times for $j = 1, \ldots, m$, with $T_{1:m} \leq T_{2:m} \leq \ldots \leq T_{m:m}$. We assume that these component failure times are independent and identically distributed. The system's signature is the m-vector q with j-th component

$$q_j = P(T_S = T_{j:m}) \tag{1}$$

so q_j is the probability that the system failure occurs at the moment of the j-th component failure. Assume that $\sum_{j=1}^{m} q_j = 1$; this assumption implies that the system functions if all components function, has failed if all components have failed, and that system failure can only occur at times of component failures. The signature provides a qualitative description of the system structure that can be used in reliability quantification [14]. The survival function of the system failure time can be derived by

$$P(T_S > t) = \sum_{j=1}^{m} q_j P(T_{j:m} > t) \tag{2}$$

If the failure time distribution for the components is known and has cumulative distribution function (CDF) $F(t)$, then

$$P(T_{j:m} > t) = \sum_{r=m-j+1}^{m} \binom{m}{r} [1 - F(t)]^r [F(t)]^{m-r} \tag{3}$$

The expected value of the system failure time can be derived as function of the expected values of the order statistics $T_{j:m}$, by

$$E(T_S) = \sum_{j=1}^{m} q_j E(T_{j:m}) \tag{4}$$

[2] This assumption could be relaxed but is reasonable for many real-world systems.

[3] This assumption could be relaxed but for coherent systems it would lead to the trivial cases of systems that either always fail or always function.

From these expressions it is clear that the system structure is fully taken into account through the signature and is separated from the information about the random failure times of the components. This enables e.g. straightforward comparison of the reliability of two systems with the same single type of components if their signatures are stochastically ordered [14]. Consider two systems, each with m components and all failure times of the $2m$ components assumed to be *iid*. Let the signature of system A be q^a and of system B be q^b, and let their failure times be T^a and T^b, respectively. If

$$\sum_{j=r}^{m} q_j^a \geq \sum_{j=r}^{m} q_j^b \tag{5}$$

for all $r = 1, \ldots, m$, then

$$P(T^a > t) \geq P(T^b > t) \tag{6}$$

for all $t > 0$. Such a comparison is even possible if the two systems do not have the same number of components, as one can increase the length of a system signature in a way that does not affect the corresponding system's failure time distribution [14], so one can always make the two systems' signatures of the same length. Consider a system with m components, signature $q = (q_1, q_2, \ldots, q_m)$ and failure time T_S. For ease of notation, define $q_0 = q_{m+1} = 0$. Now define a signature q^* as the vector with $m+1$ components

$$q_j^* = \frac{j-1}{m+1} q_{j-1} + \frac{m+1-j}{m+1} q_j \tag{7}$$

for $j = 1, \ldots, m+1$, and define a random failure time T_S^* with probability distribution defined by the survival function

$$P(T_S^* > t) = \sum_{j=1}^{m+1} q_j^* P(T_{j:m+1} > t) \tag{8}$$

with

$$P(T_{j:m+1} > t) = \sum_{r=m-j+2}^{m+1} \binom{m+1}{r} [1 - F(t)]^r [F(t)]^{m+1-r} \tag{9}$$

Then the probability distributions of T_S and T_S^* are identical, so

$$P(T_S > t) = P(T_S^* > t) \tag{10}$$

for all $t > 0$. Note that the signatures q and q^* represent systems with m and $m+1$ components with *iid* failure times and CDF $F(t)$, respectively. Note further that q^* may not actually correspond to a physical system structure, but applying this extension (consecutively) enables two systems with different numbers of components to be compared.

Many systems' structures do not have corresponding signatures which are stochastically ordered. For example, the signatures $(\frac{1}{4}, \frac{1}{4}, \frac{1}{2}, 0)$ and $(0, \frac{2}{3}, \frac{1}{3}, 0)$, which relate to basic system structures, are not stochastically ordered. An attractive way to compare such systems' failure times, say T^a and T^b, is by considering the event $T^a < T^b$. To get more detailed insight into the difference between the two systems' failure times, one can consider the event $T^a < T^b + \delta$ as function of δ [4]. This way to compare two systems' failure times does not require the failure times of the components in one system to be *iid* with the failure times of the components in the other system. Let system A consist of m_a components with *iid* failure times, and system B of m_b components with *iid* failure times, with components of the different systems being of different types and their random failure times assumed to be fully independent, which means that any information about the failure times of components of the type used in system A does not contain any information about the failure times of components of the type used in system B. The ordered random failure times of the components in system A and of those in system B are denoted by $T^a_{1:m_a} \leq T^a_{2:m_a} \leq \ldots \leq T^a_{m_a:m_a}$ and $T^b_{1:m_b} \leq T^b_{2:m_b} \leq \ldots \leq T^b_{m_b:m_b}$, respectively. Let the failure time distribution for components in system A have CDF F_a and for components in system B have CDF F_b. Using the signatures q^a and q^b of these systems, the probability for the event $T^a < T^b$ is

$$P(T^a < T^b) = \sum_{i=1}^{m_a} \sum_{j=1}^{m_b} q_i^a q_j^b P(T^a_{i:m_a} < T^b_{j:m_b}) \tag{11}$$

with

$$P(T^a_{i:m_a} < T^b_{j:m_b}) = \int_0^\infty f_a^i(t) P(T^b_{j:m_b} > t) dt \tag{12}$$

where $f_a^i(t)$ is the probability density function (PDF) for $T^a_{i:m_a}$, which, with PDF $f_a(t)$ for the failure time of components of in system A, is equal to

$$f_a^i(t) = f_a(t) \sum_{r_a=m_a-i+1}^{m_a} \binom{m_a}{r_a} [1-F_a(t)]^{r_a-1} [F_a(t)]^{m_a-r_a-1} [r_a - m_a(1-F_a(t))] \tag{13}$$

and

$$P(T^b_{j:m_b} > t) = \sum_{r_b=m_b-j+1}^{m_b} \binom{m_b}{r_b} [1 - F_b(t)]^{r_b} [F_b(t)]^{m_b-r_b} \tag{14}$$

In Section 2 of this paper an alternative to the signature is proposed and investigated, we call it the system's survival signature. In Section 3 we discuss the difficulties in generalizing the signature to systems with multiple types of components, where the proposed survival signature appears to have a clear advantage which suggests an interesting and important new area of research. We end the paper with a concluding discussion in Section 4, where we briefly comment on computation of the system signature and survival signature, the possibility to use partial information, the generalization to theory of imprecise reliability in case the components' failure time distribution is not precisely known [5, 6], and the possible use of these concepts if only failure times from tested components are available and nonparametric predictive statistical methods are used for inference on the system reliability [2].

2 The Survival Signature

The signature was introduced to assist reliability analyses for systems with m components with *iid* failure times, with the specific role of modelling the structure of the system and separating this from the random failure times of the components. In this section, we consider an alternative to the signature which can fulfill a similar role, and which actually is closely related to the signature. Let $\Phi(l)$, for $0 = 1,\ldots,m$, denote the probability that a system functions given that *precisely* l of its components function. As in Section 1, we restrict attention again to coherent systems, for which $\Phi(l)$ is an increasing function of l, and we assume that $\Phi(0) = 0$ and $\Phi(m) = 1$. There are $\binom{m}{l}$ state vectors $\underline{x}$ with precisely l components $x_i = 1$, so with $\sum_{i=1}^{m} x_i = l$; we denote the set of these state vectors by S_l. Due to the *iid* assumption for the failure times of the m components, all these state vectors are equally likely to occur, hence

$$\Phi(l) = \binom{m}{l}^{-1} \sum_{\underline{x} \in S_l} \phi(\underline{x}) \tag{15}$$

As the function $\Phi(l)$ is by its definition related to survival of the system, and, as we will see later, it is close in nature to the system signature, we call it the system survival signature.

Let $C_t \in \{0,1,\ldots,m\}$ denote the number of components in the system that function at time $t > 0$. If the probability distribution of the component failure time has CDF $F(t)$, then for $l \in \{0,1,\ldots,m\}$

$$P(C_t = l) = \binom{m}{l} [F(t)]^{m-l} [1 - F(t)]^{l} \tag{16}$$

It follows easily that

$$P(T_S > t) = \sum_{l=0}^{m} \Phi(l) P(C_t = l) \tag{17}$$

The terms in the right-hand side of equation (17) explicitly have different roles, with the term $\Phi(l)$ taking the structure of the system into account, that is how the system's functioning depends on the functioning of its components, and the term $P(C_t = l)$ taking the random failure times of the components into account. Taking these two crucial aspects for determining the survival function for the system failure time into account separately in this way is similar in nature to the use of system signatures as discussed in Section 1.

Equations (2) and (17) imply

$$\Phi(l) = \sum_{j=m-l+1}^{m} q_j \tag{18}$$

which is easily verified by

$$\sum_{j=1}^{m} \sum_{r=m-j+1}^{m} q_j \binom{m}{r} [1 - F(t)]^r [F(t)]^{m-r} =$$
$$\sum_{r=1}^{m} \sum_{j=m-r+1}^{m} q_j \binom{m}{r} [1 - F(t)]^r [F(t)]^{m-r} \tag{19}$$

Equation (18) is logical when considering that the right-hand side is the probability that the system failure time occurs at the moment of the $(m-l+1)$-th ordered component failure time or later. The moment of the $(m-l+1)$-th ordered component failure time is exactly the moment at which the number of functioning components in the system decreases from l to $l-1$, hence the system would have functioned with l components functioning.

In Section 1 a straightforward comparison was given of the failure times of two systems A and B. Let us denote the survival signatures of these systems by $\Phi^a(l)$ and $\Phi^b(l)$, respectively, and assume that both systems consist of m components and that all these $2m$ components are of the same type, so have *iid* random failure times. The comparison in Section 1 was based on the stochastic ordering of the systems' signatures, if indeed these are stochastically ordered. Due to the relation (18) between the survival signature $\Phi(l)$ and the signature q, this comparison can also be formulated as follows: If

$$\Phi^a(l) \geq \Phi^b(l) \tag{20}$$

for all $l = 1, \ldots, m$, then

$$P(T^a > t) \geq P(T^b > t) \tag{21}$$

for all $t > 0$.

As explained in Section 1, the possibility to extend a signature in a way that retains the same system failure time distribution is an advantage for comparison of different system structures. The survival signature $\Phi(l)$ can be similarly extended as is shown next, by defining explicitly the survival signature $\Phi^*(l)$ that relates to a system with $m+1$ components which has the same failure time distribution as a system with m components and survival signature $\Phi(l)$ (throughout the superscript * indicates the system with $m+1$ components, and all components considered are assumed to have *iid* failure times). For $l = 1, \ldots, m+1$, let

$$\Phi^*(l) = \Phi(l-1) + \frac{m-l-1}{m+1} q_{m-l-1} \tag{22}$$

and from (18), (22) and $\Phi(0) = 0$ we have

$$\Phi^*(1) = \frac{m}{m+1} q_m = \frac{m}{m+1} \Phi(1) \tag{23}$$

and

$$\Phi^*(m+1) = \Phi(m) = 1 \tag{24}$$

and

$$\Phi(l+1) = \Phi(l) + q_{m-l} \tag{25}$$

Furthermore, it is easy to prove that

$$P(C_t = l) = \frac{m+1-l}{m+1} P(C_t^* = l) + \frac{l+1}{m+1} P(C_t^* = l+1) \tag{26}$$

The failure time T_S^* of the extended system with $m+1$ components and survival signature $\Phi^*(l)$ has the same survival function, and hence the same probability distribution, as the failure time T_S of the original system with m components and survival signature $\Phi(l)$. This is proven as follows

$$\begin{aligned}
P(T_S > t) &= \sum_{l=1}^{m} \Phi(l) P(C_t = l) \\
&= \sum_{l=1}^{m} \Phi(l) \left[\frac{m+1-l}{m+1} P(C_t^* = l) + \frac{l+1}{m+1} P(C_t^* = l+1) \right] \\
&= \Phi^*(1) P(C_t^* = 1) + \sum_{l=1}^{m-1} \left[\Phi(l+1) \frac{m-l}{m+1} + \Phi(l) \frac{l+1}{m+1} \right] P(C_t^* = l+1) \\
&\quad + \Phi^*(m+1) P(C_t^* = m+1) \\
&= \Phi^*(1) P(C_t^* = 1) \\
&\quad + \sum_{l=1}^{m-1} \Phi^*(l+1) P(C_t^* = l+1) + \Phi^*(m+1) P(C_t^* = m+1) \\
&= \sum_{l=1}^{m+1} \Phi^*(l) P(C_t^* = l) = P(T_S^* > t)
\end{aligned} \tag{27}$$

Comparison of the failure times of two systems A and B, each with a single type of components but these being different for the two systems, with the use of signatures, was given in Equation (11). This comparison is also possible with the use of the survival signatures, which we denote by $\Phi^a(l_a)$ and $\Phi^b(l_b)$ for systems A and B, respectively. The result is as follows

$$P(T^a < T^b) = \int_0^\infty f_S^a(t) P(T^b > t) dt \tag{28}$$

with $f_S^a(t)$ the PDF of T^a, given by

$$f_S^a(t) = f_a(t) \sum_{l_a=0}^{m_a} \Phi^a(l_a) \binom{m_a}{l_a} [1-F_a(t)]^{l_a-1} [F_a(t)]^{m_a-l_a-1} [l_a - m_a(1-F_a(t))] \tag{29}$$

and

$$P(T^b > t) = \sum_{l_b=0}^{m_b} \Phi^b(l_b) \binom{m_b}{l_b} [1 - F_b(t)]^{l_b} [F_b(t)]^{m_b-l_b} \tag{30}$$

Using relation (18) and change of order of summation as in (19), it is easy to show that (28) is actually the same formula as (11), so there is no computational difference in the use of either the signature or the survival signature for such comparison of two systems with each a single type of components. We can conclude that the method using the survival signature as presented in this section is very similar in nature to the method using signatures for systems with components with *iid* failure times. In Section 3 we consider the generalization of the signature and the survival signature to the very important case of reliability inferences for systems with multiple types of components.

3 Systems with Multiple Types of Component

Most practical systems and networks for which the reliability is investigated consist of multiple types of components. Therefore, a main challenge is generalization of the theory of signatures to such systems. Although an obvious challenge for research, little if any mention of it has been made in the literature. We will consider whether or not it is feasible to generalize the standard concept of the signature to systems with multiple types of components, and we will also consider this for the survival signature.

We consider a system with $K \geq 2$ types of components, with m_k components of type $k \in \{1, 2, \ldots, K\}$ and $\sum_{k=1}^{K} m_k = m$. We assume that the random failure times of components of the same type are *iid*, while full independence is assumed for the random failure times of components of different types. Due to the arbitrary ordering of the components in the state vector, we can group components of the same type together, so we use state vector $\underline{x} = (\underline{x}^1, \underline{x}^2, \ldots, \underline{x}^K)$ with $\underline{x}^k = (x_1^k, x_2^k, \ldots, x_{m_k}^k)$ the sub-vector representing the states of the components of type k. We denote the ordered random failure times of the m_k components of type k by $T_{j_k:m_k}^k$, for ease of presentation we assume that ties between failure times have probability zero.

System signatures were introduced explicitly for systems with a single type of components, which are assumed to have *iid* failure times. To generalize the signature approach to multiple types of components, it will be required to take into account, at the moment of system failure T_S which coincides with the failure of a specific component, how many of the components of each other type have failed. The generalized signature can again be defined in quite a straightforward manner, namely by defining

$$q_k(j_k) = P(T_S = T^k_{j_k:m_k}) \tag{31}$$

for $k = 1, \ldots, K$ and $j_k = 1, \ldots, m_k$, so the total signature can be defined as

$$q = (q_1(1), \ldots, q_1(m_1), q_2(1), \ldots, q_2(m_2), \ldots, q_K(1), \ldots, q_K(m_K)) \tag{32}$$

With this definition, the survival function of the system's failure time is

$$P(T_S > t) = \sum_{k=1}^{K} \sum_{j_k=1}^{m_k} q_k(j_k) P(T^k_{j_k:m_k} > t) \tag{33}$$

However, deriving this generalized signature is complex and actually depends on the failure time probability distributions of the different types of components, hence this method does not any longer achieve the separation of the system structure and the failure time distributions as was the case for a single type of components in Section 1. To illustrate this, we consider the calculation of q for the case with $K = 2$ types of components in the system. For ease of notation, let $T^2_{0:m_2} = 0$ and $T^2_{m_2+1:m_2} = \infty$. Calculation of $q_1(j_1)$ is possible by

$$\begin{aligned} q_1(j_1) &= P(T_S = T^1_{j_1:m_1}) = \\ &\sum_{j_2=0}^{m_2} \left[P(T_S = T^1_{j_1:m_1} \mid T^2_{j_2:m_2} < T^1_{j_1:m_1} < T^2_{j_2+1:m_2}) \right. \\ &\qquad \left. \times P(T^2_{j_2:m_2} < T^1_{j_1:m_1} < T^2_{j_2+1:m_2}) \right] \end{aligned} \tag{34}$$

Derivation of the terms $P(T^2_{j_2:m_2} < T^1_{j_1:m_1} < T^2_{j_2+1:m_2})$ involves the failure time distributions of both component types. It is possible to define the generalized signature instead by the first term in this sum, so the conditional probability of $T_S = T^1_{j_1:m_1}$ given the number of components of type 2 that are functioning at time $T^1_{j_1:m_1}$, but this does not simplify things as the probabilities $P(T^2_{j_2:m_2} < T^1_{j_1:m_1} < T^2_{j_2+1:m_2})$ will still be required for all $j_1 \in \{1, \ldots, m_1\}$ and $j_2 \in \{1, \ldots, m_2\}$, and as these probabilities involve order statistics from different probability distributions this is far from straightforward. Of course, for the general case with a system consisting of

$K \geq 2$ types of components, the arguments are the same but the complexity increases as function of K. Calculating the system reliability via this generalized signature involves the calculation of m terms $q_k(j_k)$, each of which requires

$$\prod_{\substack{l=1\\l\neq j}}^{K}(m_l+1) \tag{35}$$

probabilities for orderings of order statistics from different probability distributions to be derived. This quickly becomes infeasible, which is probably the reason why such a generalization of the signature has not been addressed in detail in the literature. It may also explain why the signature, although a popular topic in the reliability research literature, does not appear to have made a substantial impact on practical reliability analyses.

We will now investigate if the survival signature, as presented in Section 2, may be better suited for the generalization to systems with multiple types of components. Let $\Phi(l_1, l_2, \ldots, l_K)$, for $l_k = 0, 1, \ldots, m_k$, denote the probability that a system functions given that *precisely* l_k of its components of type k function, for each $k \in \{1, 2, \ldots, K\}$; again we call this function $\Phi(l_1, l_2, \ldots, l_K)$ the system's survival signature, it will be clear from the context whether or not there are multiple types of components. There are $\binom{m_k}{l_k}$ state vectors $\underline{x}^k$ with precisely l_k of its m_k components $x_i^k = 1$, so with $\sum_{i=1}^{m_k} x_i^k = l_k$; we denote the set of these state vectors for components of type k by S_l^k. Furthermore, let $S_{l_1,\ldots,l_K}$ denote the set of all state vectors for the whole system for which $\sum_{i=1}^{m_k} x_i^k = l_k$, $k = 1, 2, \ldots, K$. Due to the *iid* assumption for the failure times of the m_k components of type k, all the state vectors $\underline{x}^k \in S_l^k$ are equally likely to occur, hence

$$\Phi(l_1, \ldots, l_K) = \left[\prod_{k=1}^{K}\binom{m_k}{l_k}^{-1}\right] \times \sum_{\underline{x}\in S_{l_1,\ldots,l_K}} \phi(\underline{x}) \tag{36}$$

Let $C_t^k \in \{0, 1, \ldots, m_k\}$ denote the number of components of type k in the system that function at time $t > 0$. If the probability distribution for the failure time of components of type k is known and has CDF $F_k(t)$, then for $l_k \in \{0, 1, \ldots, m_k\}$, $k = 1, 2, \ldots, K$,

$$\begin{aligned} P(\bigcap_{k=1,\ldots,K} \{C_t^k = l_k\}) &= \prod_{k=1}^{K} P(C_t^k = l_k) \\ &= \prod_{k=1}^{K} \left(\binom{m_k}{l_k}[F_k(t)]^{m_k-l_k}[1-F_k(t)]^{l_k}\right) \end{aligned} \tag{37}$$

The probability that the system functions at time $t > 0$ is

$$P(T_S > t) = \sum_{l_1=0}^{m_1} \cdots \sum_{l_K=0}^{m_K} \Phi(l_1, \ldots, l_K) P(\bigcap_{k=1}^{K} \{C_t^k = l_k\}) =$$

$$\sum_{l_1=0}^{m_1} \cdots \sum_{l_K=0}^{m_K} \left[\Phi(l_1, \ldots, l_K) \prod_{k=1}^{K} P(C_t^k = l_k) \right] =$$

$$\sum_{l_1=0}^{m_1} \cdots \sum_{l_K=0}^{m_K} \left[\Phi(l_1, \ldots, l_K) \prod_{k=1}^{K} \left(\binom{m_k}{l_k} [F_k(t)]^{m_k - l_k} [1 - F_k(t)]^{l_k} \right) \right] \quad (38)$$

Calculation of (38) is not straightforward but it is far easier than calculation using the generalized signature in (33). Of course, the survival signature $\Phi(l_1, \ldots, l_K)$ needs to be calculated for all $\prod_{k=1}^{K}(m_k+1)$ different $(l_1, \ldots, l_K)$, but this information must be distracted from the system anyhow and is only required to be calculated once for any system, similar to the (survival) signature for systems with a single type of components. The main advantage of (38) is that again the information about system structure is fully separated from the information about the components' failure times, and the inclusion of the failure time distributions is straightforward due to the assumed independence of failure times of components of different types. The difficulty in (33) of having to find probabilities of rankings of order statistics from different probability distributions is now avoided, which leads to a very substantial reduction and indeed simplification of the computational effort.

We can conclude that the survival signature, which is proposed in this paper and which is very closely related to the classical signature in case of systems with components with *iid* failure times, seems to provide an attractive way to generalize the concept of signature to systems with multiple types of components. It should be emphasized that the survival signature provides all that is needed to calculate the survival function for the system's failure time, and as this fully determines the failure time's probability distribution all further inferences of interest can be addressed using the survival signature. While we have assumed independence of the failure times of components of different types, the proposed approach in this paper can also be used if these failure times are dependent, in which case the joint probability distribution for these failure times must of course be used in Equation (37). This would still maintain the main feature of the use of the proposed survival signature in Equation (38), namely the explicit separation of the factors taking into account the information about the system structure and the information about the component failure times. Also with the less attractive generalization of the classical signature it is possible to deal with dependent failure times for components of different types, but it is likely to substantially complicate the computation of probabilities on the orderings of order statistics for failure times of components of different types.

Theoretical properties of the survival signature for systems with multiple types of components are an important topic for future research. This should include analysis of possibilities to extend such a signature while keeping the corresponding systems' failure times distributions the same, which could possibly be used with some adapted form of stochastic monotonicity (on the subvectors relating to components of the same type) for comparison of failure times of systems that share the same multiple types of components. It seems possible to compare the failure times of different systems with multiple types of components using the survival signature along the lines as presented in Section 2 for systems with components with *iid* failure times, but this should also be developed in detail. The first results for this survival signature, as presented in this paper, are very promising, particularly due to the possibility to use the survival signature for systems with multiple types of components, so such further research will be of interest.

We briefly illustrate the use of the survival signature for a system with $K = 2$ types of components, types 1 and 2, as presented in Figure 1 (where the types of the components, 1 or 2, are as indicated).

With $m_1 = m_2 = 3$ components of each type, the survival signature $\Phi(l_1, l_2)$ must be specified for all $l_1, l_2 \in \{0, 1, 2, 3\}$; this is given in Table 1. To illustrate its derivation, let us consider $\Phi(1, 2)$ and $\Phi(2, 2)$ in detail. The state vector is $\underline{x} = (x_1^1, x_2^1, x_3^1, x_1^2, x_2^2, x_3^2)$, where we order the three components of type 1 from left to right in Figure 1, and similar for the three components of type 2. To calculate $\Phi(1, 2)$, we consider all such vectors $\underline{x}$ with $x_1^1 + x_2^1 + x_3^1 = 1$ and $x_1^2 + x_2^2 + x_3^2 = 2$, so precisely 1 component of type 1 and 2 components of type 2 function. There are 9 such vectors, for only one of these, namely $(1, 0, 0, 1, 0, 1)$, the system functions. Due to the *iid* assumption for the failure times of components of the same type, and independence between components of different types, all these 9 vectors have equal probability to occur, hence $\Phi(1, 2) = 1/9$. To calculate $\Phi(2, 2)$ we need to check all 9 vectors $\underline{x}$ with $x_1^1 + x_2^1 + x_3^1 = 2$ and $x_1^2 + x_2^2 + x_3^2 = 2$. For 4 of these vectors the system functions, namely $(1, 1, 0, 1, 0, 1)$, $(1, 1, 0, 0, 1, 1)$, $(1, 0, 1, 1, 1, 0)$ and $(1, 0, 1, 1, 0, 1)$, so $\Phi(2, 2) = 4/9$.

We consider two cases with regard to the failure time distributions for the components. In Case A, we assume that the failure times of components of type 1 have an Exponential distribution with expected value 1, so with

$$f_1(t) = e^{-t} \quad \text{and} \quad F_1(t) = 1 - e^{-t} \tag{39}$$

and that the failure times of components of type 2 have a Weibull distribution with shape parameter 2 and scale parameter 1, so with

$$f_2(t) = 2te^{-t^2} \quad \text{and} \quad F_2(t) = 1 - e^{-t^2} \tag{40}$$

In Case B, these same probability distributions are used but for the other components type than in Case A, so the failure times of components of type

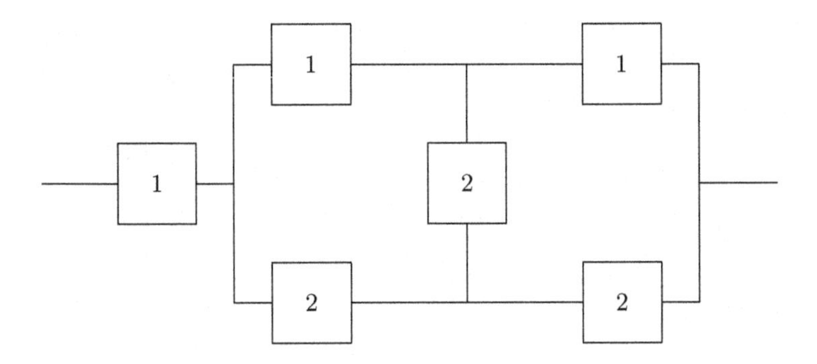

Fig. 1 System with 2 types of components

1 have the above Weibull distribution while the failure times of components of type 2 have the above Exponential distribution.

The survival functions for the failure time of this system, for both Cases A and B, are calculated using Equation (38) and are presented in Figure 2. Type 1 components are a bit more critical in this system, due to the left-most component in Figure 1. The Exponential distribution makes early failures more likely than the Weibull distribution used in this example, which leads initially to lower survival function for Case A than for Case B. It is intesting that these two survival functions cross, it would have been hard to predict this without the detailed computations.

Whilst presenting these survival functions is in itself not of major interest without an explicit practical problem being considered, the fact that the computations based on Equation (38) are straightforward indicates that the survival signature can also be used for larger and more complicated systems. Of course, deriving the survival signature itself is not easy for larger systems, this is briefly addressed in the following section.

Table 1 Survival signature of system in Figure 1

l_1	l_2	$\Phi(l_1,l_2)$	l_1	l_2	$\Phi(l_1,l_2)$
0	0	0	2	0	0
0	1	0	2	1	0
0	2	0	2	2	4/9
0	3	0	2	3	6/9
1	0	0	3	0	1
1	1	0	3	1	1
1	2	1/9	3	2	1
1	3	3/9	3	3	1

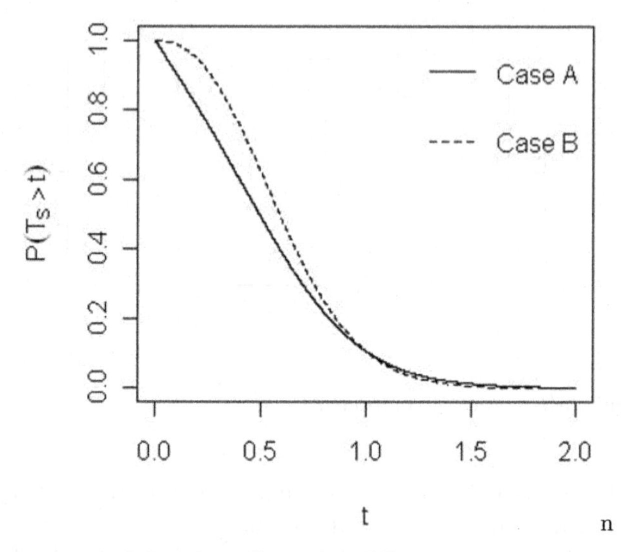

Fig. 2 System survival functions Cases A and B

4 Discussion

Computation of the signature of a system with components with *iid* failure times is difficult unless the number of components is small or the system structure is quite trivial [10, 14], and the same holds for the survival signature introduced in this paper. However, if derivation of the exact survival function for the system's failure time is required then it is unavoidable that all details of the system's structure must be taken into account, hence computation of the signature or the survival signature is necessary. For specific inferences of interest, e.g. if one wants to assess whether or not the system's reliability at a specific time t exceeds a specific value, computation of the exact (survival) signature may not be required. If one has partial information about the signature, then optimal bounds for $P(T_S > t)$ can be derived easily using the most 'pessimistic' and 'optimistic' signatures that are possible given the partial information [1]. Partial information on the survival signature is also quite straightforward to use, due to the fact that $\Phi(l)$ is increasing in l so corresponding bounds for $P(T_S > t)$ are easy to derive. This also holds for the case with multiple types of components, as $\Phi(l_1, \ldots, l_K)$ is increasing in each l_k for $k \in \{1, \ldots, K\}$. Due to the far more complex nature of the generalization of the classical signature to systems with multiple types of components, it is likely that they would be less suited for dealing with partial information, it seems of little interest to investigate this further. If such bounds for $P(T_S > t)$ are already conclusive for a specific inference, then there is no need for further computations which might reduce the effort considerably. In many applications one may not know the precise probability distribution for the components of a system. One way to deal with lack of such exact knowledge is the use of a set of probability distributions, in line with the theory of imprecise reliability [6] which uses lower and upper probabilities [5] for uncertainty quantification. Generalizing the use of (survival) signatures in order to use sets of probability distributions for the components' failure times is difficult as derivation of the optimal corresponding bounds for the survival function for the system's failure time involves multiple related constrained optimisation problems. One may also wish to use a statistical framework for inference on the survival function for the components' failure times. A particularly attractive way to do this is by using nonparametric predictive inference (NPI) [2], as it actually leads to relatively straightforward calculation of the optimal bounds for the survival function for the system's failure time, because the multiple optimisation problems involved can all be solved individually (which is rather trivial in that setting) and their optima can be attained simultaneously [4]. Currently, the generalization to systems with multiple types of components, as presented in this paper using the survival signature, is being investigated within the NPI framework. We expect to report exciting results from this research in the near future, which will generalize recent results on NPI for system reliability [3, 8, 11] to more general systems.

With a suitable generalization of the signature to systems with multiple types of components, as we believe the survival signature to be, there are many challenges for future research. For example, one may wish to decide on optimal testing in order to demonstrate a required level of system reliability, possibly taking costs and time required for testing, and corresponding constraints, into account [13]. It will also be of interest to consider possible system failure due to competing risks, where the NPI approach provides interesting new opportunities to consider unobserved or even unknown failure modes [7, 12]. Of course, the main challenges will result from the application of the new theory to large-scale real-world systems, which we expect to be more feasible with the new results presented in this paper.

References

1. Al-nefaiee, A.H., Coolen, F.P.A.: Nonparametric prediction of system failure time using partially known signatures. In: Proceedings Statistical Models and Methods for Reliability and Survival Analysis and their Validation, Bordeaux (to appear, July 2012)
2. Coolen, F.P.A.: Nonparametric predictive inference. In: Lovric (ed.) International Encyclopedia of Statistical Science, pp. 968–970. Springer, Berlin (2011)
3. Coolen, F.P.A., Aboalkhair, A.M., MacPhee, I.M.: Diversity in system reliability following component testing. The Journal of the Safety and Reliability Society 30(4), 78–93 (2010)
4. Coolen, F.P.A., Al-nefaiee, A.H.: Nonparametric predictive inference for failure times of systems with exchangeable components. Journal of Risk and Reliability (to appear, 2012)

5. Coolen, F.P.A., Troffaes, M.C.M., Augustin, T.: Imprecise probability. In: Lovric (ed.) International Encyclopedia of Statistical Science, pp. 645–648. Springer, Berlin (2011)
6. Coolen, F.P.A., Utkin, L.V.: Imprecise reliability. In: Lovric (ed.) International Encyclopedia of Statistical Science, pp. 649–650. Springer, Berlin (2011)
7. Coolen-Maturi, T., Coolen, F.P.A.: Unobserved, re-defined, unknown or removed failure modes in competing risks. Journal of Risk and Reliability 225, 461–474 (2011)
8. Coolen-Schrijner, P., Coolen, F.P.A., MacPhee, I.M.: Nonparametric predictive inference for systems reliability with redundancy allocation. Journal of Risk and Reliability 222, 463–476 (2008)
9. De Finetti, B.: Theory of Probability. Wiley, New York (1974)
10. Eryilmaz, S.: Review of recent advances in reliability of consecutive k-out-of-n and related systems. Journal of Risk and Reliability 224, 225–237 (2010)
11. MacPhee, I.M., Coolen, F.P.A., Aboalkhair, A.M.: Nonparametric predictive system reliability with redundancy allocation following component testing. Journal of Risk and Reliability 223, 181–188 (2009)
12. Maturi, T.A., Coolen-Schrijner, P., Coolen, F.P.A.: Nonparametric predictive inference for competing risks. Journal of Risk and Reliability 224, 11–26 (2010)
13. Rahrouh, M., Coolen, F.P.A., Coolen-Schrijner, P.: Bayesian reliability demonstration for systems with redundancy. Journal of Risk and Reliability 220, 137–145 (2006)
14. Samaniego, F.J.: System Signatures and their Applications in Engineering Reliability. Springer, New York (2007)

Effective Oblivious Transfer Using a Probabilistic Encryption

Alexander Frolov

Abstract. Some novel effective 1-out-of-2, 1-out-of-n, $n-1$-out-of-n, and m-out-of-n interactive and non-interactive oblivious transfer protocols (OT protocols) using a probabilistic encryption are presented. Their key information is adapted from corresponding Bellare – Rivest fractional OT protocols and the encryption is carried out on ElGamal. They can be realized in a multiplicative as well as an additive group of prime order. It is shown that due to usage of different encryption keys this implementation can be simplified in such a way that single randomizer is sufficient for all encryptions. The proposal allows to increase the information rate by $2n/(n+1)$ times and to reduce by the same factor the computational complexity of the second round phase of interactive and of the communication phase of non-interactive m-out-of-n OT protocols explored probabilistic encryption. These propositions have potential applications in all cryptographic protocols based on the m-out-of-n oblivious transfer using probabilistic encryption including generalized oblivious transfer, in particular in electronic commerce.

1 Introduction

The notion of oblivious transfer (OT) has been introduced by M. Rabin [1] to help solve the problem of "exchanging secrets," studied by M. Blum [2]. So the simplest case of OT is the transmission of one bit with the probability 1/2. Corresponding 1/2 fractional OT protocol is the protocol in which one party A can transfer a message to the other party B in such a way that:

– B receives the message with probability 1/2,
– A is oblivious as to whether the transfer was successful or not, that is A does not know whether B received the message.

In [3] using ElGamal encryption [4], the 1-out-of-2 OT protocol is established. In this protocol one party A can transfer two messages to the other party B in such a way that:

Alexander Frolov
National Research University Moscow Power Engineering Institute, 111250. Moscow, Krasnokazarmennaya, Russian Federation
e-mail: abfrolov@gmail.com

W. Zamojski et al. (Eds.): Complex Systems and Dependability, AISC 170, pp. 131–147.
springerlink.com

– B receives only one message of your choice,
– A does not know which of two messages B received.

The bibliography of initial steps of the OT protocols development can be found in [5] and in other related publications. The next most essential step was made by M. Bellare and R.L. Rivest [5]. They developed the theory of fractional oblivious transfer and proposed the interactive and non-interactive m/n fractional OT protocols in which one party A can transfer a message to the other party B in such a way that:

– B receives the message with the probability m/n,
– A is oblivious as to whether the transfer was successful or not, that is A does not know whether B received the message.

In [6] the basic notion of 1-out-of-2 OT protocol was extended to 1-out-of-n OT protocol.

The Bellare – Rivest $1/n$ fractional OT and $(n-1)/n$ fractional OT protocols generalize the 1/2 fractional OT protocols, but their other m/n OT protocols $(1<m<n-1)$ are based on the original polynomial scheme. All these protocols involve n ElGamal encryptions. In [7] this scheme is applied to calculate the public keys of m-out-of-n oblivious transfer protocols (m-out-of-n OT protocols) with repeated use of ElGamal. In such protocols, one party A can transfer n messages to the other party B in such a way that:

– B receives combination of m messages of your choice,
– A does not know which of possible messages combinations B received.

In [8] on the basis of other matrix scheme, the interactive and non-interactive m-out-of-n OT protocols which repeatedly use the Nyberg–Rueppel digital signature with message recovery [9,10] have been described.

In [11] the notion of OT was extended as a generalized oblivious transfer (GOT). In GOT protocols the retrieval restrictions are defined as a decreasing monotone collection of subsets of the given message set $\boldsymbol{U}$. In these protocols, one party A can transfer all messages of the set $\boldsymbol{U}$ to the other party B in such a way that:

– B receives all messages from only one of these subsets of your choice,
– A does not know which messages B received.

In [12] there are described GOT protocols that repeatedly involve m-out-of-n OT protocols.

The OT protocols considered it this chapter start with the *Global set-up phase* when the group G of large prime order q with intractable discrete logarithm and Diffie–Hellman problems, its generator b, its element U with unknown for both parties discrete logarithm $\log_b U$ are declared.

The interactive OT protocols involve the *First round phase* of receiver's public key set-upping and its verifiable transmission to the sender and the *Second round phase* of oblivious transmission m of n messages from the sender to the receiver.

The non-interactive OT protocols instead of the first round phase involve the *Receiver's public key set-up and certification phase* when receiver's public key is set-upped and sent to trusted center for verifying, certification and publishing. The second round phase is renamed to *Communication phase* with the same functionality.

As it has been pointed out in [5] non-interactive successive oblivious transfer are not independent: the receiver's secret key remains unchanged in successive runs of protocol.

During the second round phase of interactive protocols as well as the communication phase of non-interactive m-out-of-n OT protocols, n ElGamal encryptions or digital signatures with message recovery are involved repeatedly. In each case the encryption or signature use a new randomizer. In general, the re-use of randomizer in the probabilistic encryption schemes or digital signatures could lead to the disclosure of the message or the secret key. It will happen, when one use such schemes with the same encryption key or digital signature key. But within a single session of OT protocol, encryptions or signatures carried out repeatedly using principally different secret keys. Therefore, the re-use of randomizer is safe. As a consequence, the comlexity of sender's calculation of randomizers is reduced by n times. In the probabilistic encryption systems the amount of transmitted information as well as the number of exponentiation is reduced which makes the protocol more efficient.

The following notation is used.

The OT protocol is a traditional interactive oblivious transfer protocol;

the NIOT protocol is a traditional non-interactive oblivious transfer protocol;

the EOT and ENIOT protocols are contracted (effective) variants of the OT and NIOT protocols resp.;

the SUNIOT and SUENIOT protocols are the single use NIOT and ENIOT protocols.

In this chapter, the m/n fractional OT protocols [5] are transformed into the m-out-of-n OT protocols and the possibility of their simplifying due use of single randomizer in separate run of protocol is founded and estimated.

Let e_{OT} be the number of items passed on the second round phase or the communication phase in the performance of traditional OT protocol and e_{EOT} be the number of items passed on these phases in the performance of the corresponding simplified (effective) OT protocol. Then the ratio $\rho_{\mathrm{EOT}} = \dfrac{e_{\mathrm{OT}}}{e_{\mathrm{EOT}}}$ expresses the increase in the information rate of these phases. Simultaneously, this ratio characterizes the computational efficiency of these phases. Indeed, the numbers e_{OT} and e_{EOT} are the numbers of exponentiations during their performance. The information rate and computational complexity of the first round phase as well as of the key-set-up phase remain the same.

The chapter is organized as follows. In section 2 the traditional 1-out-of-2 OT protocols and their proposed contracted variants are discussed and the values

$\rho_{\text{1-out-of-2 EOT}}$, $\rho_{\text{1-out-of-2 ENIOT}}$, and $\rho_{\text{1-out-of-2 SUENIOT}}$ are estimated. In section 3 the important partial variants of the m-out-of-n protocols are considered and the values $\rho_{\text{1-out-of-}n\text{ EOT}}$, $\rho_{\text{1-out-of-}n\text{ ENIOT}}$, and $\rho_{\text{1-out-of-}n\text{ SUENIOT}}$, $\rho_{n\text{-1-out-of-}n\text{ EOT}}$, $\rho_{n\text{-1-out-of-}n\text{ ENIOT}}$, and $\rho_{n\text{-1-out-of-}n\text{ SUENIOT}}$ are estimated. Section 4 is devoted to common m-out-of-n protocols involving Bellare – Rivest polynomial scheme and the values $\rho_{m\text{-out-of-}n\text{ EOT}}$, $\rho_{m\text{-out-of-}n\text{ ENIOT}}$, and $\rho_{m\text{-out-of-}n\text{ SUENIOT}}$ are estimated. In section 5 we recall the concept of GOT protocol introduced in [12] to show how one can improve its effectiveness implementing m-out of-n EOT protocols. In section 6 the EOT and ENIOT protocols on the non-supersingular elliptic curve of large order are presented. In conclusion (section 7) the field of application of proposed protocols is characterized and one open question is discussed.

Considering the protocols, we omit the descriptions of the global set-up phase. We only recall the key set-up procedures and prove the safety of ElGamal encryptions implementations simplifying. The protocols in sections 2, 3, and 4 are described in terms of multiplicative group G but each of them can be transformed as a protocol in the additive group replacing multiplication and exponentiations by additions and scalar multiplications. Protocols in section 6 are given in the additive group.

We also suppose that the transmitted messages are represented as the elements of the group G.

Remark. To simplify the calculations, one can represent messages as binary codes of limited length k—1 and use the invertible mapping $\psi: G \to F_2^k$ [13]. But we will not use this method because it is not directly connected with the novelty of our proposal.

2 Effective 1-Out-of-2 OT- and NIOT Protocols

The traditional interactive and non-interactive 1-out-of-2 OT protocols are described as in original papers [3] and in textbooks [13, 14].

The functionality of 1-out-of-2 OT protocol has been described in section 1. The key set-up procedure is the following. The receiver B chooses the secret key x, $0<x<q$, and permutation π on the set $\{1, 2\}$ such that $\pi(1)=i$, where i is an index of chosen message m_i, then B calculates $\beta_{\pi(1)} = b^x$ and $\beta_{\pi(2)} = Ub^{-x}$, the receiver public key is (β_1, β_2). The receiver B knows the discrete logarithm of only element β_i and does not possesses any information on $\log_b \beta_{3-i}$. The checking procedure performed by the sender A or the trusted center T consists of verification of the predicate $\beta_1\beta_2=U$.

The second round phase of 1-out-of-2 OT- and the communication phase of 1-out-of-2 NIOT protocols are the following:

A chooses at random two distinct numbers $0<y_1$, $y_2<q-1$, and sends to B two pairs of group G elements

$$(c_1, c_2) = (b^{y_1}, b^{y_2})$$

and

$$(\alpha_1, \alpha_2) = (m_1 \beta_1^{y_1}, m_2 \beta_2^{y_2}).$$

This information corresponds to two ElGamal cryptograms

$$(b^{y_1}, m_1 \beta_1^{y_1}) \text{ and } (b^{y_2}, m_2 \beta_2^{y_2}).$$

B calculates

$$\alpha_i c_i^{-x} = \alpha_i b^{-y_i x} = m_i \beta_i^{y_i} b^{-y_i x} = m_i \beta_i^{y_i} \beta_i^{-y_i} = m_i.$$

Note that the randomly selected numbers y_1 and y_2 could be the same and equal to the number y: $y_1 = y_2 = y$.

Claim 2.1. *The reuse of randomizer in the* 1-out-of-2 *OT- and 1-out-of-2 NIOT* protocol*s is safe. The problem of deriving of the second message by receiver and Diffie — Hellman problem are polynomially equivalent.*

Proof. Let the randomizers be the same: $y_1 = y_2 = y$. Note that knowing of $b^y =$ and $b^{x'} = Ub^{-x} = \beta_{3-i}$ is not sufficient to calculate $\beta_{3-i}{}^y = b^{x'y}$, since in this case solving of the Diffie — Hellman problem or discrete logarithm problem is required. Moreover, the additional knowing of m_i does not allow to calculate m_{3-i}, because different secret keys x and x' of ElGamal cryptosystem have been used for the calculation of β_i and β_{3-i}. So B cannot compute m_{3-i} and randomizer reuse in separate performances as of the 1-out-of-2 OT- and the 1-out-of-2 NIOT protocols is safe for the sender: A can be sure that only one message has been received. At the same time the recipient, keeping x and i secret, still convinced that A can not distinguish which of two elements is an element β_1.

In this situation, party B knows elements x, b, $\beta_i = b^x$, $\beta_{3-i} = b^{x'}$, $\beta_1 \neq \beta_2$, $c = b^y$, $d_i = m_i \beta_i^y = m_i b^{xy}$, $m_{3-i} \beta_{3-i}^y$ and $U = \beta_1 \beta_2$. Let B2.1 be an algorithm for finding of m_{3-i} under these conditions. Then it allows to solve the Diffie — Hellman problem: there are given b, b^y, b^x, find b^{xy}. The solution is the following: take an arbitrary number z, $1 < z < q$, $\beta_1 = b^z \neq b^x$, put $\beta_2 = b^x$, compute $U = \beta_1 \beta_2$, and choose arbitrary elements d_1 and d_2 of group G, $1 < d_1$, $d_2 < q$, supposing that $d_1 = m_1 \beta_1^y = m_1 b^{xy}$, $d_2 = m_2 \beta_2^y$. Then using b^y and applying the breaking algorithm B2.1 one could get m_2 and later compute $\beta_2^y = b^{xy}$. On the other hand, if one could compute $\beta_{3-i}^y = b^{x'y}$ then m_{3-i} could be computed. Hence the problem of the second message deriving and Diffie — Hellman problem are polynomially equivalent. This completes the proof.

It follows that the second round phase of the 1-out-of-2 OT protocol as well as the communication phase of the 1-out-of-2 NIOT protocol can be simplified as follows:

A checks that $C = \beta_1 \beta_2$, chooses random number $0 < y < q-1$, calculates and sends to *B* the element $c = b^y$ and the pair of group *G* elements

$$(\alpha_1, \alpha_2) = (m_1 \beta_1^y, m_2 \beta_2^y).$$

This information corresponds to two ElGamal cryptograms

$$(b^y, m_1\beta_1^{\ y}) \text{ and } (b^y, m_2\beta_2^{\ y}).$$

B calculates

$$\alpha_i c^{-x} = \alpha_i b^{-yx} = m_i \beta_i^y b^{-yx} = m_i \beta_i^y \beta_i^{-y} = m_i.$$

Corollary 2.1. *The effectiveness of the second round phase of* 1-*out-of-*2 *EOT- and of the communication phase of* 1-*out-of-*2 *ENIOT* protocol *is estimated as following*: $\rho_{\text{1-out-of-2 EOT}} = \rho_{\text{1-out-of-2 ENIOT}} = 4/3$, and $\rho_{\text{1-out-of-2 SUENIOT}} = 2$.[1]

3 Effective 1-Out-of-*n* and *n*−1 of *n* Protocols

The 1-Out-of-n Protocols

The 1-out-of-*n* OT protocol can be obtained using key-information of the 1/*n* fractional OT protocol [5]. The receiver's key set-up procedure is the following.

The receiver *B* chooses at random the secret key (x, i), $0<x<q\text{-}1$, $i\in\{0, 1, n-1\}$ where *i* is an index of chosen message m_i, calculates $\beta_i=b^x$ and $\beta_j=\beta_i U^{\,j-i}$ for $j\neq i$. The receiver public key is $(\beta_0, \beta_1, \ldots, \beta_{n-1})$. The receiver *B* knows the discrete logarithm $x=x_i$ of element β_i only and does not possesses any information on $\log_b\beta_j=x_j$, $j\neq i$. The checking procedure performed by the sender *A* or the trusted center *T* consists of verification of the predicates $\beta_j=\beta_0 U^j$ for all $j=0,\ldots,n-1$.

The second round phase of the 1-out-of-*n* OT protocol and the communication phase of the 1-out-of-*n* NIOT protocol is the following:

A chooses at random *n* distinct numbers $0<y_0, y_1,\ldots,y_{n-1}<q$, and sends to *B* two *n*-tuples of group *G* elements

$$(c_0, c_1, \ldots, c_{n-1}) = (b^{y_0}, b^{y_1}, \ldots, b^{y_{n-1}})$$

and

$$(\alpha_0, \alpha_1, \ldots, \alpha_{n-1}) = (m_0\beta_0^{\ y_0}, m_1\beta_1^{\ y_1}, \ldots, m_{n-1}\beta_{n-1}^{\ \ y_{n-1}}).$$

This information corresponds to *n* ElGamal cryptograms

$$(b^{y_j}, m_j\beta_j^{\ y_j}), j=0,\ldots,n-1.$$

B calculates

$$\alpha_i c_i^{-x} = \alpha_i b^{-y_i x} = m_i \beta_i^{y_i} b^{-y_i x} = m_i \beta_i^{y_i} \beta_i^{-y_i} = m_i.$$

Note that the randomly selected numbers $y_1, y_2, \ldots, y_n$ could be the same and equal to the number *y*: $y_1=y_2=\ldots=y_n=y$.

[1] Here and below estimating the single using non-interactive protocols, we take into account that the sender public key $c=b^y$ could be computed and published in advance.

Claim 3.1. *The multiple reuse of randomizer in the 1-out-of-n OT and 1-out-of- n NIOT protocols is safe. The problem of deriving of the second message by receiver and Diffie — Hellman problem are polynomially equivalent.*

Proof. Let the randomizers be the same: $y_1 = y_2 = \ldots = y_n = y$. Note that knowing of b^y and $b^{x_j} = \beta_j$, $j \neq i$, is not sufficient to calculate $\beta_j^{\ y} = b^{x_j y}$ since in this case solving of the Diffie — Hellman problem or discrete logarithm problem is required. Moreover, the additional knowledge of m_i, does not allow calculate m_j since different secret keys x_i and x_j of ElGamal cryptosystem have been used for the calculation of β_i and β_j. So B cannot compute m_j and randomizer reuse in separate performances as the 1-out-of-n OT and the 1-out-of-n NIOT protocols is safe for the sender: A can be sure that only one message has been received. At the same time the recipient, keeping x and i secret, still convinced that A cannot distinguish which of n elements is an element β_1.

In this situation, there are known elements x_i, b, $\beta_i = b^{x_i}$, $c = b^y$, $\beta_j = b^{x_j}$, $m_i\beta_i^y = m_i b^{x_i y}$, $m_j\beta_j^y = m_j b^{x_j y}$, and U. Let B3.1 be an algorithm for finding of m_j under these conditions. Then it allows to solve the Diffie — Hellman problem: there are given b, b^y, b^x, find b^{xy}. The solution is the following: take an arbitrary number z, $1<z<q$, $b^z \neq b^x$, pick $\beta_0 = b^x$, compute $\beta_1 = b^z$, $U = b^z / b^x$, and choose arbitrary nonzero elements d_0 and d_1 of group G, $1< d_0, d_1<q$, supposing that $d_0 = m_0\,\beta_0^{\,y} = m_0\, b^{x\,y}$, $d_1 = m_1\beta_1^{\,y}$. Then using b^y and applying the breaking algorithm B3.1 one could get m_1 and later compute $\beta_1^{\,y} = = b^{xy}$. On the other hand, if one could compute $\beta_j^y = b^{x'y}$ then m_j could be computed. Hence the problem of the second message deriving and Diffie — Hellman problem are polynomially equivalent. This completes the proof.

It follows that the second round phase of the 1-out-of-n OT protocol as well as the communication phase of the 1-out-of-n NIOT protocol can be simplified as follows:

A chooses at random number $0<y<q$, calculates and sends to B element $c = b^y$ and n-tuple of group G elements

$$(\alpha_0,\alpha_1,\ldots,\alpha_{n-1}) = (m_0\beta_0^{\ y}, m_1\beta_1^{\ y}, \ldots, m_{n-1}\beta_{n-1}^{\ \ y}).$$

This information corresponds to n ElGamal cryptograms

$$(b^y, m_j\beta_j^{\ y}), j=0,\ldots,n-1.$$

B calculates

$$\alpha_i c^{-x} = \alpha_i b^{-yx} = m_i\beta_i^y b^{-yx} = m_i\beta_i^y\beta_i^{-y} = m_i.$$

Corollary 3.1. *The effectiveness of the second round phase of the 1-out-of-n EOT and of the communication phase of the 1-out-of-n n ENIOT* protocols *is estimated as following*:

$$\rho_{\text{1-out-of-}n\text{ EOT}} = \rho_{\text{1-out-of-}n\text{ ENIOT}} = 2n/(n+1), \textit{ and } \rho_{\text{1-out-of-}n\text{ SUENIOT}} = 2.$$

The n—1-Out-of-n Protocols

The n–1-out-of-n OT protocol can be obtained using key-information of the $(n-1)/n$ fractional OT protocol [5]. The receiver's key set-up procedure is the following.

The receiver B chooses at random n distinct numbers $x_1, x_2,\ldots,x_n$ such that $0<x_i<q$, $i\in\{1,\ldots, n\}$, $x_1+x_2+\ldots+x_n =q$, and a permutation π on the set $\{1,\ldots, n\}$ such that $\pi(i) =j_i$, $i=1,\ldots,n-1$, where j_i are indices of chosen messages m_{j_i}, calculates $\beta_{\pi(i)} = b^{x_i}$, $i=1,\ldots, n-1$, $\beta_{\pi(n)} = Ub^{x_n}$. The receiver's public key is $(\beta_1, \beta_2, \ldots, \beta_n)$. The receiver B knows the discrete logarithms x_i of elements $\beta_{\pi(i)}= \beta_{j_i}$, $i=1, \ldots, n-1$, and does non possesses any information on $\log_b\beta_{\pi(n)}$. The checking procedure performed by the sender A or the trusted center T consists of verification of the predicate $\Pi_{i=1}^{n}\beta_i = U$.

The second round phase of the $n-1$-out-of-n OT protocol and of the communication phase of the $n-1$-out-of-n NIOT protocol is the following:

A chooses at random n distinct numbers $0<y_1, y_2,\ldots,y_n<q$, and sends to B two n-tuples of group G elements

$$(c_1,c_2,\ldots,c_{n-1})=(b^{y_1},b^{y_2},\ldots,b^{y_n})$$

and

$$(\alpha_1,\alpha_2,\ldots,\alpha_n)=(m_1\beta_1^{\,y_1},m_2\beta_2^{\,y_2},\ldots,m_n\beta_n^{\,y_n}).$$

This information corresponds to n ElGamal cryptograms

$$(b^{y_i},m_i\beta_i^{y_i}),\ i=1,\ldots,n.$$

For $i=1,\ldots, n-1$, B calculates

$$\alpha_{\pi(i)}c_{\pi(i)}^{-x_i} = \alpha_{\pi(i)}b^{-y_{\pi(i)}x_i} = m_{\pi(i)}\beta_{\pi(i)}^{y_{\pi(i)}}b^{-y_{\pi(i)}x_i} =$$
$$= m_{\pi(i)}\beta_{\pi(i)}^{y_{\pi(i)}}\beta_{\pi(i)}^{-y_{\pi(i)}} = m_{\pi(i)} = m_{j_i}.$$

Note that the randomly selected numbers $y_1, y_2, \ldots, y_n$ could be the same and equal to the number y: $y_1 =y_2 =\ldots=y_n=y$.

Claim 3.2. *The multiple reuse of randomizer in n–1-out-of-n OT and n–1-out-of-n NIOT protocol is safe. The problem of deriving of the n-th message by receiver and Diffie — Hellman problem are polynomially equivalent.*

Proof. Let the randomizers be the same: $y_1 =y_2 =\ldots=y_n =y$. Note that knowing of b^y and $b^x = \beta_{\pi(n)}$, is not sufficient to calculate $\beta_{\pi(n)}^{\ y} = b^{xy}$ since in this case solving of the Diffie—Hellman problem or discrete logarithm problem is required. Moreover, the additional knowledge of all messages $m_{\pi(i)}$, $i\neq n$, does not allow to calculate $m_{\pi(n)}$, since different secret keys x_i and x of ElGamal cryptosystem have been used for the calculation of $\beta_{\pi(i)}$ and $\beta_{\pi(n)}$. So randomizer reuse in separate performances of the $n-1$-out-of-n OT protocol as well as $n-1$-out-of-n NIOT protocol is safe for the sender: A can be sure that only $n-1$ messages have been received. At the same time the recipient, keeping $(x_1, x_2, \ldots, x_{n-1})$ and π secret, still convinced that A cannot distinguish which of n elements is an element $\beta_{\pi(n)}$.

In this situation, for i=1,2, ..., $n-1$, there are known the elements x_i, $b, d_{\pi(i)} = m_{\pi(i)}\beta^y_{\pi(i)}$, $\beta_{\pi(i)}=b^{x_i}$, $c = b^y$, $\beta_{\pi(n)}=b^x$, and U. Let B3.2 be an algorithm for finding of $m_{\pi(n)}$ under these conditions. Then it allows to solve the Diffie—Hellman problem: there are given b, b^y, b^x, find b^{xy}. The solution is the following: take distinct arbitrary numbers $x_1, x_2, \ldots, x_{n-1}$ such that $0<x_i<q$, for $i=1,\ldots,n-1$, calculate $\beta_i^y = (b^y)^{x_i} = b^{x_i y}$, compute $U = b^x \Pi_{i=1}^{n-1}\beta_i$. Hence $\beta_n = b^x = Ub^{x_n}$, $x_1+x_2+\ldots+x_{n-1}+x_n=q$. Then choose distinct elements $d_1, \ldots, d_{n-1}$, d_n, supposing that for all $i=1,\ldots,n$ $1<d_i<q$, $d_i=m_i\beta_i^y$. Now applying breaking algorithm B3.2 and using b^y one can compute m_n and later $\beta_n^y = b^{xy}$. On the other hand, if the receiver could compute $\beta_{\pi(n)}{}^y = b^{xy}$ then $m_{\pi(n)}$ could be computed. Hence the problem of the n-th message deriving and Diffie—Hellman problem are polynomially equivalent. This completes the proof.

It follows that the second round phase of the $n-1$-out-of-n OT protocol as well as communication phase of the $n-1$-out-of-n NIOT protocol can be simplified as follows:

A chooses at random number $0<y<q$, calculates and send to B element $c = b^y$ and n-tuple of group G elements

$$(\alpha_1,\alpha_2,\ldots,\alpha_n)=(m_1\beta_1{}^y, m_2\beta_2{}^y,\ldots,m_n\beta_n{}^y).$$

This information corresponds to n ElGamal cryptograms

$$(b^y, m_i\beta_i{}^y),\ i=1,\ldots,n.$$

For $i=1,\ldots,n-1$, B calculates

$$\alpha_{\pi(i)}c^{-x_i} = \alpha_{\pi(i)}b^{-yx_i} = m_{\pi(i)}\beta^y_{\pi(i)}b^{-yx_i} =$$
$$= m_{\pi(i)}\beta^y_{\pi(i)}\beta^{-y}_{\pi(i)} = m_{\pi(i)} = m_{j_i}.$$

Corollary 3.2. *The effectiveness of the second round phase of the $n-1$-out-of-n EOT and of the communication phase of the $n-1$-out-of-n ENIOT protocols is estimated as following*:

$$\rho_{n-1\text{-out-of-}n\text{ EOT}} = \rho_{n-1\text{-out-of-}n\text{ ENIOT}} = 2n/(n+1), \textit{ and } \rho_{n-1\text{-out-of-}n\text{ SUENIOT}} = 2.$$

4 Effective *m*-Out-of-*n* Protocols

The Global set-up phase of this protocol is completed by letting $\alpha_0=1\in Z_q$, and fixing of n distinct elements $\alpha_1,\ldots,\alpha_n$ of $Z_q\setminus\{\alpha_0\}$. All these elements are public.

The m-out-of-n OT protocol can be obtained using key-information of the m/n fractional OT protocol [5].

The receiver's key set-up procedure is the following.

The receiver's public key is a vector $(\beta_1, \beta_2, \ldots, \beta_n, W_0, W_1, \ldots, W_m)\in G^{n+m+1}$.

To compute it, the receiver B chooses at random a size m subset of $[n]=\{1,2,\ldots,n\}$ specifying an injective map $\pi:[m]\rightarrow[n]$, where $\pi(1),\ldots, \pi(m)$ are the m chosen indices. Later he chooses elements $x_{\pi(1)},\ldots, x_{\pi(m)}\in Z_n$ at random and sets $\beta_{\pi(i)}=b^{x_{\pi(i)}}\in G$ for $i=1,\ldots,m$. This specifies m elements of $\beta_1, \beta_2,\ldots,\beta_n$, in such a way that receiver B knows their discrete logarithms. The other $n-m$ elements have to be specified in such a way that B doesn't know and cannot compute their discrete logarithms. To compute these $n-m$ elements, B first compute elements $W_0, W_1,\ldots,W_m$ as follows.

The receiver B defines $m+1$ by $m+1$ matrix

$$\mathbf{A}=\begin{bmatrix} \alpha_0^0 & \alpha_0^1 & \cdots & \alpha_0^m \\ \alpha_{\pi(1)}^0 & \alpha_{\pi(1)}^1 & \cdots & \alpha_{\pi(1)}^m \\ \vdots & \vdots & \vdots & \vdots \\ \alpha_{\pi(m)}^0 & \alpha_{\pi(m)}^1 & \cdots & \alpha_{\pi(m)}^m \end{bmatrix}.$$

This is Vandermonde matrix over the field Z_q. It is invertible. B computes its inverse

$$\mathbf{B}=\mathbf{A}^{-1}=\begin{bmatrix} \beta_{0,0} & \beta_{0,1} & \cdots & \beta_{0,m} \\ \beta_{1,0} & \beta_{1,1} & \cdots & \beta_{1,m} \\ \vdots & \vdots & \vdots & \vdots \\ \beta_{m,0} & \beta_{m,1} & \cdots & \beta_{m,n} \end{bmatrix}.$$

B now sets

$$W_0=U^{\beta_{0,0}}\cdot\Pi_i^m\beta_{\pi(i)}^{\beta_{0,1}},$$

$$W_1=U^{\beta_{1,0}}\cdot\Pi_i^m\beta_{\pi(i)}^{\beta_{1,1}},$$

$$\vdots$$

$$W_m=U^{\beta_{m,0}}\cdot\Pi_i^m\beta_{\pi(i)}^{\beta_{m,1}}.$$

Finally, B specifies the remaining elements of the public key:

$$\beta_i=\Pi_{j=0}^m W_j^{\alpha_i^j}$$

for all $i\in[m]$ that are not in the range of π.

The verification procedure performed by the sender A or the trusted center T is the following:

1) check that the public key consists of $m+n+1$ elements of G;
2) for all $i=1,\ldots,n$ check the predicates

$$U = \Pi_{j=0}^{m} W_j;$$

$$\beta_i = \Pi_{j=0}^{m} W_j^{\alpha_i^j}.$$

The second round phase of the m-out-of-n OT protocol and the communication phase of the m-out-of-n NIOT protocol are the following:

A chooses at random n distinct numbers $0<y_1, y_2,\ldots,y_n<q$, and sends to B two n-tuples of group G elements

$$(c_1,c_2,\ldots,c_{n-1}) = (b^{y_1}, b^{y_2},\ldots,b^{y_n})$$

and

$$(\alpha_1,\alpha_2,\ldots,\alpha_n)=(m_1\beta_1^{y_1}, m_2\beta_2^{y_2},\ldots,m_n\beta_n^{y_n}).$$

This information corresponds to n ElGamal cryptograms

$$(b^{y_i}, m_i\beta_i^{y_i}),\ i=1,\ldots,n.$$

For $i=1,\ldots,n-1$, B calculates

$$\alpha_{\pi(i)}c_{\pi(i)}^{-x_i} = \alpha_{\pi(i)}b^{-y_{\pi(i)}x_i} = m_{\pi(i)}\beta_{\pi(i)}^{y_{\pi(i)}}b^{-y_{\pi(i)}x_i} =$$
$$= m_{\pi(i)}\beta_{\pi(i)}^{y_{\pi(i)}}\beta_{\pi(i)}^{-y_{\pi(i)}} = m_{\pi(i)} = m_{j_i}.$$

Note that the randomly selected numbers $y_1, y_2, \ldots, y_n$ could be the same and equal to the number y: $y_1=y_2=\ldots=y_n=y$.

Claim 4.1. *The multiple reuse of randomizer in the m-out-of-n OT and the m-out-of-n NIOT protocols is safe.*

Proof. Let the randomizers be the same: $y_1=y_2=\ldots=y_n=y$. Note that knowing of b^y and $b^{x_i}=\beta_i$, i is not in the range of π, is not sufficient to calculate $\beta_i^y=b^{x_i y}$ since in this case solving of the Diffie — Hellman or discrete logarithm problems is required. Moreover, the additional knowledge of all messages $m_{\pi(i)}$ does not allow to calculate m_i, since different secret keys $x_{\pi(i)}$ and x_i of ElGamal cryptosystem have been used for the calculation of $\beta_{\pi(i)}$ and β_i. So randomizer reuse in separate performances as of the m-out-of-n OT protocol and the m-out-of-n NIOT protocols is safe for the sender: A can be sure that only m messages have been received. At the same time the recipient, keeping $(x_1, x_2, \ldots, x_{m-1}, x_m)$ and π

secret, still convinced that A cannot distinguish which of n elements are the elements $\beta_{\pi(I,)}$. $i=1, \ldots, m$. This completes the proof.

It follows that the second round phase of the *m-out-of-n* OT protocol as well as the communication phase of the *m*-out-of-*n* NIOT protocol can be simplified as follows:

A chooses at random the number $0<y<q$, calculates and sends to B element $c=b^y$ and n-tuple of group G elements

$$(\alpha_1,\alpha_2,\ldots,\alpha_n)=(m_1\beta_1^{\ y},m_2\beta_2^{\ y},\ldots,m_n\beta_n^{\ y}).$$

This information corresponds to n ElGamal cryptograms

$$(b^y,m_i\beta_i^{\ y}),\ i=1,\ldots,n.$$

For $i=1,\ldots, n-1$, B calculates

$$\alpha_{\pi(i)}c^{-x_i}=\alpha_{\pi(i)}b^{-yx_i}=m_{\pi(i)}\beta_{\pi(i)}^{y}b^{-yx_i}=$$
$$=m_{\pi(i)}\beta_{\pi(i)}^{y}\beta_{\pi(i)}^{-y}=m_{\pi(i)}=m_{j_i}.$$

Corollary 4.1. *The effectiveness of the second round phase of the m-out-of-n EOT and of the communication phase of the m-out-of-n ENIOT protocols are estimated as following*:

$$\rho_{m\text{-out-of-}n\ \text{EOT}}=\rho_{m\text{-out-of-}n\ \text{ENIOT}}=(2n/(n+1),\ \textit{and}\ \rho_{m\text{-out-of-}n\ \text{SUENIOT}}=2.$$

The following question remains open: let B4.1 be breaking algorithm for computing of m_i, i is not in the range of π under the conditions when B knows elements $x_{\pi(i)}$, $m_{\pi(i)}(\beta_{\pi(i)}{}^{y})$, $\beta_{\pi(i)}=b^{x_{\pi(i)}}$ for $i=1,\ldots, m$, $c=b^y$, and U. Is it possible to solve Diffie – Hellman problem using this algorithm?

Remark that the simplest variant of generalized OT protocol can be organized using the logic of *m*-out-of-*n* OT protocol. This special GOT protocol is the at-most-*m*-out-of-*n* OT protocol that allows sending n messages from which the receiver could get at most m. This protocol can be implemented as the *m-out-of-m+n* OT protocol if transferred m messages $m_{n+1},\ldots,m_{n+m}$ are randomly chosen group G elements. Receiver obtains m out of $m+n$ messages including m' informative messages and $m-m'$ of these auxiliary codes which could be returned to sender on the level of commercial relations.

For these protocol, $\rho_{\text{at-most-}m\text{-out-of-}n\ \text{EOT}}=\rho_{\text{at-most-}m\text{-out-of-}n\ \text{ENIOT}}=2(n+m)/(n+m+1)$, $\rho_{\text{at-most-}m\text{-out-of-}n\ \text{SUENIOT}}=2$.

5 The Effective GOT Protocols

In this section, we recall the GOT protocol introduced in [12] to show how one can improve its effectiveness implementing *m*-out of-*n* EOT protocols. Let ***U*** be the collection of messages M_i embedded in the field F, $|U|=n$, and $\boldsymbol{A}$ be the monotone decreasing collection of subsets of the set U, that B is allowed to retrieve. The collection $\boldsymbol{A}$ is represented by the collection $\boldsymbol{A}^0$ of maximal elements of $\boldsymbol{A}$. Without loss of generality one can assume that collection $\boldsymbol{A}^0$ consists of the subsets of the same cardinality m, $m<n$.

Let Γ be monotone increasing closure of $\boldsymbol{A}^0$

$$\Gamma=\{\boldsymbol{C}\subseteq\boldsymbol{U}:\exists\boldsymbol{B}\in\boldsymbol{A}^0,\boldsymbol{B}\subseteq\boldsymbol{C}\},$$

and Σ be a secret sharing scheme realizing Γ. Let $s\in F$ be a secret random value selected by the party A and let s_i be the corresponding share of M_i. The sender A and the receiver B engage in m-out-of-n OT protocol for the following set of pairs of values:

$$W=\{<M_i+x_i, s_i>: i=1,\dots,n\}.$$

Here, $x_i\in F$, $i=1,\dots,n$ are random and independent field elements selected by A. Getting the set of k pairs

$$<M_{i_j}+x_{i_j}, s_i>, j=1,\dots,k,$$

where $\{M_{i_1},\dots,M_{i_k}\}\in\boldsymbol{A}^0$; party B recover the secret s and sends it to the sender A. If this value is correct, A sends to B the complete set of random shifts, $\{x_1,\dots,x_n\}$. Then using $x_{i_j},\dots,x_{i_k}$ and $M_{i_1}+x_{i_1},\dots,M_{i_k}+x_{i_k}$ the receiver B can calculate $\{M_{i_1},\dots,M_{i_k}\}$. Implementing m-out-of-n EOT protocol one can improve the effectiveness of considered GOT protocol.

As we can see, interactive GOT protocol involves four phases: the *First round phase* of receiver's public key set-upping and its verifiable transmission to the sender; the *Second round phase* of oblivious transmission m of n messages from the sender to the receiver; the *Third round phase* of secret s recovering and transmitting to the sender, the *Fourth round phase* of secret verifying and sending of random and independent field elements to the receiver for recovering of m allowed messages.

Corollary 5.1. *The effectiveness of the second round phase of EGOT protocol is estimated as following*: $\rho_{EGOT}=2n/(n+1)$.

6 The Elliptic Curve EOT and NIEOT Protocols

In this section, we describe protocols defined in additive group with infeasible discrete logarithm and Diffie–Hellman problems. These groups can be obtained as subgroups of Koblitz's curves [15].

Global set-up phase. Take the non-supersingular elliptic curve

$$EC: Y^2+XY=X^3+X^2+1$$

over the field $GF(2^{163})$ in the polynomial basis

$$\lambda^{162},\ldots,\lambda^2,\lambda,1$$

generated by the root λ of irreducible polynomial

$$X^{163}+X^7+X^6+X^3+1.$$

The order of the elliptic curve group equals

N=11692013098647223345629483507196896696658237148126=
2×5846006549323611672814741753598448348329118574063.

The point

P=(1f0c047bba3cc88cc681c3fa2ae92612b01d563ba,
527e27fbc1b349f822f52352039e0613ce54ec26c)

is the point of order N. Hence the order of point

Q=2P=
=(2b0bc3a5c0e9fa0d3f9f4d897169f986d22dbf6fc,
32e817608e3ab30c2559ac3c8d0ef606522ae0e86)

is a prime number

q=5846006549323611672814741753598448348329118574063.

So we can choose the additive group $<Q>$ of large prime order q with the generator Q.

Let the element U, $U\in<Q>$ be the elliptic curve point with unknown for anyone discrete logs.

Terminating the Global set-up phase, we let $\alpha_0=1\in Z_q$, and fix n distinct elements $\alpha_1,\ldots,\alpha_n$ of $Z_q\backslash\{\alpha_0\}$. For example, $\alpha_1=2$, $\alpha_3=3,\ldots,\alpha_n=n+1$. (We assume that $n<q-1$). All these elements are public.

The receiver's key set-up procedure is the following.

The receiver's public key is a vector $(\beta_1,\beta_2,\ldots,\beta_n,W_0,W_1,\ldots,W_m)\in EC(GF(2^{163}))^{n+m+1}$. To compute it, the receiver B chooses at random a size m subset of $[n]=\{1,2,\ldots,n\}$ specifying an injective map $\pi:[m]\to[n]$, where $\pi(1),\ldots,\pi(m)$ are the m chosen indices. Later B chooses elements $x_{\pi\{1\}},\ldots,x_{\pi\{m\}}\in Z_n$ at random and sets $\beta_{\pi\{i\}}=x_{\pi(i)}Q\in G$ for $i=1,\ldots,m$. This specifies m elements out of the $\beta_1,\beta_2,\ldots,\beta_n$ in such a way that the receiver B knows their discrete logarithms. The other $n-m$ elements have to be specified in such a way that B doesn't know, and can't compute their discrete logs. To compute these $n-m$ elements, B first computes elements W_0, W_0, $W_1,\ldots,W_m$ as follows: he defines $m+1$ by $m+1$ Vandermonde matrix **A** and computes its inverse **B**.

Then B sets

$$W_0 = \beta_{0,0}U + \sum_i^m \beta_{0,i}\beta_{\pi(i)},$$
$$W_1 = \beta_{1,0}U + \sum_i^m \beta_{1,i}\beta_{\pi(i)},$$
$$\vdots$$
$$W_m = \beta_{m,0}U + \sum_i^m \beta_{m,i}\beta_{\pi(i)}$$

Finally, B specifies the remaining elements of the public key:

$$\beta_i = \sum_{j=0}^m \alpha_i^j W_j$$

for all $i \in [m]$ that are not in the range of π.

The verification procedure performed by the sender A or the trusted center T is the following.

1) Check that public key consists of $m+n+1$ elements of $G=\langle Q \rangle$.
2) Check the predicates

$$U = \sum_{j=0}^m W_j;$$
$$\beta_i = \sum_{j=1}^m \alpha_i^j W_j \text{ for all } i=1,\ldots,n.$$

Suppose that all messages m_i, $i=1,\ldots, n$ are embedded into the points M_i (using for example algorithms from [16]). The second round phase of the *m-out-of-n* OT protocol as well as the communication phase of the m-out-of-n NIOT protocol can be performed as follows:

A chooses at random the number $0<y<q$, calculates and sends to B element $c= yQ$ and n-tuple of group G elements

$$(\alpha_1,\alpha_2,\ldots,\alpha_n)=(M_1 + y\beta_1, M_2 + y\beta_2,\ldots, M_n + y\beta_n).$$

This information corresponds to n ElGamal cryptograms

$$(yQ, M_i + y\beta_i),\ i=1,\ldots, n.$$

For $i=1,\ldots, m$, B calculates

$$\alpha_{\pi(i)} - x_i c = \alpha_{\pi(i)} - yx_iQ = M_{\pi(i)} + y\beta_{\pi(i)} - yx_iQ =$$
$$= M_{\pi(i)} + y\beta_{\pi(i)} - y\beta_{\pi(i)} = M_{\pi(i)} = M_{j_i}.$$

7 Conclusion Remarks

In this chapter, some novel effective 1-out-of-2, 1-out-of-*n*, *n*–1-out-of-*n*, and *m*-out-of-*n* interactive and non-interactive oblivious transfer protocols (OT protocols) using a probabilistic encryption have been presented. Their key information is adapted from corresponding Bellare – Rivest fractional OT protocols and the encryption is carried out on ElGamal. They can be realized both in a multiplicative and in an additive group of prime order. It has been shown that through the use of different encryption keys, this implementation can be simplified using a single randomizer for all encryptions in separate session. The proposal allows to increase by $2n/(n+1)$ times the information rate and to reduce by the same factor the computational complexity of the second round phase of interactive and of the communication phase of non-interactive *m*-out-of-*n* OT protocols explored probabilistic encryption. To validate the proposed approach in the development of application systems following [8], one can consider a straightforward application of the proposed protocols in electronic commerce. Using the protocol in section 6 as an example we can construct an online video shop. In this example, party *A* is the Internet merchant who wants to sell videos over the Internet, while the clients can get what they want without revealing which ones they have selected. Let the collection of videos contains 1000 samples and client want to get some of them obliviously. Then on the commutative phase of EOT protocol 1001 exponentiations have to be involved by sender and the same number of items (elliptic curve points) have to be sent from *A* to *B*, i.e. 326 326 bits, instead of 2000 exponentiations and 652000 bits in the case of implementing of OT protocol. The propositions of this chapter have potential applications in all cryptographic protocols based on the *m*-out-of-*n* oblivious transfer using probabilistic encryption including generalized oblivious transfer, in particular in electronic commerce. The most important is application of OT protocols for multiparty computations [17, 18, 19], in particular for oblivious polynomial evaluation [20, 21]. The following question remains open: are the problem of deriving of the extra message by receiver participated in *m*-out-of-*n* OT protocol and Diffie–Hellman problem polynomially equivalent?

Acknowledgement. This research has been supported by Russian Foundation of Basic Research, project, 11-01-00792-a.

References

[1] Rabin, M.O.: How to exchange secrets by oblivious transfer. Technical Report TR-81, Aiken Computation Laboratory, Harvard University (1981)
[2] Blum, M.: How to exchange (secret) keys. Trans. Computer Systems 1, 175–193 (1983)
[3] Even, S., Goldreich, O., Lempel, A.: A randomized protocol for signing contracts. Communications of the ACM 28, 637–647 (1985)
[4] ElGamal, T.: A public-key cryptosystem and a signature scheme based on discrete logarithms. IEEE Trans. Inform. Theory IT-31(4), 469–472 (1985)

[5] Bellare, M., Rivest, R.L.: Translucent cryptography – an alternative to key escrow, and its implementation via fractional oblivious transfer. MIT/LCS Technical Report 683 (1990)
[6] Brasard, G., Crépeau, C., Robert, J.M.: Oblivious transfer and intersecting codes. IEEE Transaction of Information Theory, Special Issue on Coding and Complexity 42, 1769–1780 (1996)
[7] Mamontov, A.I., Frolov, A.B.: On one scheme for oblivious transfer of combinations of messages. MPEI Bulletin 3, 113–119 (2005) (in Russian)
[8] Mu, Y., Zhang, J., Varadharajan, V.: *m* out of *n* Oblivious Transfer. In: Batten, L.M., Seberry, J. (eds.) ACISP 2002. LNCS, vol. 2384, pp. 395–405. Springer, Heidelberg (2002)
[9] Nyberg, K., Rueppel, R.A.: A new signature scheme based on the DSA giving message recovery. In: 1st ACM Conference on Computer and Communications Security, Fairfax, Virginia, pp. 58–61 (1993)
[10] Nyberg, K., Rueppel, R.A.: Message recovery for signature schemes based on the discrete logarithm problem, pp. 182–193. Springer (1994)
[11] Ishai, Y., Kushelevitz, E.: Private simultaneous messages protocols with applications. In: Proc. of ISTCS 1997, pp. 174–184. IEEE Computer Society (1997)
[12] Tassa, T.: Generalized oblivious transfer by secret sharing. Designs, Codes and Cryptography 58, 1:11–1:21 (2011)
[13] Koblith, N.: A Course in number theory and cryptography. Springer, New York (1994)
[14] Salomaa, A.: Public-key cryptography. Springer, New York (1990)
[15] Koblitz, N.: Constructing Elliptic Curve Cryptosystems in Characteristic 2. In: Menezes, A., Vanstone, S.A. (eds.) CRYPTO 1990. LNCS, vol. 537, pp. 156–167. Springer, Heidelberg (1991)
[16] Rosing, M.: Implementing elliptic curve cryptography. Manning Publications Co., Greenwich (1998)
[17] Yao, A.C.: Protocols for secure computation. In: Proc. of IEEE Foundation of Computer Science (FOCS), pp. 160–164 (1982)
[18] Goldreich, O., Vainish, R.: How to Solve Any Protocol Probleman Efficiency Improvement. In: Pomerance, C. (ed.) CRYPTO 1987. LNCS, vol. 293, pp. 73–86. Springer, Heidelberg (1988)
[19] Killian, J.: Founding cryptography on oblivious transfer. In: Proc. of the 20th Annual ACM Symposium on Theory of Computing (STOC), pp. 20–31 (1988)
[20] Noar, M., Pinkas, B.: Oblivious polynomial evaluation. In: Proc. of the 31st Annual ACM Symposium on Theory of Computing (STOC), pp. 245–254 (1999)
[21] Noar, M., Pinkas, B.: Computationally secure oblivious transfer. Journal of Cryptology 18, 1–35 (2005)

Gap-and-IMECA-Based Assessment of I&C Systems Cyber Security

Vyacheslav Kharchenko, Anton Andrashov, Vladimir Sklyar,
Andriy Kovalenko, and Olexandr Siora

Abstract. This chapter presents an approach to cyber security assessment, which is based on Gap Analysis (GA) and Intrusion Modes and Effects Criticality Analysis (IMECA) techniques, applicable to complex Instrumentation and Control (I&C) systems, including safety-critical FPGA-based I&C systems. Elements of the GA-and-IMECA procedure of assessment are proposed. As an example, the proposed approach and technique are considered in the context of assessing the cyber security properties of FPGA-based I&C systems, taking into account vulnerabilities of products and discrepancies of appropriate processes.

1 Introduction

1.1 Motivation

I&C systems are complex systems that consist of both hardware and software components, which continuously interact with each other in order to perform their intended functions. One of the development and operation problems of modern I&C systems for critical application is the reliable assessment and assurance of the two main system attributes, namely safety and cyber security. The assessment of cyber security, which also influences the safety of I&C systems and other

Vyacheslav Kharchenko
National Aerospace University "KhAI", Ukraine, 61070, Kharkov, 17, Chkalov str.
e-mail: v.kharchenko@khai.edu

Andriy Kovalenko · Vyacheslav Kharchenko
Centre for Safety Infrastructure-Oriented Research and Analysis, Ukraine, 61085, Kharkov, 37, Astronomicheskaya str.
e-mail: a.kovalenko@csis.org.ua

Anton Andrashov · Vladimir Sklyar · Olexandr Siora
Research and Production Corporation "Radiy", Ukraine, 25006, Kirovograd, 29, Geroyev Stalingrada str.
e-mail: {a.andrashov,v.sklyar,marketing}@radiy.com

W. Zamojski et al. (Eds.): Complex Systems and Dependability, AISC 170, pp. 149–164.
springerlink.com

controlled applications, is a very important, complicated, and challenging problem. During the assessment, it is necessary to take into account a set of various features and factors, their interrelations and interactions. Modern realities require improving I&C systems security, both in terms of requirements and their implementation. Moreover, assurance of cyber security for critical I&C systems is a requirement of national and international regulatory documents, as well as actual practice in safety engineering [1].

The Field Programmable Gate Arrays (FPGA) technology is now being widely used worldwide in process industries, and increasingly in I&C systems for various safety and security critical domains, such as Nuclear Power Plants (NPPs), onboard computer-based systems, electronic medical systems, etc. [2,3]. The application of FPGA technology allows developers to implement the required functions in a convenient and reliable way.

There are several challenging problems in the area of cyber security assurance for complex FPGA-based I&C systems, including the following: consideration of all possible vulnerabilities that can appear in the final product due to process discrepancies, which were present at earlier stages of the product life cycle, prioritization of such vulnerabilities according to their criticality and severity, determination of both sufficient and cost-effective countermeasures either to eliminate the identified (or potential) vulnerabilities or to make the vulnerabilities difficult to exploit by an adversary. In our opinion, the accurate evaluation of the actual level of the vulnerabilities' criticality and severity (and security of the system in whole) is one of the main challenges. Inaccurate estimation can cause additional efforts, costs and may present undesirable level of risk. In the framework of this chapter, I&C safety is considered as an attribute of high importance. Security is an attribute, which affects safety [4].

One of the possible ways to consider all possible security vulnerabilities for complex I&C systems is using a process-product approach. Such an approach requires performance assessments not only for products (components of the I&C system received at different life cycle stages), but for all the processes within the product life cycle. Application of process-product approach is inevitable in case of FPGA-based I&C systems, due to FPGA's dual nature: it consists of both hardware and software, with its inherent complexity. Such a process-product approach should also be considered in FPGA-specific regulatory documents that would address issues such as system safety assessment, design life cycle, verification and validation, configuration management, documentation requirements, etc. in order to identify all possible discrepancies. Each discrepancy can potentially lead to the introduction of security vulnerabilities (or breaches) into the final product, during the implementation of life cycle processes.

1.2 Work Related Analysis

Authors in [5] describe security-related gaps, unique to commercial embedded system design only. Importance and uniqueness of the embedded security challenges, an enumeration of security requirements, concepts, and design challenges are presented. Though, the paper is limited to security processing requirements and architecture, illustrated with a popular secure sockets layer protocol, and processing workload example.

Paper [6] introduces the concepts of designing secure hardware in embedded systems. The major classes of attacks and the mindset of potential attackers are presented, as well as examples of previous hardware attacks are discussed. Typical product development cycle and recommends ways to incorporate security, risk assessment, and policies into the process are presented.

Failure Mode, Effects and Criticality Analysis (FMECA) is an extension of standard formalized technique called Failure Mode and Effects Analysis initially intended for the systems reliability analysis devoted to the specification of failure modes, their sources, causes, criticality, and influence on system's operability [7]. "Failure modes" means the ways, or modes, in which something within an I&C system might fail. Failures are any errors or defects in a form of deviations from normal operation, which can affect the user of I&C system, and can be potential (that can happen in future) or actual (that have already happened). "Effects analysis" refers to studying the consequences of those failures. In addition, FMECA extends FMEA (Failure Modes and Effects Analysis) by including a criticality analysis, which is used to chart the probability of failure modes against the severity of their consequences.

In the FMEA-technique, all possible failures are prioritized according to consequences severity, frequency and detectability. Such technique is used during design stages in order to avoid failures in a system being developed. During certain consequent stages it can also be used for the purpose of process control. The overall purpose of FMEA-techniques is to take actions to eliminate or reduce possible failures.

There are a lot of FMECA technique modifications related to various components, including software (SFMECA), to various levels of I&C hierarchy (HFMECA), to various processes, including design (DFMECA) and others [8,9]. In general cases, Concept and Event Modes and Effect Criticality Analysis may be considered. These modifications are not used to assess I&C security.

IMECA (Intrusion Modes and Effects Criticality Analysis) is a modification to FMECA-technique that takes into account possible intrusions into the system [10]. During the assessment of I&C systems, IMECA can be used in addition to standardized FMECA for safety-related domains, because each vulnerability can become a failure in a case of intrusion into such systems [11,12].

The objectives of this chapter are to customize the IMECA-technique and to develop an applicable approach to assessment the level of I&C systems cyber security. The rest of the paper is structured as follows: Section 2 describes the underlying concepts of the gap-and-IMECA-based approach, as well as its application to assessment of safety-critical I&C systems. Section 3 provides a methodological-level interpretation of the proposed approach in the context of cyber security of FPGA-based I&C systems.

2 General Gap-and-IMECA-Based Approach to Assessment of Safety-Critical I&C Systems

2.1 Conception of Gap-and-IMECA-Based Approach

Here, as one of possible solutions for I&C systems assessment problem, we propose an approach, which is based on IMECA technique.

One of the fundamental concepts behind the idea of the approach is the concept of gap. Before providing a definition for gap, we propose the taxonomy of the main notions used in the chapter. Such taxonomy covers the notions of process, product, intrusion, discrepancy, gap, anomaly, vulnerability and attack (see Fig. 1). We outlined clearly some important attributes of a process, product and intrusion, as well as their interrelations. Also, the proposed taxonomy allows tracing a case of non-ideal process in product development along with possible consequences of process implementation.

The main notions in Fig. 1 are process, product, and intrusion. Processes are being implemented through the development stages of I&C system life cycle model in order to produce products. Also, products can be vulnerable to intrusions of various types that can affect the product. Results of implementation of the processes (i.e., all the set of processes that led to the creation of the product) can have effects on possible consequential changes in such processes. Each process comprises activities, and, in a case of "non-ideal" process, some of them can contain discrepancies.

So, now we can define gap as a set of discrepancies of any single process (which can consist of a set of sub-processes) within the life cycle of I&C system that can introduce some anomalies in a product and/or cannot reveal (and eliminate) existing anomalies in a product. In particular, such anomalies can be caused by imperfection of product specification (or even representation), implementation, verification, and/or other non-compliances.

In terms of cyber security, some of the anomalies can be vulnerabilities of the product. Vulnerabilities, in turn, can be exploited by an adversary during intrusion into the product to implement an attack in order to introduce some unintended functionality into the product.

Direct relation between vulnerabilities and unintended functionality in Fig. 1 denotes some possible situation, which is not covered by the scope of this chapter; such a situation may occur in the presence of hardware Trojans within the components of the product, and, hence, requires additional comprehensive analysis.

Hence, we propose a process-based approach to GA, because "non-ideal" processes, which contain discrepancies, can produce various problems in the corresponding products, and the following statements are true:

1. Presence of gaps in $Process_j$ results in anomalies in $Product_i$ even if $Product_{i-1}$ is "ideal".

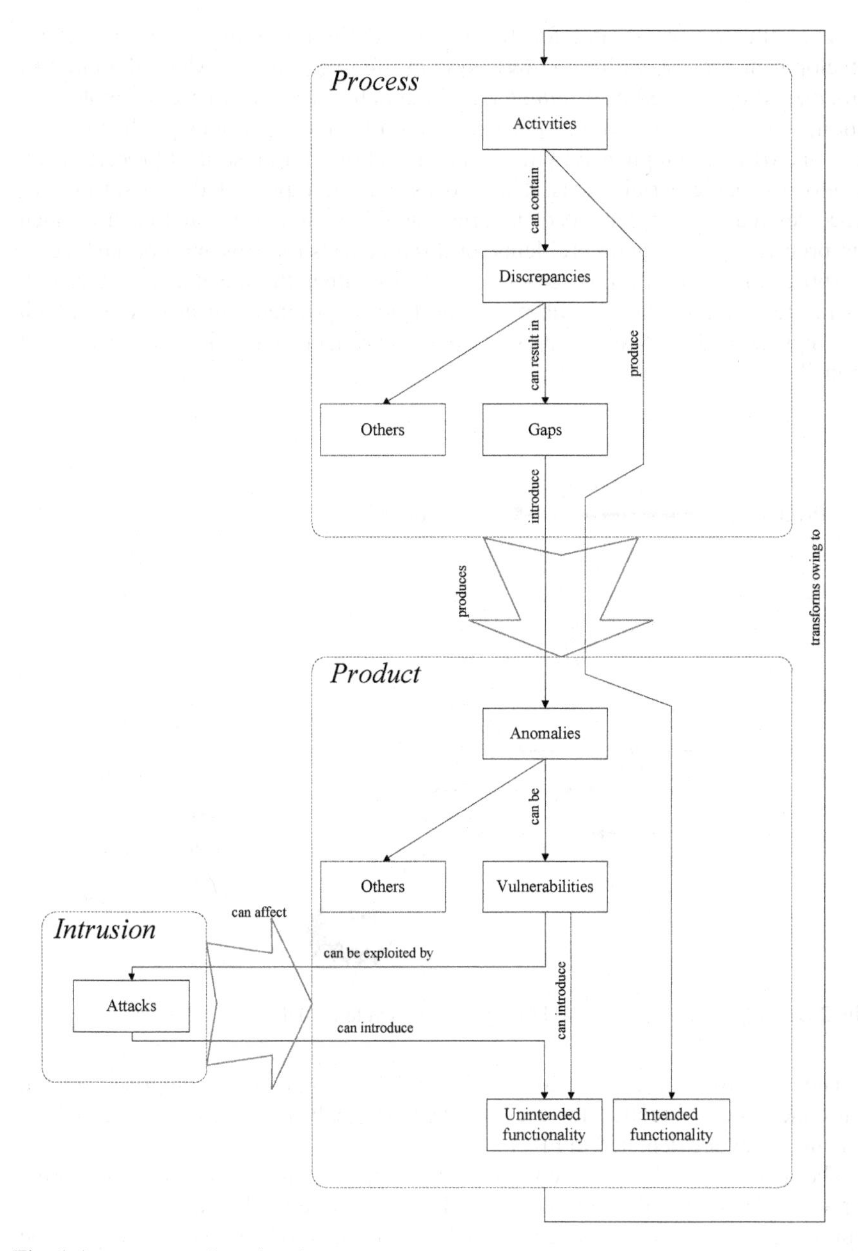

Fig. 1 A taxonomy of used notions

2. Presence of anomalies within *Product*$_{i-1}$ can be eliminated by "ideal" *Process*$_j$ in many cases. This may be true in case of verification and validation processes, however, it does not apply to design processes. For example, anomaly in the technical specification is not eliminated by an "ideal" direct translation process (since it may not include verification).

As an illustrative example for the proposed definition of gap, let us consider a development process within the I&C system life cycle model, where the input of $Process_j$ is represented by $Product_{i-1}$, and the output (result of process implementation) – is $Product_i$ (see Fig. 2). The transition from the previous product (*i-1*) to next one (*i*) is accomplished by the implementation of a prescribed process (*j*) by developers, using certain tools. This process can be represented as a set of sub-processes that are implemented in serial and/or parallel ways, and each of such sub-processes may contain problems (or discrepancies towards appropriate “ideal” sub-process) due to various reasons caused by either the developer or the tool. Therefore, the problems in sub-processes lead to problems in processes, which are implemented in order to produce a new product and can result in product anomalies.

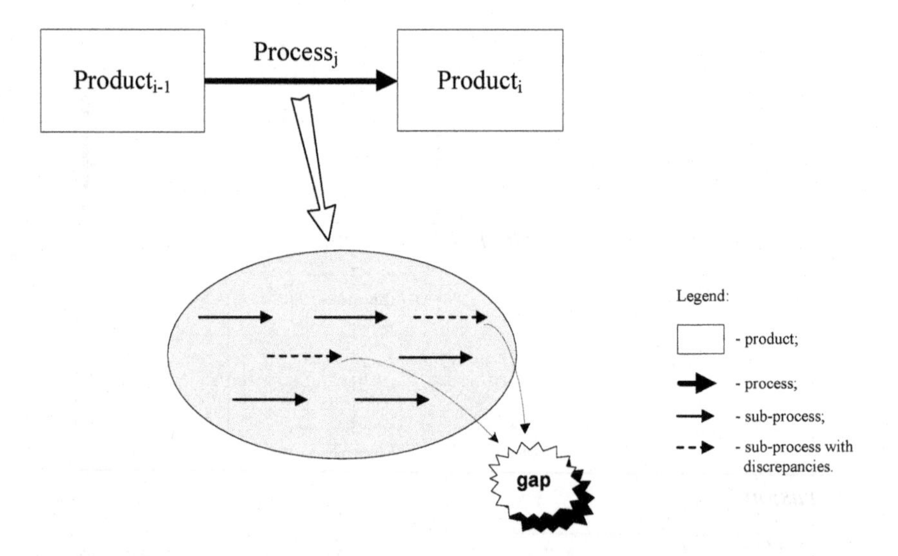

Fig. 2 Development process in the I&C system life cycle model

The activities, required to implement the approach, comprise several consequent steps intended for a comprehensive analysis and assessment of I&C systems. They are depicted in Fig. 3.

The key idea of assessment is in the application of the process-product approach. Therefore, the life cycle model of I&C systems should include detailed representation of life cycle processes and appropriate products. Then, it is possible to identify problems (or discrepancies) within the model, i.e. gaps. In general, such gaps may reflect various aspects of the I&C system, depending on what system properties are assessed (for example, safety and security).

Hence, depending on the I&C system aspects under assessment, each gap should be represented in a form of a formal description; such formal description should be made for a set of discrepancies identified within the gap. The IMECA technique is the most convenient, in our opinion, to perform such description: each identified gap can be represented by a single local IMECA table and each discrepancy inside the gap can be represented by a single row in that local IMECA table. In this way, complete traceability of life cycle processes, appropriate products and inherent properties of corresponding discrepancies can be achieved. As a result, the number of local IMECA tables would correspond to the number of identified gaps, and the number of rows within each local IMECA table would correspond to the number of identified discrepancies within the appropriate gap (see an example in subsection 3.2).

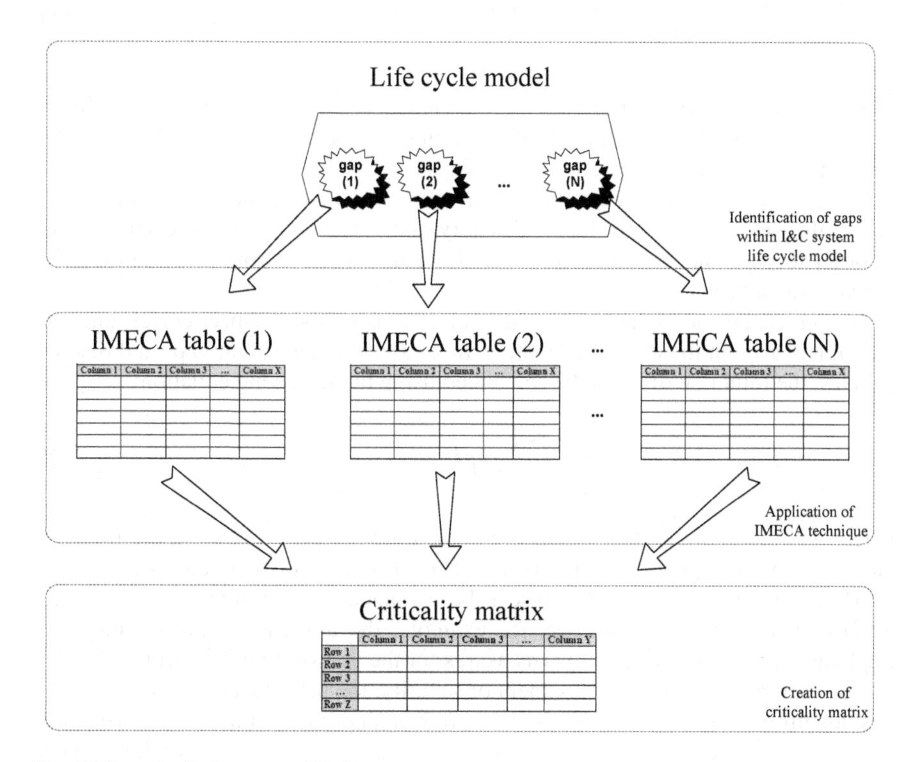

Fig. 3 The principal stages of I&C system assessment

After completing the appropriate columns, for example on the basis of expert assessment, for all local IMECA tables, each gap being represented by a set of discrepancies with appropriate numerical values. Data within each row of local IMECA tables reveal, in explicit form, the weaknesses of the I&C system aspect under assessment: for example, in terms of safety – system faults and failures, in terms of security – intrusion probability and severity.

Further, in order to implement the approach, the following cases are possible, depending on the scope of the assessment:

1. Assessment of the I&C system as a whole. Then, a set of particular IMECA tables (which represent all the identified gaps by a set of discrepancies) should be

integrated into the single global IMECA table that reflects the whole system. In this case, each row of the global IMECA table forms the basis for creating a global criticality matrix.

2. Assessment of particular (sub-)systems within the I&C system. In this case, it is possible to create an appropriate set of local criticality matrixes that correspond to certain (sub-)systems, based on a set of local IMECA tables.

Integration of local criticality matrixes into a global one is carried out in accordance with the following rule:

$$e_{yz}^{G} = \bigcup_{k=1}^{n} e_{yz}^{L_k}, \tag{1}$$

where e^G is an element of the global criticality matrix, e^{L_k} is the corresponding element of the k-th local criticality matrix, and n is the total number of local criticality matrixes (equal to total number of gaps).

Moreover, the scales for the numerical values of a discrepancy (for example, its probability and severity) for local criticality matrixes can be set to the same value in order to eliminate the necessity of additional analysis during the creation of a global criticality matrix.

In both cases, the highest risk of the selected assessment aspect corresponds to the highest row in the criticality matrix. In a case of independent gaps and discrepancies, the total risk of R can be calculated using the following equation:

$$R = \sum_{i=1}^{n} \sum_{j=1}^{m} p_{ij} D_{ij}, \tag{2}$$

where n is the total number of gaps, m is the total number of rows in the IMECA table, p is the occurrence probability, and D is the corresponding damage.

Moreover, the criticality matrix can be extended to be K-dimensional (where K>2) that allows us to consider, for example, the amount of time required to implement the appropriate countermeasures for the assessed I&C system.

For example, during the assessment of security, the prioritization of vulnerabilities identified on the basis of process-product approach, should be performed according to their criticality and severity, representing their corresponding stages in the cyber security assurance of the given I&C system. The main goal of this step is to identify the most critical security problems within the given set. Prioritization may require the creation of a criticality matrix, where each vulnerability is represented within single rows. In such cases, it is possible to manage the security risks of the whole I&C system via changing the positions of the appropriate rows within the matrix (the smallest row number in the matrix corresponds to the smallest risk of occurrence).

During the performance of GA, the identification of discrepancies (and the corresponding vulnerabilities in case of security assessment), can be implemented via separate detection/analysis of problems caused by human factors, techniques and tools, taking into account the influence of the development environment.

Then, after all identified vulnerabilities are prioritized, it is possible to assure security of the I&C system by implementing of appropriate countermeasures. Such countermeasures should be selected on the basis of their effectiveness (also, in context of assured coverage), technical feasibility, and cost-effectiveness. But there is an inevitable trade-off between a set of identified vulnerabilities and a minimal number of appropriate countermeasures, which allows us to eliminate vulnerabilities or to make them difficult to be exploited by an adversary. The problem of choosing such appropriate countermeasures is an optimization problem and is still challenging.

2.2 Example of Proposed Approach Application

As an illustrative example for the proposed approach, consider a typical development process for a VHDL code, implemented by a developer (see Fig 4a).

The input to the process is represented by a technical specification document (containing the comprehensive description of the object being developed), and the result is the VHDL code (development object). In such a case the possible discrepancies can be caused by design faults, developer's errors, and/or errors in appropriate procedures intended for the developer. Moreover, during the subsequent stages of the overall development process, existing problems in the product can be either eliminated or multiplied. Then, it is possible to represent the identified set of the process' discrepancies (or single gap) in a form of IMECA-based table, where each row corresponds to a discrepancy within the process.

Such a complex gap can be eliminated, for example, via the implementation of another development process (see Fig. 4b), which includes three entities: technical specifications, an Event-B tool model (a form of technical specification representation in terms of a tool that is understandable to developer and can automatically be translated into a VHDL code), and the VHDL code itself.

Transitions from previous entities to the next are accomplished by the execution of certain processes, namely: formal notations development process (implemented by the developer, and consisting of translation of technical specifications into a model, in terms of internal instructions of the Event-B tool, allowing the developer to mathematically prove the correctness of the resulting notation) and the translation process (implemented by special add-ons of the Event-B tool, and consisting of generating the final VHDL code on the basis of the derived model) [13].

Discrepancies in such processes can be caused by the applied tools only, since the formal notations development process is followed by the model in Event-B tool that is mathematically verifiable. Discrepancies of the translation process (or discrepancies of its sub-processes) can be caused by the Event-B tool, for example, in a case, when such tool is not fully tested or certified.

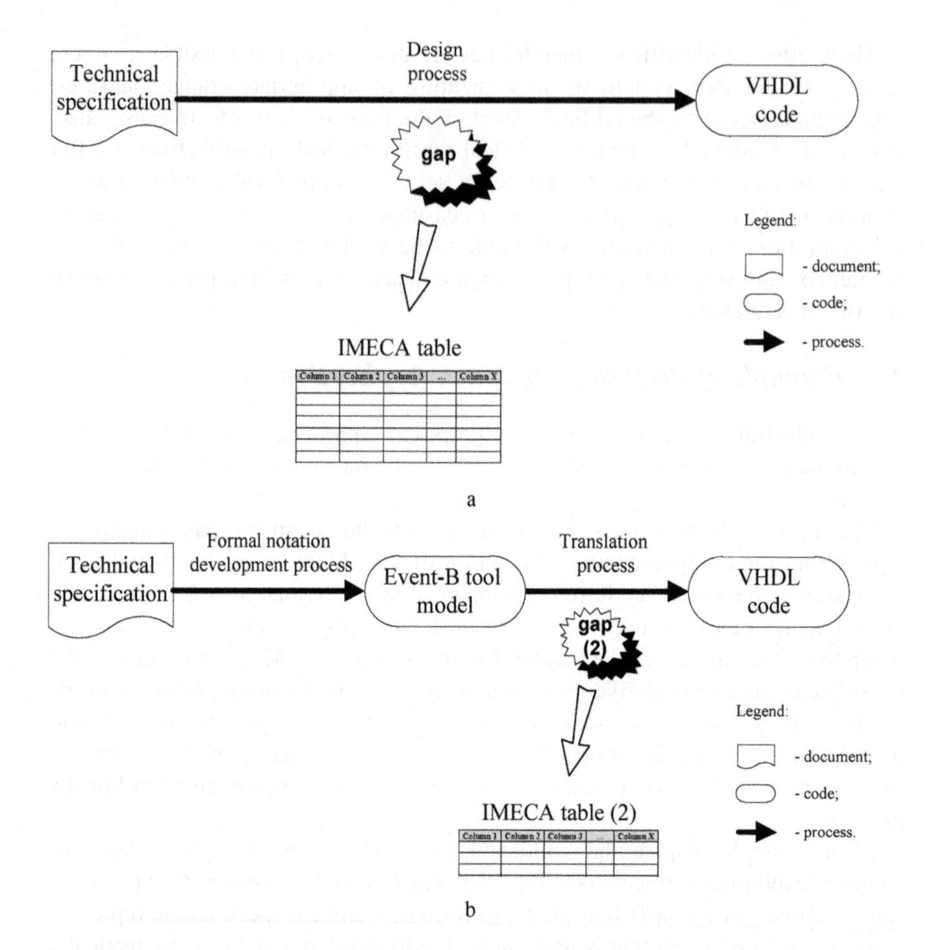

Fig. 4 Development processes for VHDL code

In this way, it is possible to state that we can identify the only existing gap. Moreover, such a gap can be eliminated if certified tools are applied. Thus, in the case given in Equation (2), the risk factor R is reduced due to the reductions in the values of parameters n (from 2 to 1), m, and p_{ij}.

3 Assessment of FPGA-Based I&C Systems Cyber Security

3.1 Life Cycle Model of FPGA-Based I&C System

Basis of modern critical I&C systems is usually formed by FPGA chips, which are used in various hardware components. Vulnerabilities of FPGA technology can unintentionally arise or can be introduced by an adversary during different stages of FPGA chip life cycle. A model of FPGA-based I&C system life cycle is depicted in Fig. 5, and includes:

1) stages implemented by FPGA chip vendor:

– a stage of FPGA chip design (Stage 1);

– a stage of FPGA chip manufacturing (Stage 2)

– a stage of FPGA chip packaging and testing (Stage 3);

2) stages implemented by I&C system developer:

– a stage of FPGA electronic design (which describes I&C system's logic) development for integration into FPGA chip (Stage 4);

– a stage of FPGA electronic design implementation and testing (Stage 5);

3) a stage implemented by user of I&C system:

– a stage of operation of FPGA-based I&C system at intended location (Stage 6).

There are factors that can contribute to intended or unintended introduction of vulnerabilities into FPGA-based I&C system during implementation of various processes for the following life cycle stages:

– use of malicious tools (EDA tools or CAD tools) during either FPGA chip designing by a vendor or during FPGA electronic design development by an I&C system developer;

– use of compromised devices during integration of developed FPGA electronic design into FPGA chip by an I&C system developer;

– use of IP-cores from third-party vendors during development of FPGA electronic design by an I&C system developer;

– the presence of adversaries (insiders) in development teams.

Some vendors of FPGA chips do not have own manufacturing capacity: in such a case, after implementation of design processes for FPGA chip, that includes application of appropriate tools, they place orders for chip manufacturing among appropriate foundries. Such foundries can introduce additional vulnerabilities into FPGA chips by stealing or modifying FPGA design. Moreover, supply chain of manufactured FPGA chips to developer of I&C system is usually traceable and can be audited that, however, does not reduce its importance from point of view of cyber security assurance problem for FPGA-based I&C systems.

Most of life cycle stages of FPGA chip and FPGA-based I&C system are implemented using software tools. Such tools are usually used, for example, during design of printed circuit boards for FPGA chips, in development of FPGA electronic designs, during simulations, etc. Hence, developers of tools for design automation, in turn, can introduce new vulnerabilities into FPGA-based I&C systems being developed.

Some vulnerabilities can be introduced into FPGA-based I&C systems by their designers via using of IP-cores in FPGA electronic design. IP-core is completed functional description intended for integration into FPGA electronic design, which is being developed. IP-cores can be either in a form of modules for hardware description languages or in a form of compiled netlists. IP-cores are used by designers to save their resources and time. IP-cores can be produced by FPGA chip vendor or third-party vendors, and, in order to assure cyber security of FPGA-based I&C system, it is necessary to facilitate safe distribution and integration of such IP-cores by designers of I&C systems.

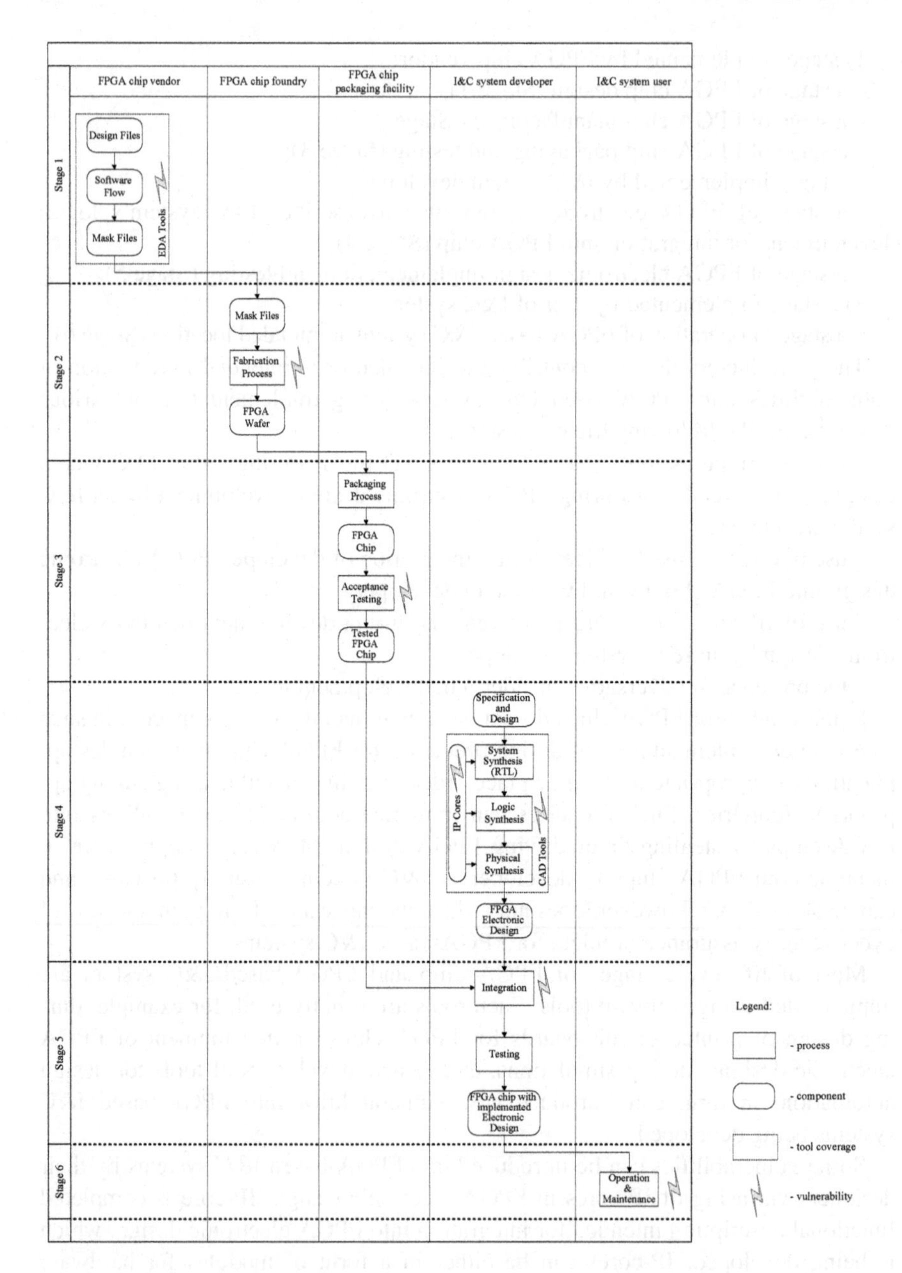

Fig. 5 Life cycle model of FPGA-based I&C system

3.2 *Gap-and-IMECA-Based Assessment of FPGA-Based I&C System*

So, proposed gap-and-IMECA-based approach, as applied to cyber security assessment, can be expressed in the following activities sequence:

Step 1. Identification of security gaps lists for all the components (or modules) of I&C system, being assessed, during each life cycle stage. Such lists should include both process gaps (in terms of discrepancies) and product cyber security gaps (in terms of vulnerabilities).

Step 2. Determination of an appropriate set of vulnerabilities for each identified process gap, security gap and possible scenarios to exploit the vulnerabilities. So, for each identified discrepancy or vulnerability, there should be created local IMECA table that reflects: attack mode, attack nature, attack cause, occurrence probability, effect severity, type of effects, and countermeasures.

Step 3. Performance of GA on the basis of IMECA-technique: each gap (identified during Step 1) being represented by one or several rows in a local IMECA table, where the number of such rows corresponds to the number of appropriate discrepancies or vulnerabilities identified during Step 2. GA should be performed in order to reveal appropriate cyber security risks.

Step 4. Assessment of appropriate columns (occurrence probability and effect severity) in each particular IMECA table, for example, on the basis of expert evaluation. Then, each row of such a local IMECA table represents security weaknesses, which should be analyzed further (during Step 6) in context of the whole I&C system.

Step 5. Analysis of cyber security risks of I&C system components during different stages: each row in local IMECA tables forms the basis for creation of security criticality matrix, which reveals the weaknesses of appropriate components in a visual form. The highest cyber security risk corresponds to the highest row in security criticality matrix.

In order to illustrate IMECA-based assessment, we present results for attacks modes possible during operation and maintenance stage of FPGA-based I&C system (see Table 1).

Criticality matrix is depicted in Fig. 6a. Each of the numbers inside the matrix represents an appropriate row number of IMECA table. From cyber security assurance point of view, the possible way of risk reduction is in decreasing of attacks' occurrence probability, since related damage is constant. Fig. 6b represents worst-case criticality diagonal for the matrix; acceptable values of risks are below the diagonal. Cases of probability, decreasing for rows 2, 3, and 5 are denoted by dotted lines with arrows: the problem is in decreasing of the probability by the degree sufficient to move row of IMECA table below the criticality diagonal. Such decreasing of the probability can be achieved, for example, by implementation of certain process countermeasures.

Table 1 Results of IMECA for FPGA attacks

Row number	Gap in stage of	Attack mode	Attack nature	Attack cause	Occurrence probability	Effect severity	Type of effects	Countermeasures
1	Operation	Black Box Attack	Active	Simple logic of electronic design	Very low	Very low	Reverse engineering of logic by adversary	Complication of electronic design logic
2	Operation	Read-back Attack	Active	Absence of chip security bit and/or availability of physical access to chip interface	Moderate	High	Obtaining of secret information by adversary	The use of security bit. Application of physical security controls
3	Operation	Cloning Attack	Active	Storing of decoded configuration	Moderate	High	Obtaining of configuration data by adversary	Checking of chip's internal ID before powering up an electronic design. Encoding of configuration file. Storing of configuration file within FPGA chip (requires internal power source)
4	Operation	Physical Attack	Active	Absence of monitoring of parameters (voltage, temperature, clock) of environment and chip	Low	Moderate	Obtaining of information concerning patented algorithms by adversary	Decreasing memory retention effect. Monitoring of parameters (voltage, temperature, clock) of environment and chip
5	Operation	Side-Channel Attack	Active	Correlation of measurable parameters with its function	High	High	Leak of undesirable information	Addition of random noise in measurable parameters (or masking of information by random values). Decrease of difference in power consumption. Changing of electronic design logic

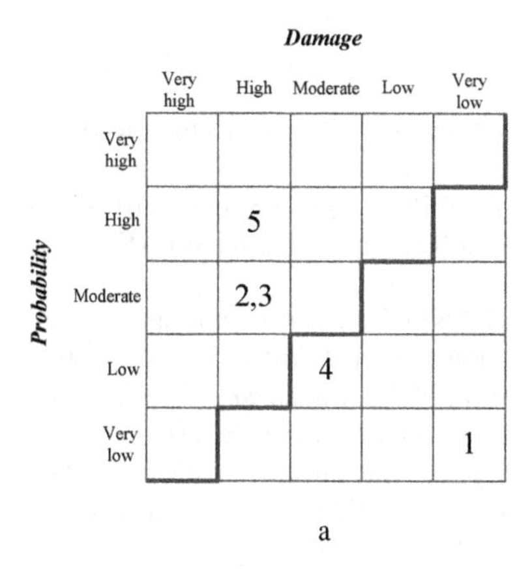

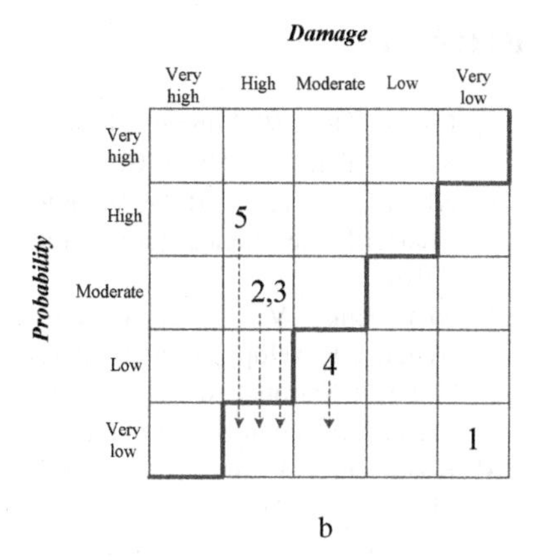

Fig. 6 Criticality matrixes

4 Conclusion

A problem of I&C systems assessment is still challenging due to the fact that such systems consist of interconnected complex components with different functions and different nature. The majority of modern I&C systems, including safety-critical I&C systems, are being FPGA-based, hence, it is impossible to perform their assessment without consideration of all the special features for all the technologies used. In this chapter we discussed some problems related to assessment of various aspects of I&C systems, including FPGA-based systems.

To assure cyber security of modern complex I&C systems, as well as to decrease a probability of vulnerabilities exploitation and appearance of security breaches, a cyber security assessment approach is proposed. This approach implies identification of all possible discrepancies, on the basis of product and life cycle processes, and their assessment via application of IMECA technique.

The proposed approach is based on both gap conception and IMECA technique. Such an approach is applicable in assessment of various aspects of I&C systems, since it considers process-product model to reveal all the process discrepancies that can potentially result in product anomalies.

Application of the proposed approach and technique was illustrated by an example of cyber security assessment for some FPGA-based I&C system. Gap-and-IMECA-based technique was applied in development of a company standard in Research and Production Corporation Radiy that is harmonized with international standards. This standard is used during implementation of development and verification activities for safety-critical I&C systems for nuclear power plants [3].

Next steps of research and development activities may be connected with creation and implementation of tool-based support for the proposed approach, taking into account results of qualitative and quantitative assessment.

References

[1] IEC 61508:2010, Functional Safety of Electrical/Electronic/Programmable Electronic Safety-related Systems (2010)
[2] NUREG/CR-7006, Review Guidelines for Field-Programmable Gate Arrays in Nuclear Power Plant Safety Systems, U.S. Nuclear Regulatory Commission (February 2010)
[3] Kharchenko, V., Sklyar, V. (eds.): FPGA-based NPP Instrumentation and Control Systems: Development and Safety Assessment, Research and Production Corporation "Radiy", National Aerospace University named after N.E. Zhukovsky "KhAI", State Scientific Technical Center on Nuclear and Radiation Safety, 188 p (2008)
[4] Kharchenko, V. (ed.): Critical Infrastructures Safety: Mathematical and Engineering Methods of Analysis and Assurance, Department of Education and Science of Ukraine, National aerospace university named after N. Zhukovsky "KhAI", 641 p (2011)
[5] Ravi, S., Raghunathan, A., Kocher, P.: Security in Embedded Systems: Design Challenges. ACM Transactions on Embedded Computing Systems 3(3), 461–491 (2004)
[6] Grand, J.: Practical Secure Hardware Design for Embedded Systems. In: Proc. of the 2004 Embedded Systems Conference, San Francisco, California, March 29-April 1 (2004)
[7] IEC 812, Analysis Techniques for System Reliability – Procedure for Failure Modes and Effects Analysis (FMEA). International Electrotechnical Commission, Geneva (1985)
[8] Lutz, R., Helmer, G., Moseman, M., Statezni, D., Tockey, S.: Safety Analysis of Requirements for a Product Family. In: Proc. 3rd Int'l Conf. on Requirements Engineering (ICRE 1998), pp. 24–31 (1998)
[9] Elyasi Komari, I., Kharchenko, V., Babeshko, E., Gorbenko, A., Siora, A.: Extended Dependability Analysis of Information and Control Systems by FME(C)A-technique: Models. In: Procedures, Application, DepCoS – RELCOMEX 2009, pp. 25–32 (2009)
[10] Gorbenko, A., Kharchenko, V., Tarasyuk, O., Furmanov, A.: F(I)MEA-Technique of Web Services Analysis and Dependability Ensuring. In: Butler, M., Jones, C.B., Romanovsky, A., Troubitsyna, E. (eds.) Rigorous Development of Complex Fault-Tolerant Systems. LNCS, vol. 4157, pp. 153–167. Springer, Heidelberg (2006)
[11] Avizienis, A., Laprie, J.-C., Randell, B., Landwehr, C.: Basic Concepts and Taxonomy of De-pendable and Secure Computing. IEEE Transactions on Dependable and Secure Computing 1(1), 11–33 (2004)
[12] Babeshko, E., Kharchenko, V., Gorbenko, A.: Applying F(I)MEA-technique for SCADA-based Industrial Control Systems Dependability Assessment and Ensuring. In: DepCoS-RELCOMEX 2008, pp. 309–315 (2008), doi:10.1109/DepCoS-RELCOMEX.2008.23
[13] Abrial, J.-R.: Modeling in Event-B, 612 p. Cambridge University Press (2010)

Approach to Methods of Network Services Exploitation

Katarzyna Nowak and Jacek Mazurkiewicz

Abstract. The chapter is focused on the methods of network services exploitation. The approach is based on two streams of data: dependability factors and the features defined by the type of business service realized. The dependability means the combination of the reliability and functional parameters of the network. We try to analyze two types of sophisticated systems: discrete transport systems and the computer networks. The proposed method is based on modeling and simulating of the system behavior. This way it is possible to operate with large and complex networks described by various - not only classic – distributions and set of parameters. The results are converted to the unified system description and generic model. The model can be used as a source to create different measures – also for the economic quality of the network systems. The presented problem is practically essential for defining and organization of network services exploitation.

1 Introduction

The contemporary network systems are created as very sophisticated products of human idea characterized by the complex structure. On the other hand the systems combine two types of resources: technical (engineering stuff) and information (algorithms, processes and management procedures). The systems are human-controlled and computer-aided devices. The reliability parameters of the system resources are very screwed-up – so the exploitation analysis of contemporary systems needs adequate models and calculation methods [24, 29].

During more than 60 years the reliability theory was altered from the reliability of single and separated objects (elements) considered only two states ("efficient work", failure) to the contemporary dependability of systems or even the

Katarzyna Nowak · Jacek Mazurkiewicz
Institute of Computer Engineering, Control and Robotics, Wroclaw University of Technology, ul. Wybrzeze Wyspianskiego 27, 50-370 Wroclaw, Poland
e-mail: `Katarzyna.M.Nowak@pwr.wroc.pl`, `Jacek.Mazurkiewicz@pwr.wroc.pl`

W. Zamojski et al. (Eds.): Complex Systems and Dependability, AISC 170, pp. 165–178.
springerlink.com

dependability of service nets. The indicated development of the reliability theory is the consequence of expanding the event sets taken into consideration for the reliability models. The present system dependability theory considers not only classical reliable events (failures or repairs) but tries to combine all types of the faults generated by the system resources (hardware, algorithms, human-factor) and the environmental features which may disturb the operable state (attacks – for example). The main goal of the system exploitation analysis is to convert the discussion focused on the reliability function of elements (or structures created by the element sets) into the task performance or efficiency estimation. The tasks are realised according to the system services [20, 21]. The classical models used for reliability analysis are mainly based on Markov or Semi-Markov processes [2] which are idealized, it is hard to reconcile them with practice and is insufficient in general. We suggest the Monte Carlo simulation [30] for proper reliability and functional parameters calculation. No restriction on the system structure and on a kind of distribution is the main advantage of the method [31].

We call the approach as the functional-reliability models of network system exploitation. The computer systems analysis is the root for our elaboration but we believe it is useful for modelling of the wider spectrum of systems which realise tasks based on fully or partially available resources. We think about a discrete transport system or power management systems for example.

The computer and software equipment allows making the exploitation analysis more sophisticated. The simulation technique is the real chance to operate with large systems – where the number of elements is significant. The elements can be described by different sets of features. We can observe – in parallel – large number of events in quite long time-periods. This way we can collect data sets to very detailed presentation of the system life. Based on the data we are able to elaborate the formal theoretical approach to the network system exploitation. [21, 30, 31] Of course it is necessary to eliminate all these features which are very system-depend and not enough generic.

2 Transportation Systems

2.1 Traffic Modeling

Modeling traffic flow for design, planning and management of transportation systems in urban and highway area has been addressed since the 1950s mostly by the civil engineering community. The following definitions and concepts of traffic simulation modeling can be found in works such as Gartner et al. [11]. Depending on the level of detail in modeling the granularity of traffic flow, traffic models are broadly divided into two categories: macroscopic and microscopic models. According to Gartner et al. [11], a macroscopic model describes the traffic flow as a fluid process with aggregate variables, such as flow and density. The state of the system is then simulated using analytical relationships between average variables such as traffic density, traffic volume, and average speed. On the other hand, a microscopic model reproduces interaction of punctual elements (vehicles, road segments, intersections, etc) in the traffic network. Each vehicle in the system is emulated according to its individual characteristics (length, speed, acceleration, etc.). Traffic is then simulated, using processing logic and models describing vehicle driving behavior, such as car-following and lane-changing models. Those models reproduce driver-driver and driver-road interactions. Despite its great accuracy level, for many years this highly detailed modeling was considered a computationally intensive approach. Since the last twenty years, with the improvements in processing speed, this microscopic approach becomes more attractive. In fact, Ben-Akiva et al. [3], Barcelo et al. [1] and Liu et al. [17] claim that using microscopic approach is essential to track the real-time traffic state and then, to define strategy to decrease congestion in urban transportation networks. For the control of congestion, they explain that the models must accurately capture the full dynamics of time dependant traffic phenomena and must also track vehicles' reactions when exposed to *Intelligent Transportation Systems* (*ITS*). From the latter assertions, in order to control traffic congestion in internal transportation networks it appears that the microscopic modeling will be more appropriate. A common definition of congestion is the apparition of a delay above the minimum travel time needed to traverse a transportation network. As stated in Taylor et al. [25], this notion is context-specific; and complex because a delay may always appear in dynamic transport system, but this delay must exceed a threshold value in order to be considered.

2.2 Microscopic Analysis

Few works have considered the traffic behavior when studying outdoors vehicle-based internal transport operational problems. In the surface mining environment, pickup and delivery operations involve a fleet of trucks transporting materials from excavation stations to dumping stations, through a designed shared road network. At pickup stations, shovels are continuously digging during a shift according to a pre-assigned mining production plan. Trucks are moving in a cyclical manner between shovels (pickup stations), and dumping areas (delivery stations). A truck cycle time is defined as the time spent by a truck to accomplish an affected mission that consists of travelling to a specific shovel, being serviced by the shovel and hauling material to a specific dumping area. Burt and Caccetta [6] state that mine productivity is very sensitive to truck dispatching decisions which are closely related to the truck cycle time. Thus several papers have studied and proposed algorithms and software to resolve this problematic issue. In fact, this critical decision consists of finding, according to the real environment, to which best shovel a truck must be affected. Such decision has to be generated continuously during a shift, whenever a truck finished dumping at a delivery station. Despite the several proposed dispatching software, recent articles by Krzyzanowska [16] formally criticize the simplistic assumption behind those software which tend to provide dispatching decisions with the objective to optimize a truck cycle times previously calculated. Generally speaking, those software systems based the optimisation process on the past period collected data of trucks cycle times and assume that for the next period trucks will spend on average the same time to accomplish missions. But in the reality of mining operation, the duration of truck travel time appears to be very sensitive to the variable traffic state and road conditions. Burt and Caccetta [6] and Krzyzanowska [16], point out the unresolved problematic of truck bunching and platoon formation in mining road network which apparently induce lower productivity.

2.3 Commodity Movement

Similarly to material transportation in mining operation, several papers (Ioannou [14], Vis [26]) have provided methods for improving container terminal complex operations. In such applications, three types of handling operations are defined: vessel operations, receiving/delivery operations and container handling and storage operations in the stack yards. As we are interested by internal transportation systems, our review concerns the papers dealing with the container handling and storage operations in the stack yards. Generally speaking, vessels bring inbound containers to be picked up by internal trucks and distributed to the respective stocks in the yard. Once discharged, vessels have to leave with on board outbound containers which also are delivered by internal trucks from the storage yard. For this purpose, trucks are moving through a terminal internal road network. In order to decrease the vessel turnaround time, which is the most important performance measure of container terminals, it is important to perform those operations as quickly as possible. In fact according to [3], this movement of containers between quay sides and storage yards appears to greatly affect the productivity of containership's journey. Vis and Koster [26] gives an extended review of numerous research papers, providing algorithms to solve this complex routing and scheduling problem. They criticize the lack of consistency of the simplistic assumptions made to solve the proposed models within the real-world highly stochastic environment. The ignored traffic situation in the complex seaport internal transportation network is strongly criticized in recent papers [4], [11]. For example, in [3], a travel time of a container internal truck is modeled as a static mean time of travel, based on the distance and the truck average speed. Duinkerken et al. [7], put a uniform distribution between zero and 30% of the nominal travel time formulation, aiming to assimilate the complexity of traffic. More accurate work to solve this issue is the one provided recently by Liu, Chu and Recker [17]. They integrate a traffic model to the internal service model and reported the effectiveness of this integration which allows analyzing the tractor traffic flow in a port container terminal. Conscious about the critical problem of congestion in the road network inside a terminal, a quantitative measure of congestion to be added as a controllable decision variable had been developed. For this purpose, they considered the road system inside the terminal as a directed network and they measured flows on arcs in units of trucks travelling per unit time. Those two last works appear as providing the leader approach in term of consideration of congestion and traffic in container terminals; however, their approach is ultimately macroscopic. As we have lately discussed, even if this macroscopic approach allows analyzing the traffic behavior, the highly detailed microscopic model is more efficient for an effective real-time traffic monitoring and control.

2.4 Real System Description

The analyzed transportation system is a simplified case of the *Polish Post*. The business service [27], [29], [31] provided the *Polish Post* is the delivery of mails. The system consists of a set of nodes placed in different geographical locations. We have the headquarter (*HQ*) located in the central part of Poland and two kinds of nodes could be distinguished: central nodes (*CN*) and ordinary nodes (*ON*). The single central node with the set of ordinary nodes corresponding with it creates the sub-system.

There are bidirectional routes between nodes. Mails are distributed among ordinary nodes by trucks, whereas between central nodes by trucks, railway or by plane. The mail distribution could be understood by tracing the delivery of some mail from point *A* to point *B*. At first the mail is transported to the nearest ordinary node *A*. Different mails are collected in ordinary nodes, packed in larger units called containers and then transported by trucks scheduled according to management architecture decision to the nearest central node. In central node containers are repacked and delivered to appropriate (according to delivery address of each mail) central node. In the second – closest to the destination place - central node the mail is again repacked and delivered in a container to destination ordinary node.

The headquarter collects all data about the actual situation in whole transportation system and makes the necessary decisions as the reaction for the temporary needs. The headquarter is not in use in transportation action – if we think about the loading, unloading processes, etc. The central nodes aggregate the data from the single region of the country. And finally the ordinary nodes control the local situation to the end-user. The scale of necessary actions depends on the actual needs. In the Polish Post there are 14 central nodes and more than 300 ordinary nodes. There more than one million mails going through one central node within 24 hours. It gives a very large system to be modeled and simulated.

The process of any system modeling requires defining the level of details. Increasing the system details causes the simulation becoming useless due to the computational complexity and a large number of required parameter values to be given.

2.5 Formal Model

The model can be described as follows [27]:

$$DTS = \langle TM, BS, TI, CM, TT \rangle \tag{1}$$

where: *CM* – client model, *BS* – business service, *TM* – task model,
TI – technical and human infrastructure, *TT* – vehicles' time-table.

Task Model (TM): We can discuss several kinds of a commodity transported in the system. Single kind commodity is placed in a unified container, and containers are transported by vehicles. The commodities are addressed and there are no other parameters describing them.

Business Service (*BS*): It is a set of services based on business logic that can be loaded and repeatedly used for concrete business handling process. *Business Service* can be seen as a set of service components and tasks that are used to provide service in accordance with business logic for this process. Each service component in *DTS* consists of a task of delivering a container from a source node to the destination one.

Technical Infrastructure (TI) consists of: Nodes, Routes, Vehicles, Maintenance Crews.

Nodes: We have single central node in the each part of the sub-system and the set of central nodes in the whole system. The central node is the destination of all commodities taken from other – ordinary nodes. The central node is also the global generator of commodities driven to the nodes of the system. The generation of containers is described by Poisson process. In case of central node there are separate processes for each ordinary node. Whereas, for ordinary nodes there is one process, since commodities are transported from ordinary nodes to the central node or in other direction. Ordinary nodes are described by intensity of container generation (routed to central node) and central node is described be a table of intensities of containers for each ordinary node. Moreover the length between each two nodes is given.

Vehicles: We assumed that all vehicles are of the same type and are described by following functional and reliability parameters: mean speed of a journey, capacity – number of containers which can be loaded, reliability function and time of vehicle maintenance. Central node is the base place for vehicles. They start from the central node and the central node is the destination of their travel. The temporary state of each vehicle is characterized by following data: vehicle state, distance travelled from the beginning of the route, capacity of the commodity. The vehicle running to the end of the route is able to take different kinds of commodity (located in unified containers, each container includes single-kind commodity). The vehicle hauling a commodity is always fully loaded or taking the last part of the commodity if it is less than its capacity.

Routes: Each route describes possible trip of vehicles. The set of routes we can describe as series of nodes:

$$R = \langle c, v_1, \ldots, v_n, c \rangle \quad v_i \in ON \quad and \quad c = CN \tag{2}$$

Maintenance Crews: Maintenance crews are identical and indistinguishable. The crews are not combined to any node, are not combined to any route, they operate in the whole system and are described only by the number of them. The temporary state of maintenance crew is characterized by: number of crews which are not involved into maintenance procedures and queue of vehicle waiting for the maintenance.

Client Model (CM): The service realised by the clients of the transport system is sending mails from a source node to a destination one. *Client Model* consist of a set of clients (*C*). Each client is allocated in one of nodes of the transport system:

$$allocation : C \rightarrow No\,. \tag{3}$$

A client allocated in an ordinary node is generating containers (since, we have decided to monitor containers not separate mails during simulation) according to the Poisson process with destination address set to ordinary nodes. In the central node, there is a set of clients, one for each ordinary node. Each client generates containers by a separate Poisson process and is described by intensity of container generation:

$$intensity : C \to R_+ . \tag{4}$$

The central node is the destination address for all containers generated in ordinary nodes.

Time-Table (TT): Vehicles operate according to the time-table exactly as city buses or intercity coaches. The *Time-Table* consists of a set of routes (sequence of nodes starting and ending in the central node, times of approaching each node in the route and the recommended size of a vehicle [28].

The number of used vehicles or the capacity of vehicles does not depend on temporary situation described by number of transportation tasks or by the task amount for example. It means that it is possible to realize the journey by completely empty vehicle or the vehicle cannot load the available amount of commodity (the vehicle is to small). *Time-Table* is a fixed element of the system in observable time horizon, but it is possible to use different time-tables for different seasons or months of the year. Each day a given time-table is realized, it means that at a time given by the time table a vehicle, selected randomly from vehicles available in the central node, starts from central node and is loaded with containers addressed to each ordinary nodes included in a given route. This is done in a proportional way. Next, after approaching given node (it takes some time according to vehicle speed - random process and road length) and the vehicle is waiting in an input queue if there is any other vehicle being loaded/unload at the same time. There is only one handling point in each node. The time of loading/unloading vehicle is described by a random distribution. The containers addressed to given node are unloaded and empty space in the vehicle is filled by containers addressed to a central node. The operation is repeated in each node on the route and finally the vehicle is approaching the central node when is fully unloaded and after it is available for the next route. The process of vehicle operation could be stopped at any moment due to a failure (described by a random process). After the failure, the vehicle waits for a maintenance crew (if there are no available due to repairing other vehicles), is being repaired (random time) and after it continues its journey [29].

3 Complex Information Systems

Few years from now desktop applications were the most popular and used, but with increasing population of the Internet and with growing possibilities of portability of applications, many systems become an online services. Trends related with that fact (i.e. cloud computing [10], service oriented architecture [18]) became a standard in some online services (i.e. document exchange, software development and code refactoring, online banking, tax-payment systems). Hence need for those systems are still growing more and more complex and high-tech technology is required. One thing was not change during the time – user needs. Complexity of the system resulted in defining some of the Information System as a *Complex Information System* (*CIS*) – systems with extensive infrastructure aimed to satisfy user needs in case of a service. To meet these needs, Information System should be created in as an optimal technical infrastructure that (with combination of business point of view) will provide an appropriate level of the system goals. Taking these aspects into consideration we focus on service and user requirements – functional and dependability. Since the client expects from the system that it will provide some service in an infrastructure located on a provider side and with a suitable configuration, therefore user expect to receive a solution for the task that was send to the system as a request. For this reasons, we can model *CIS* as a 4-tuple:

$$CIS =< Z,HS,M,K > \tag{5}$$

where: Z – tasks, HS – technical infrastructure (hardware, software, links),
K – chronicle of the system (understood as time functions of the system),
M – clients.

As mentioned since we propose to analyze *CIS* systems on a basis of their service, we can modify (5) and define *Business Service* (*BS*) as a set of business logic, that can be loaded and repeatedly used for concrete business handling process (i.e. online banking, flight booking, etc.).

$$CISB =< Z, BS,HS,M,K > \tag{6}$$

Taks (*Z*) are considered as an input data specified by the clients in case of business service usage (i.e. selection of page and subpages related with various possibilities and scenarios).

Business service (BS) can be seen as a set of *Service Components* (SC_i) and tasks that are used to provide service in accordance with business logic for this process. Each *Service Component* (SC_i) is a service located on defined host (server) that determined service behavior, possibilities and requirements (i.e. authentication, data base service, web service, etc.). Since service component is not a physical unit, it can be locates on any machine, therefore one host can have more than one service component.

Technical infrastructure (HS) is defined as a set of network devices and links built to provide network service with respect to TCP/IP aspects. Each device is described by unique ID and some parameters (i.e. host performance, software, operation system).

Chronicle of the system (*K)* is the time function on each level of abstraction.

Clients (*M*) consists of a set of users where each user is defined by its allocation (host), number of users of a given profile (i.e. similar type of behavior), set of activities (a sequence of task calls - name of task and a name of service component) and inter-activity delay time.

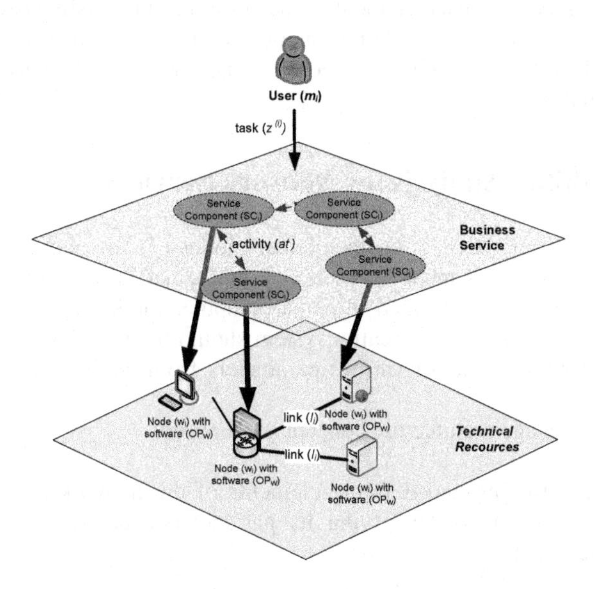

Fig. 1 Business service oriented information system – levels of abstraction

4 Abstract Approach for Network Services

Taking into consideration that both systems described above providing a service in a sense of user request accomplishment, we can note that we can speak about one common approach. An approach of describing and analyzing network services as a general network, but still in a sense of its proposes. In this method, the key point of the view is the *Task* (*T*) given to the systems, its specification and its time. Since results of task are conditioned by the scenario, choreography within a service must be defined and known.

Moreover resources used to realize this choreography must be dependable. It is worth to note that, on a one network of *Technical Infrastructure* (*TI*) more than one choreography can be realized and more than one configuration of the service is possible (i.e. reconfiguration of the system is possible). Task is realized as an input to the *Business Service* (*BS*), therefore its choreography is based on predefined service components located in network nodes. As mentioned, specifying the task and its parameters is a *user* role (*M*) and the time functions on each level of abstraction - C*hronicle of the System* (*K*). Both systems *DTS* (1) and *CIS* (5) can be represented as a 4-tuple:

$$ANS = < Z, BS, TI, M, K > \tag{7}$$

where: *ANS* – Abstract Network Services

The unified description can guarantee the required level of abstraction for the analysis we are going to provide (Fig. 1). For example the discussion at the field of queue theory ought to be realized forgetting the technical details which are useless. But if we try to transform the results into the economic matters the more sophisticated features for each element of the system can be easily provided.

It requires only the proper definition of functions and their parameters. In general the abstract approach is a tool to tune the level of network system exploitation analysis.

5 Dependability Analysis for Network Services

Proposed abstract approach of mathematical model (7) for network services is given (in section 4). This model is seen as a mathematical description of all the system aspects easily transformed to some computer model for further analysis. To gain results that are close to reality system should be described as precise as it is possible and with respect of various parameters from the lowest to the highest level of the system.

Therefore tree stage architecture concept is proposed:

1. *Monitoring* – finding and defining elements of the network (nodes, devices, vehicles, routes, links), specifying its parameters/characteristics, faults and failures tracing, etc.;

2. *Modelling* – describing elements (service components and resources) as an input for analysis tools, representing collected data in a description language of selected format for computer processing;

3. *Analysis* – processing of computer representation of the system, analysis and visualization of its behaviour and inference based on the results.

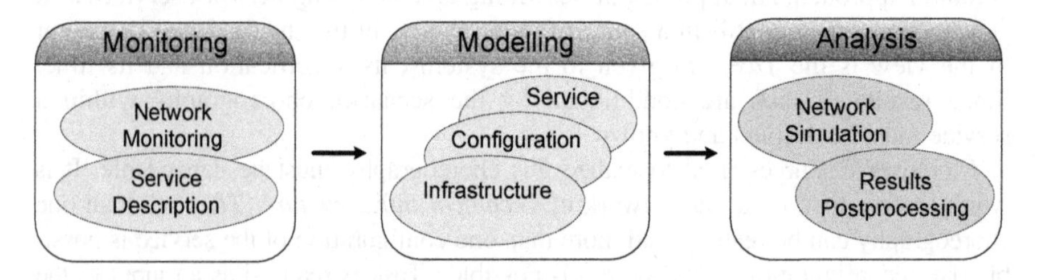

Fig. 2 Method for data collecting and data processing – draft architecture

The presented architecture (Fig. 2) can be combined with several techniques and methods, depending of a level of abstraction and analysis purposes. For this reasons as a monitoring method, some know tools can be used (i.e. ping command) or more advances concept can be adapted. For example, multi-agent monitoring can be used [15, 21, 23]. Data gathered in this way can be collected as one common format (text file, database, etc.).

Research prove that using standards as a basis of this representation is one of the best techniques for system modelling [4, 21, 27]. Description languages proposed as a standard in that matter seem to be suitable (i.e. WS-CDL [13], BPEL [19]) and easily transformable to any analysis tool. Surely various analysis methods can be used, i.e. formal analysis [8], chains [8] and network simulation [22]. Depending on results and their representation post-processing method can be chosen: Matlab charts, table representations, etc.

The presented approach requires the event-driven simulation. Which is based on a idea of event, which could is described by time of event occurring and type of an event. The simulation is done by analyzing a queue of event (sorted by time of event occurring) while updating the states of system elements according to rules related to a proper type of event. Due to a presence of randomness in the system model the analysis of it has to be done based on Monte-Carlo approach [9]. What requires a large number of repeated simulations.

The processing of events is done in objects representing the network system elements. The objects are working in parallel. The life cycle of each object consists of waiting for an event directed to this object and then execution of tasks required to perform the event. These tasks includes the changes of internal state of the object (for example when vehicle approaches the node it is unloaded, i.e. the number of hauled containers decreases) and sometimes creating a new event (for example the event vehicle starts from the node generates new event vehicle reached the node – next node in the trip).

The random number generator is used to deal with random events, i.e. failures. It is worth to notice that the current analyzed event not only generates a new event but also could change time of some future events (i.e. time of approaching the node is changed when failure happens before). The time of a new event is defined by the sum of current time (moment of execution of the current event) and the duration of a given task (for example vehicle repair). Only times of starting a given route (event vehicle starts from the central node) are predefined (according to the time table). Duration of all other tasks is defined by system elements states or are given by random processes.

Moreover each object representing a node has additional process (working in parallel) which is responsible for generating the amount of commodity. The life cycle of this process is very simple: waiting a random time, generating a commodity with a given destination address and storing a commodity in the store house (implemented as a queue) of a given node.

The event-simulation program could be written in a general purpose programming language (like C++), in a fast prototyping environment (like Matlab) or a special purpose discrete-event simulation kernel. One of such kernels, is the Scalable Simulation Framework (*SSF*) [28] which is a used for *SSFNet* [29][30] computer network simulator. *SSF* is an object-oriented *API* - a collection of class interfaces with prototype implementations. It is available in C++ and Java. *SSFAPI* defines just five base classes: Entity, inChannel, outChannel, Process, and

Event. The communication between entities and delivery of events is done by channels (channel mappings connects entities).

As it was mentioned a presence of randomness in the system model, the Monte-Carlo approach is used. The original *SSF* was not designed for this purpose so some changes in *SSF* core were done to allow to restart the simulation from time zero several times within one run of simulation program.

The statistical analysis of the system behavior requires a very large number of simulation repetition, therefore the time performance of developed simulator is very important.

6 Conclusions

We have presented a unified, abstract, formal model for modelling of network systems exploitation problems. The approach has been created based on *Discrete Transport System* (*DTS*) and *Complex Information System* (*CIS*). The proposed approach allows performing various dependability analysis of the network systems, for example:

- determine what will cause a "local" change in the system,
- make experiments in case of increasing volume of commodity (containers, amount of bytes) per day incoming to system,
- identify weak point of the system by comparing few its configuration,
- better understand how the system behaves,
- foresee changes caused by human resource influence.

Based on the results of simulation it is possible to create different metrics to analyse the system in case of reliability, functional and economic case.

The metric could be analysed as a function of different essential functional and reliability parameters of network services system. Also the system could be analyse in case of some critical situation (like for example a few day tie-up [30]).

The presented approach - based on two streams of data: dependability factors and the features defined by the type of business service realized - makes a starting point for practical tool for defining an organization of network systems maintenance.

It is possible to operate with large and complex networks described by various - not only classic – distributions and set of parameters. The model can be used as a source to create different measures – also for the economic quality of the network systems. The presented problem is practically essential for defining and organization of network services exploitation.

Acknowledgment. Work reported in this paper was sponsored by a grant No. N N509 496238, (years: 2010-2013) from the Polish Ministry of Science and Higher Education.

References

[1] Barcelo, J., Codina, E., Casas, J., Ferrer, J.L., Garcia, D.: Microscopic Traffic Simulation: a Tool for the Design, Analysis And Evaluation Of Intelligent Transport Systems. Journal of Intelligent and Robotic Systems: Theory and Applications 41, 173–203 (2005)

[2] Barlow, R., Proschan, F.: Mathematical Theory of Reliability. Society for Industrial and Applied Mathematics, Philadelphia (1996)

[3] Ben-Akiva, M., Cuneo, D., Hasan, M., Jha, M., Yang, Q.: Evaluation of Freeway Control Using a Microscopic Simulation Laboratory. Transportation Research, Part C (Emerging Technologies) 11C, 29–50 (2003)

[4] Birta, L., Arbez, G.: Modelling and Simulation: Exploring Dynamic System Behaviour. Springer, London (2007)

[5] Bonabeau, E.: Agent-based modelling: methods and techniques for simulating human systems. Presented at. Proc. Natl. Acad. Sci. (2002)

[6] Burt, C.N., Caccetta, L.: Match Factor for Heterogeneous Truck and Loader Fleets. International Journal of Mining, Reclamation and Environment 21, 262–270 (2007)

[7] Duinkerken, M.B., Dekker, R., Kurstjens, S.T.G.L., Ottjes, J.A., Dellaert, N.P.: Comparing Transportation Systems for Inter-Terminal Transport at the Maasvlakte Container Terminals. OR Spectrum 28, 469–493 (2006)

[8] Duflos, S., Diallo, A.A., Le Grand, G.: An Overlay Simulator for Interdependent Critical Information Infrastructures. In: Dependability of Computer Systems. DepCoS-RELCOMEX 2007, pp. 27–34 (2007)

[9] Fishman, G.: Monte Carlo: Concepts, Algorithms, and Applications. Springer (1996)

[10] Gao, Y., Freeh, V.W., Madey, G.R.: Conceptual Framework for Agent-based Modelling and Simulation. In: Proceedings of NAACSOS Conference, Pittsburgh (2003)

[11] Gartner, N., Messer, C.J., Rathi, A.K.: Traffic Flow Theory and Characteristics. In: Board, T.R. (ed.) University of Texas at Austin, Texas (1998)

[12] Gold, N., Knight, C., Mohan, A., Munro, M.: Understanding service-oriented software. IEEE Software 21, 71–77 (2004)

[13] Hongli, Y., Xiangpeng, Z., Zongyan, Q., Geguang, P., Shuling, W.: A Formal Model for Web Service Choreography Description Language (WS-CDL). In: Proc. of ICWS 2006. IEEE Computer Society (2006)

[14] Ioannou, P.A.: Intelligent Freight Transportation. Taylor and Francis Group, Carolina (2008)

[15] Jennings, N.R.: On Agent-Based Software Engineering. In: Artificial Intelligence, vol. 117, pp. 277–296. Elsevier Press (2000)

[16] Krzyzanowska, J.: The Impact of Mixed Fleet Hauling on Mining Operations at Venetia Mine. Journal of The South African Institute of Mining and Metallurgy 107, 215–224 (2007)

[17] Liu, H., Chu, L., Recker, W.: Performance Evaluation of ITS Strategies Using Microscopic Simulation. In: Proceedings of the 7th International IEEE Conference on Intelligent Transportation Systems, pp. 255–270 (2004)

[18] Mascal, C.M., North, M.J.: Tutorial on agent-based modelling and simulation. In: Winter Simulation Conference (2005)

[19] Mayer, P., Lubke, D.: Towards a BPEL Unit Testing Framework. In: International Symposium on Software Testing and Analysis, pp. 33–42. ACM (2006)

[20] Mellouli, S., Moulin, B., Mineau, G.W.: Laying Down the Foundations of an Agent Modelling Methodology for Fault-Tolerant Multi-agent Systems. In: ESAW, pp. 275–293 (2003)
[21] Michalska, K., Mazurkiewicz, J.: Functional and Dependability Approach to Transport Services Using Modelling Language. In: Jędrzejowicz, P., Nguyen, N.T., Hoang, K. (eds.) ICCCI 2011, Part II. LNCS (LNAI), vol. 6923, pp. 180–190. Springer, Heidelberg (2011)
[22] Nicol, D., Liu, J., Liljenstam, M., Guanhua, Y.: Simulation of Large Scale Networks Using SSF. In: Proceedings of the 2003 Winter Simulation Conference, vol. 1, pp. 650–657 (2003)
[23] Nowak, K., Mazurkiewicz, J.: Multiagent Modeling and XML-Like Description of Discrete Transport System. In: Kabashkin, I. (ed.) Transport and Telecommunication. Transport and Telecommunication Institute, Riga, vol. 12(4), pp. 14–26 (2011)
[24] Sanso, B., Milot, L.: Performability of a Congested Urban-Transportation Network when Accident Information is Available. Transportation Science 33(1), 10–21 (1999)
[25] Taylor, M.A.P., Woolley, J.E., Zito, R.: Integration of the Global Positioning System and Geographical Information Systems for Traffic Congestion Studies. Transportation Research, Part C (Emerging Technologies) 8C, 257–285 (2000)
[26] Vis, I.F.A.: Survey of Research in the Design and Control of Automated Guided Vehicle Systems. European Journal of Operational Research 170, 677–709 (2006)
[27] Walkowiak, T., Mazurkiewicz, J.: Analysis of Critical Situations in Discrete Transport Systems. In: Proceedings of International Conference on Dependability of Computer Systems, Brunow, Poland, June 30-July 2, pp. 364–371. IEEE Computer Society Press, Los Alamitos (2009)
[28] Walkowiak, T., Mazurkiewicz, J.: Availability of Discrete Transport System Simulated by SSF Tool. In: Proceedings of International Conference on Dependability of Computer Systems, Szklarska Poreba, Poland, pp. 430–437. IEEE Computer Society Press, Los Alamitos (2008)
[29] Walkowiak, T., Mazurkiewicz, J.: Functional Availability Analysis of Discrete Transport System Realized by SSF Simulator. In: Bubak, M., van Albada, G.D., Dongarra, J., Sloot, P.M.A. (eds.) ICCS 2008, Part I. LNCS, vol. 5101, pp. 671–678. Springer, Heidelberg (2008)
[30] Walkowiak, T., Mazurkiewicz, J.: Algorithmic Approach to Vehicle Dispatching in Discrete Transport Systems. In: Sugier, J., et al. (eds.) Technical Approach to Dependability, pp. 173–188. Oficyna Wydawnicza Politechniki Wroclawskiej, Wroclaw (2010)
[31] Walkowiak, T., Mazurkiewicz, J.: Functional Availability Analysis of Discrete Transport System Simulated by SSF Tool. International Journal of Critical Computer-Based Systems 1(1-3), 255–266 (2010)
[32] Walkowiak, T., Mazurkiewicz, J.: Soft Computing Approach to Discrete Transport System Management. In: Rutkowski, L., Scherer, R., Tadeusiewicz, R., Zadeh, L.A., Zurada, J.M. (eds.) ICAISC 2010. LNCS, vol. 6114, pp. 675–682. Springer, Heidelberg (2010)

Pattern Based Support for Site Certification

Dariusz Rogowski and Przemysław Nowak

Abstract. The work presents a methodology for building development environments of secure and reliable IT products or systems according to the newest approach called Site Certification. The methodology is based on design patterns worked out in the CCMODE project (Common Criteria compliant, Modular, Open IT security Development Environment) carried out by the Institute of Innovative Technologies EMAG. The design patterns help developers to write proper documents (evidences) according to the Site Certification requirements. This approach allows to gain a certificate for a development environment. Next, the certificate can also be used to diminish the costs of the product evaluation according to the Common Criteria standard. The work shows by examples how to accomplish the final document by using its pattern.

1 Introduction

The Common Criteria for Information Technology Security Evaluation (called for short the Common Criteria – CC) is also known as the international standard ISO/IEC 15408 [1–3]. The CC provides a set of requirements for security functionality of IT products. It also provides assurance based upon an evaluation of the IT product (called the TOE – Target of Evaluation) that is to be trusted. The evaluation results answer the question whether the IT product fulfills its security specification or not. Additionally, the standard can be used as a guide for developers, evaluators and eventually consumers of products with security features. However, the usage of this guide in its original shape is very complicated and difficult. That is why the pattern based methodology was worked out in the CCMODE project.

The methodology comprises design patterns for evidences that are needed for the evaluation process of the IT product. The structure of the pattern is consistent with the CC requirements. It comprises hints with definitions and guidelines taken from the standard. The patterns are ready to use and they enable developers to make necessary documentation of the IT product. More specific information about the patterns can be found in the following publications [4–7].

Dariusz Rogowski · Przemysław Nowak
Institute of Innovative Technologies EMAG, 40-189 Katowice, Leopolda 31, Poland
e-mail: drogowski@emag.pl, pnowak@emag.pl

W. Zamojski et al. (Eds.): Complex Systems and Dependability, AISC 170, pp. 179–193.
springerlink.com

IT products developed according to the CC requirements have their security features well done and structured. This development process is very similar to good engineering practices which are used in many development environments (hardware or software). Next the product has to be assessed by an independent evaluation institution. This institution must check not only the evidences made by developers but also the development environment itself which we can call the site audit.

The complete evaluation of the product has two steps: the TOE evaluation and the audit of the site. Every time when a new product is developed it must be assessed during this two-steps evaluation process. That makes the assessment more expensive and time consuming.

Nowadays many of IT products are made in several development and manufacturing sites, located across the world. Developers may construct multiple TOEs, each using a different set of these sites. In such cases, the time, cost and effort needed for the evaluation are soaring. These obstacles could be overcame by reuse of already assessed evidences of the sites. There is increasing demand coming from developers to avoid unnecessary evaluation efforts.

That is why the Bundesamt für Sicherheit in der Informationstechnik (BSI) has developed and validated a procedure in order to perform reusable evaluations of sites-related aspects. The procedure is called the Site Certification and leads to a TOE independent CC certificate [8]. This certificate confirms that a specific development environment fulfils the CC requirements. These requirements concern the life-cycle support of the products developed at the site. These certificates can be reused in a TOE evaluation later on.

One of the key issues of the CCMODE project methodology is the ability to support the process of implementation and management of development environments. The methodology comprises the survey which helps developers to find out whether their sites fulfill the CC requirements or not. It shows what must be done to provide basic procedural, physical and personnel security of the site. While building the site, every step must be documented in a proper and consistent manner for the needs of the later evaluation. The design patterns dedicated for the development environment facilitate the process of site documentation.

There are standard patterns which regard such aspects of the site as: configuration management, life-cycle definition, development security, delivery, flaw remediation, tools and techniques. These patterns are fully consistent with the assurance requirements of the CC standard. Next, the patterns are enhanced by application notes to make them consistent with the Site Certification procedures. This way the methodology comprises the set of means helping developers to build development environments which can be certified later.

In the work we describe the structure of the most important pattern for site security specification called the Site Security Target (SST) and the pattern for configuration management within the site. We show how the patterns can be used for preparing the final documentation for an exemplary development environment. We also answer the question whether it is worth to build and certify development environments according to the Site Certification approach.

2 State of the Art

Researches on the Site Certification approach began in 2006 and they were coordinated by BSI. BSI cooperated with such companies as ATSEC, TNO, Philips, IBM and T-Systems. The review of works was based on materials which are accessible on the website of the CC standard [9]; documents issued by BSI with the Common Criteria Development Board (CCDB); guidelines and mandatory documents which support the CC standard. Among these documents there are templates for the single evaluation report and the Site Security Target (SST). We also examined a few SSTs for already evaluated and certified sites which developed such types of products as smart cards and integrated circuits. The assessment of achievements gained in the Site Certification domain was also possible by reviewing the results of several latest International Common Criteria Conferences (ICCC).

2.1 Origin and Evolution of Site Certification Idea

The main source of the site certificate idea was a conviction that certification of a development site can be a significant benefit for developers who develop multiple products at one or more sites, particularly under the same procedures.

The results of researches made on the Site Certification approach were shown for the first time at the 7th ICCC (Spain, 2006) [10]. BSI presented the main procedures of the Site Certification process and first results of the trial usage of this process in real sites. BSI also described the structure of the SST document and gave the basic guidelines for developers how to develop the SST. Additionally, a definition of a site and its scope, basic procedures of certificates integration and splicing of different sites were shown at the conference. The set of minimum and optional security assurance requirements (SARs) from the ALC class was also presented.

The first trial usage of the Site Certification was done in two types of development environments: hardware and software [11]. The trial was conducted by BSI which was supported by the following companies: IBM and ATSEC in the hardware trial; Philips and T-Systems in the software trial. The trials results showed the following benefits of the Site Certification process:

- avoiding duplication of ALC-related work between different evaluations;
- reducing evaluation and certification costs;
- separating and maintaining a certificate for a site from a product certificate;
- reusing site certificates among different evaluation laboratories and national certification bodies.

The trials also proved that the Site Certification process was easily applicable and worked well in both environments. That is why we can assume that the process will be flexible enough to be applied in all kinds of development environments. After a few similar trials leaded by the ISCI (Information Security Certificate

Initiative) working group, the idea was transformed into a concrete Site Certification supporting document guidance [8] which was elaborated by BSI.

2.2 First Certificate, Guidelines and Templates

In a short time the first trial site evaluation and certification process was performed which was based on the CC supporting document guidance mentioned above [8]. The process was sponsored by Eurosmart which is an international non-profit association founded in 1995 and located in Brussels. Eurosmart represents 24 companies of the smart security industry for multi-sectors applications and includes: manufacturers of smart cards, semiconductors, terminals, equipment for smart cards system integrators, application developers and issuers. In the process there were involved parties connected with [12]:

- sites and developers – passport inlay manufacturer HID Global Erfurt; Infineon and NXP – manufacturers of security integrated circuits;
- evaluator – T-Systems GEI GmbH – a BSI-accredited laboratory;
- certifier and sponsor for some templates and guidelines – BSI.

The main goals to achieve were performing a site evaluation and certification and deriving a Site Security Target template. The SST template is to be a generic document including application notes which would serve as the basis for further SSTs. The trial paves the way for further Site Certifications by providing guidance documentation for developers and templates for evaluators. The Site Certification process was successfully completed in the first project and the following results were accomplished:

- first certificate for the site HID Global GmbH in Erfurt, the site is a part of the production flow for security integrated circuits; the SST document for the site (only sanitized version available [13]);
- supplement [14] to the Site Certification process document [8] – it comprises: hints for developers for proper documentation preparation; work units for evaluators (for aspects not covered in the Site Certification process manual); how to deal with shortcomings (interpretation, corrections);
- Site Security Target template (its details are given in the next chapters of the work) provided by Eurosmart and based on the Security IC Platform Protection Profile [15];
- Evaluation Technical Report (ETR) for Site Certification guidance [16] – this document describes how to write an ETR as part of the Site Certification process;
- template for the single evaluation report of the assurance class AST (Site Security Target evaluation) [17]. The AST class includes requirements used to evaluation of the SST document and it is described in details in the Site Certification process guide [8];
- template for the single evaluation report of the assurance class ALC for site evaluations [18].

In conclusion, the first usage of the Site Certification process in a real development environment was successful. The trial worked out a set of templates documents for developers and evaluators. The cost and time reduction was reached on both developers' and evaluators' side. The Site Certification was accepted by CCRA (Common Criteria Recognition Arrangement) members [12] as a part of a product evaluation according to the CC standard. Further, the process was conducted in many other sites. However, the very first sites which used the process were: a site of HID Global Ireland that assembles inlays and contactless smart cards; a site of SMARTRAC that manufactures ePassports and eID cards. The corresponding SSTs [19], [20] were developed and unfolded on the BSI website [21].

In general, the SST template is intended to be used within different sites. However, this one forces the sites to be relevant to the life cycle described in the Security IC Platform Protection Profile [15] on which the template is based. That is why the usage of the template is dedicated mostly to smart cards development environments. That can be a bit of limitation for developers who want to use the template in other development environments which manufacture different types of IT products. That constraint is eliminated in an enhanced SST design pattern which was worked out in the CCMODE project. The new patterns is not limited to a specific development environment. It is more general and it comprises some facilitating features in a form of hints and data fields. Additionally, we developed a set of patterns for evidences concerning requirements of the ALC class. The subset of these requirements has to be declared as minimal site requirements in the SST according to the Site Certification process. In the following chapters of this work we show a structure of the patterns, the main features which help the developers to fill in the pattern in proper way. We also present selected parts of the final documents which were prepared during the validation of the patterns. The validation was done within a development environment of gas sensors used in coal mines. The site was established in a laboratory called SecLab EMAG [4].

3 Benefits of the Site Certification

As mentioned above, the source of motivation for developing the Site Certification process was the necessity to significantly reduce time, money and effort during products evaluation. As a result, the process enables to reuse the evidences material in an efficient manner. The evidences are based on the CC set of requirements called ALC class. The ALC class (Life-cycle support) is an aspect of establishing control in the TOE development process. This control helps to avoid vulnerabilities in the TOE implementation. In this work the ALC class will be described later with more details. Most of the development and manufacturing operations are rather independent of the products themselves. So the site evaluation results can be reused many times in CC evaluation processes. These processes consider different IT products that are developed by this site.

The Common Criteria requires that every time when a new product is developed it must be assessed during a two-steps evaluation process: the TOE evaluation and the audit of the site. What is more, even though the new product is made in the same site, this very site must be evaluated once again. The Site Certification approach allows to perform a single evaluation for the site. Apart from the site audit, the Site Certification needs to prepare some documents (SST and ALC evidences). So the question is "How much does it cost?".

The results of the first trial usage of the Site Certification process confirmed its benefits. The conclusions were presented at the 10th International Common Criteria Conference (2009) [12] and bring the answer: "The costs for the Site Certification process are about 2.5 times higher than the costs for a CC audit process for a site". Figures 1 and 2 depict benefits of the Site Certification. The unit of measure on the Y axis is the cost for a standard CC audit process for one site. Figure 1 shows, that the Site Certification approach is more advantageous if three or more products are developed in the same site.

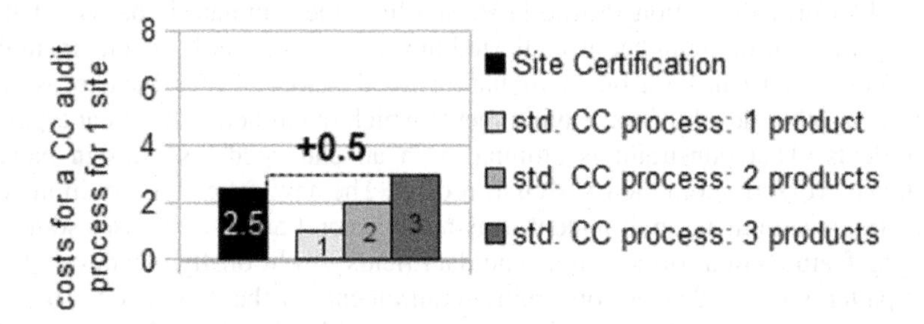

Fig. 1 Initial costs of the Site Certification and Common Criteria processes [9]

However, every two years every product must be recertified, similarly to the site. The re-audit of the site in the standard CC process costs as much as the re-audit of the site in the Site Certification process. Figure 2 depicts the following re-certifications which take place after each two-years period. For instance, after two years, two products developed in one site lead to the total number of four CC site audit processes. In that case the Site Certification process is by 0.5 unit measure cheaper.

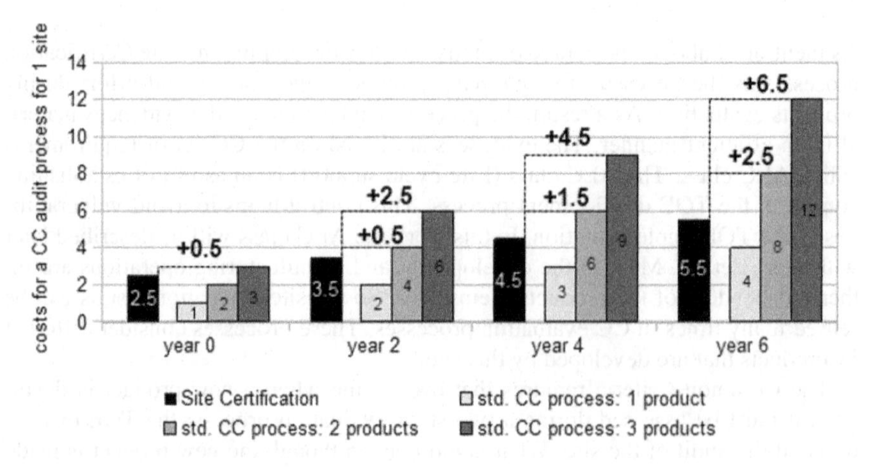

Fig. 2 Comparing the costs of the Site Certification and CC audit [9]

The benefits of the Site Certification are rising in subsequent years. Reducing the certification costs is important, but even more important is the fact that the certificate can be used many times during the evaluation of different products, and this significantly speeds up the product certification process.

Figure 3 depicts that the number of the Site Certification processes has been growing since 2009. At the same time the usage of issued certificates in standard product evaluation and certification is more and more frequent [21]. This demonstrates that the Site Certification process is so flexible and cost effective that it is very often used by developers and evaluators.

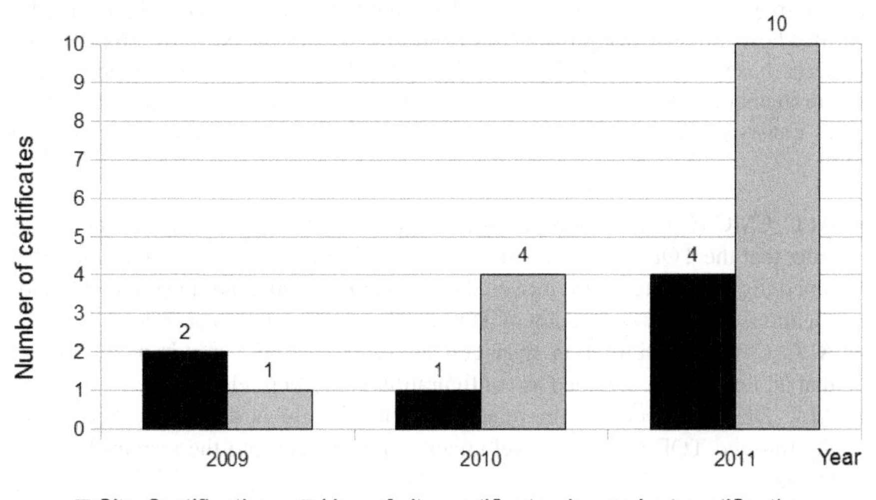

Fig. 3 Number of issued site certificates and their usage in standard CC product evaluations

The CCMODE team members took into account the Site Certification approach and they used its assumptions and requirements in the worked out methodology. We reviewed the results of BSI research, available literature and chose the best solutions to implement. The methodology comprises a set of patterns to be used by developers during the preparation of evidences for the site evaluation. The next chapter shows the results gained by BSI and co-workers and the main issues to be solved.

4 CCMODE Project in the Site Certification Context

The main goal of the CCMODE project is to work out a methodology and tools which can be used to build and manage development environments. In these environments developers can design secure IT products according to the CC requirements. The core of the methodology is a set of design patterns. The patterns are the basic tool to prepare evidence documentation needed in the evaluation processes of products security functionality. There are also many vulnerabilities in development sites which can impact the final IT product. Thus it is also important to implement security measures in development environments. These measures can mitigate the risk of tampering the product security functionality during its development phase. The measures also assure that procedures, processes and actions used in the development environment are protected enough. That is why the design patterns for evidences documentation of the development environment were additionally worked out. More about the results can be found on the CCMODE website [22].

The most important evidence document is the Site Security Target for which a pattern was worked out. The SST defines the scope of the certified site and describes how the site meets the SARs from the ALC class. Next we prepared the patterns based on the ALC class in the way which considers assumptions and requirements of the Site Certification approach. According to the CC, the ALC class consists of seven families for which the following patterns were worked out [3]:

- ALC_CMC (Configuration management capabilities) – these requirements ensure: that the TOE is correct and complete before it is sent to the consumer; that no configuration items are missed during evaluation; that no unauthorized modification, addition, or deletion of TOE configuration items was done;
- ALC_CMS (Configuration management scope) – these requirements ensure that all items to be included as configuration items were identified;
- ALC_DEL (Delivery) – the concern of this family is the secure transfer of the finished TOE from the development environment into the responsibility of the user;
- ALC_DVS (Development security) – these requirements ensure that physical, procedural, personnel, and other security measures were implemented in the development environment to protect the TOE and its parts;
- ALC_FLR (Flaw remediation) – requires identified security flaws to be tracked and corrected by the developer;
- ALC_LCD (Life-cycle definition) – application of the life-cycle model provides necessary control over the development and maintenance of the TOE and minimizes the danger of the TOE not meeting its security requirements;
- ALC_TAT (Tools and techniques) – the family includes requirements to prevent ill-defined, inconsistent or incorrect development tools from being used to develop the TOE.

The design patterns comprise application notes taken from the Site Certification guide [8]. These notes describe some adaptations that have to be done to the CC basic ALC class requirements. The adaptations allow to use the patterns in a correct and consistent manner according to the Site Certification procedures.

5 Usage of the Design Patterns

In order to evaluate a site, first a Site Security Target (SST) must be written. The SST defines the scope of the certified site and describes how the site meets the SARs from the ALC class, in particular the aspects that are of interest when re-using the site.

Next the evidences documents for ALC requirements must be written. For every site a minimum set of requirements has to be fulfilled. One is the assumption that at each site the developer uses a configuration management system that uniquely identifies all configuration items handled by that site. According to the Site Certification guide [8] the following components from the ALC class must be fulfilled: ALC_CMC.3 and ALC_CMS.3. Another assumption is that in the development environment it is necessary to use security measures to provide confidentiality and integrity of the TOE design and implementation. This requirement is described in component ALC_DVS.1. The developer must prepare documentation for all above mentioned requirements to begin the Site Certification process.

In the following chapters we describe the structure of the SST and ALC_CMC.4 patterns. We chose parts of the final documents as examples of the patterns validation results.

5.1 SST and ALC Patterns Contents

The Site Security Target pattern is one of the most important design patterns worked out in the CCMODE project. All design patterns have similar structure. They contain introductions with instructions how to fill in patterns, data fields that must be completed and endnotes with tips.

The structure of the SST pattern is based on the requirements of the AST class which is described in the Site Certification guide [8] (chapter 7 of this document). The structure of the pattern is also based on the SST template prepared for the smart security industry [15]. The AST class describes the content of the SST document and defines work units for the evaluation of the resulting document. The Site Security Target pattern consists of the following main sections (shown in Figure 4):

- SST Introduction – shows references of the SST and site, and describes the site in more detail;
- Conformance claims – describes the version of the CC that is used, and SARs which are in the scope of this site;
- Security problem definition – describes the threats and OSPs (Organisational Security Policies) that must be countered and enforced by the site;
- Security objectives for the site – shows how the site will counter the threats and enforce the OSPs;
- Extended components definition – if new SARs are needed, in this section new components (not included in CC Part 3 [3]) may be defined;
- Security requirements – provides a translation of the security objectives into a standardized language in the form of the SARs;
- Site summary specification – summarizes how the site implements the SARs.

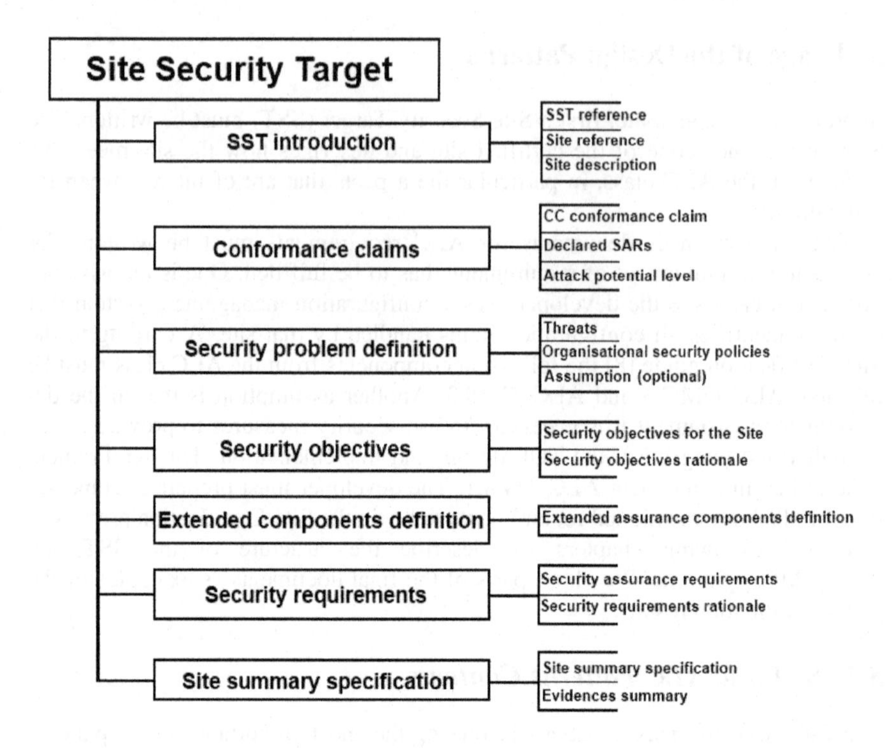

Fig. 4 The structure of the Site Security Target pattern

The structure of the Site ALC_CMC.4 pattern (Figure 5) is based on the requirements of the ALC_CMC.4 component from CC Part 3 [3], guidelines for the developer documentation [23], for evaluation reports [24], and the Site Certification guide [8].

5.2 Why Are the Patterns Helpful?

The SST pattern, similarly to the rest of the design patterns, contains many facilities for developers. Figure 6 shows an example of using the SST pattern. This is a part of "Security problem definition" – one of the main chapters. We prepared an introduction to the chapter, which can be used in any SST. The sections are consistent with the requirements of the AST_SPD family. There are data fields in square brackets that should be filled in according to endnotes. The endnote can be composed of hints, examples and references to appropriate security assurance requirements from CC part 3 [3] or site guide [8]. These data fields and footnotes are very helpful for developers but they are also a prelude to developing computer-aided application in the CCMODE project.

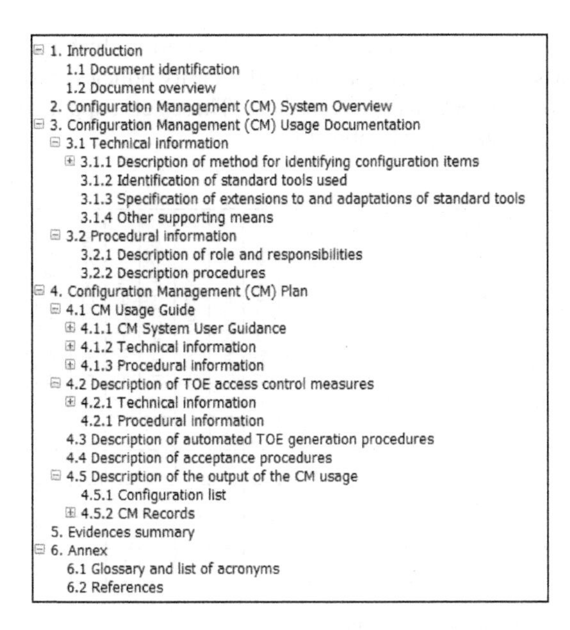

Fig. 5 The structure of the ALC_CMC pattern

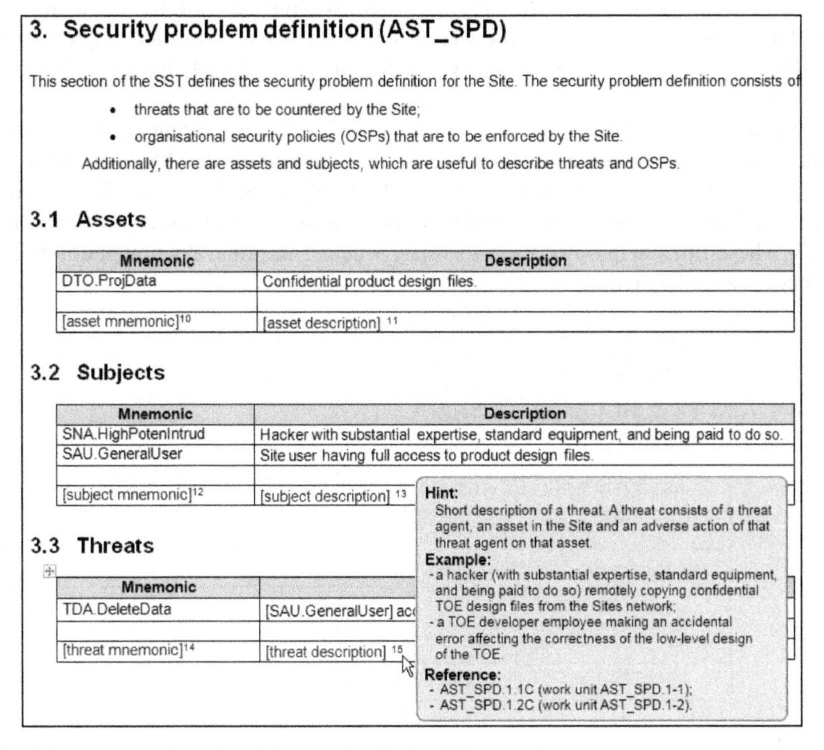

3. Security problem definition (AST_SPD)

This section of the SST defines the security problem definition for the Site. The security problem definition consists of

- threats that are to be countered by the Site;
- organisational security policies (OSPs) that are to be enforced by the Site.

Additionally, there are assets and subjects, which are useful to describe threats and OSPs.

3.1 Assets

Mnemonic	Description
DTO.ProjData	Confidential product design files.
[asset mnemonic][10]	[asset description] [11]

3.2 Subjects

Mnemonic	Description
SNA.HighPotenIntrud	Hacker with substantial expertise, standard equipment, and being paid to do so.
SAU.GeneralUser	Site user having full access to product design files.
[subject mnemonic][12]	[subject description] [13]

3.3 Threats

Mnemonic	
TDA.DeleteData	[SAU.GeneralUser] ac
[threat mnemonic][14]	[threat description] [15]

Hint:
Short description of a threat. A threat consists of a threat agent, an asset in the Site and an adverse action of that threat agent on that asset.
Example:
- a hacker (with substantial expertise, standard equipment, and being paid to do so) remotely copying confidential TOE design files from the Sites network;
- a TOE developer employee making an accidental error affecting the correctness of the low-level design of the TOE.
Reference:
- AST_SPD.1.1C (work unit AST_SPD.1-1);
- AST_SPD.1.2C (work unit AST_SPD.1-2).

Fig. 6 Security problem definition section in the SST pattern

Figure 7 shows tracing between threats and security objectives. We prepared sets of threats, assets, subjects, organisational security policies, assumptions, security objectives and connections between them. These sets support the developer in choosing the appropriate security objectives.

Security objectives / Threats	TDA.DeleteData[70]	TDA.PhysAttack	TDA.SoftAttack	TIT.MaliciousCode	TPH.HardFail	TPH.HardDamRobb	TFM.ForceMajeure
OEIT.ProtectData[71]	X						X
OEPH.PhysSafe		X				X	
OINT.TamperResistance		X				X	
OIDA.AuthRetryLimit			X				
ODEX.VirusCheck				X			
OAVB.RecovMech					X		X

Fig. 7 Tracing between threats and security objectives

The Common Criteria portal [9] provides for free an xml file with the CC standard. This allowed us to prepare a set of security assurance requirements (SARs). This collection helps to develop the next chapter of the SST "Security requirements".

The last chapter "Site summary specification" includes, among others, the evidences summary. Figure 8 depicts a part of the table with the AST requirements which are linked to other chapters and sections of the SST pattern. The requirements are also connected with work units – most granular level of evaluation work. These links help developers to verify whether the final document comprises all necessary information.

Requirement	Evidence
AST_INT.1.1C[123]	Chapter 1
AST_INT.1.2C[124]	Section 1.1
AST_INT.1.3C[125]	Section 1.2
AST_INT.1.4C[126]	Section 1.3.1
[illegible]	[illegible]
[illegible]	[illegible] 3.1, 3.2
AST_SPD.1.2C[129]	Sections 3.3, 3.1, 3.2
AST_SPD.1.3C[130]	Sections 3.4, 3.1, 3.2

All threats shall be described in terms of a threat agent, an asset, and an adverse action.

Fig. 8 Evidences summary chapter

Summary

The work presents the results of researches concerning the Site Certification approach. The approach was worked out by BSI and cooperating companies mainly coming from the smart cards industry. The developers of smart cards, integrated circuits and similar types of products demanded solutions to diminish evaluation and certification effort of their products. The evaluation costs and time are rising due to the complex life-cycle of these products and development environments made up of many sites located in different physical locations. This approach enables to reuse once evaluated requirements many times in the products evaluation process according to the Common Criteria standard.

In the CCMODE project, the methodology of the implementation and management of development environments concerning the Site Certification approach was developed [25]. The methodology is based on design patterns dedicated to evidences documentation. The work presents how to prepare necessary documentation for the development environment. The selected documents for the trial laboratory SecLab EMAG were shown as examples. It is worth to notice that all security measures and requirements described in the patterns should be first implemented in the site. Below there are main steps of the methodology to follow:

- audit of the site – gives necessary information about security needs and conditions of the site implementation;
- development of the Site Security Target which defines the logical and physical scope of the site. It presents the security problem definition (SPD) to be solved within the site. It determines minimal and optional security assurance requirements taken from the ALC class. Finally, the summary specification chapter shows how to meet the SARs in the site;
- development of the evidences documentation for ALC requirements as they are claimed in the SST;
- self evaluation of the prepared final evidences based on evidences summary chapters, hints from the patterns and guidelines of the Common Evaluation Methodology (CEM) [26];
- making decision to start the site evaluation and certification process;
- starting to use the certified site.

The work describes common properties of all patterns, their structure and hints which concern the CC requirements supplemented by application notes of the Site Certification process. The Security problem definition chapter of the SST pattern was described. An example of tracing security objectives back to threats was also presented. The tracing is a solution of a defined security problem within a site. Next the structure of ALC_CMC.4 pattern was described. It concerns the usage of configuration management system in the site.

We paid attention on evidences summary chapter introduced in all patterns. This chapter enables simple verification of the content of the final document according to the CC and Site Certification requirements. The usage of the design patterns facilitates and speeds up development of the evidences. They have data fields which are the preparation for their future use within computer aided tools. Thanks to built-in hints and guidelines, the patterns are also a kind of knowledge base about the CC standard and Site Certification accessible in one place. The trial usage of the patterns proved they are consistent with the CC standard and understandable for developers.

Eventually the documentation of the site can be passed to an independent evaluation and certification body. Then the site certificate can be reused many times during the IT product evaluation process. This possibility leads to the main advantage of the approach that reduces time and money needed in those evaluations.

References

[1] ISO/IEC 15408-1, v3.1, Information technology – Security techniques – Introduction and general model (Common Criteria Part 1) (2009)
[2] ISO/IEC 15408-2, v3.1, Information technology – Security techniques – Security functional requirements (Common Criteria Part 2) (2009)
[3] ISO/IEC 15408-3, v3.1, Information technology – Security techniques – Security assurance requirements (Common Criteria Part 3) (2009)
[4] Białas, A. (pod redakcją): Zastosowanie wzorców projektowych w konstruowaniu zabezpieczeń informatycznych zgodnych ze standardem Common Criteria. Wydawnictwo Instytutu Technik Innowacyjnych EMAG, sfinansowano ze środków UE POIG 1.3.1, Katowice (English title: Application of design patterns in the development of IT security compliant with Common Criteria) (2011)
[5] Białas, A.: Patterns Improving the Common Criteria Compliant IT Security Development Process. In: Zamojski, W., Kacprzyk, J., Mazurkiewicz, J., Sugier, J., Walkowiak, T. (eds.) Dependable Computer Systems. AISC, vol. 97, pp. 1–16. Springer, Heidelberg (2011)
[6] Bialas, A.: Patterns-based development of IT security evaluation evidences. In: The 11th International Common Criteria Conference, Antalya (2010), http://www.11iccc.org.tr/presentations.asp
[7] Białas, A.: Security-related design patterns for intelligent sensors requiring measurable assurance. Electrical Review (Przegląd Elektrotechniczny) 85(R.85)(7), 92–99 (2009) ISSN 0033-2097
[8] CCDB, Supporting Document Guidance, Site Certification. Version 1.0 Revision 1, CCDB-2007-11-001 (2007)
[9] The Common Criteria portal, http://www.commoncriteriaportal.org (accessed January 2012)
[10] Sonnenberg, F.: Site Certification Process. In: 7th ICCC, Lanzarote, Spain (2006)
[11] Borch, T.: First Trial-Use-Results of the Site Certification Process. In: 7th ICCC, Lanzarote, Spain (2006)
[12] Albertsen, H., Noller, J.: Good News & Guidelines. In: 10th ICCC, Tromso, Norway (2009)
[13] BSI, Site Security Target Lite for the Inlay Production of HID Global GmbH in Erfurt. Certification ID: BSI-DSZ-CC-S-0001, version 1.1 (2009)
[14] BSI, Guidance for Site Certification. Version 1.0 (2010)
[15] BSI, Security IC Platform Protection Profile. Version 1.0, BSI-PP-0035 (2007)

[16] BSI, Details for the structure and content of the ETR for Site Certification, ver. 1.0 (2010)
[17] BSI, Single Evaluation Report of the Assurance Class AST (Site Security Target evaluation). Version 1.0, 16th, BSI – Template_ETR-Part_AST_v1_0.doc (2010)
[18] BSI, Single Evaluation Report of the Assurance Class ALC (Life-Cycle Support). Version 1.0, 16th, BSI – Template_ETR-Part_ALC_v1_0.doc (2010)
[19] Site Security Target Lite of HID Global Ireland Teoranta in Galway Ireland. Certification ID: BSI-DSZ-CC-S-0004
[20] Site Security Target for SMT1 Smartrack Technology Ltd., Certification ID: BSI-DSZ-CC-S-0002, version 1.51 lite, September 30 (2009)
[21] BSI website, `http://www.bsi-fuer-buerger.de/EN/Topics/Certification/CertificationReports/certificationreports_node.html` (accessed on January 2012)
[22] The CCMODE project portal, `http://commoncriteria.pl` (accessed on January 2012)
[23] BSI, Guidelines for Developer Documentation according to Common Criteria Version 3.1, Bundesamt für Sicherheit in der Informationstechnik (2007)
[24] BSI, Guidelines for Evaluation Reports according to Common Criteria Version 3.1, Bundesamt für Sicherheit in der Informationstechnik ,Version 2.00 for CCv3.1 rev. 3 (2010)
[25] Nowak, P., Rogowski, D., Styczeń, I.: Certyfikacja lokalnego środowiska rozwojowego (Site Certification) jako innowacyjne podejście do oceny produktów według standardu Common Criteria. MIAG, Katowice (English title: Site Certification as innovative approach to products evaluation according Common Criteria standard) (2011)
[26] CCMB, Common Methodology for Information Technology Security Evaluation (CEM), Evaluation methodology. Version 3.1, Revision 3, CCMB-2009-07-004 (2009)

Integrating the Best 2-Opt Method to Enhance the Genetic Algorithm Execution Time in Solving the Traveler Salesman Problem

Sara Sabba and Salim Chikhi

Abstract. The traveling salesman problem (TSP) is one of the classic combinatorial optimization problem NP-complete that requires much time to find the good solution. Indeed, the genetic algorithm is a stochastic optimization algorithm; it is to find an approximate solution of a hard problem. However, genetic algorithm has a great tendency to converge to a local minimum and stay stuck in adverse solutions. To solve this problem, we study in this paper the impact of the integration of a new local optimization heuristic Best 2-opt with the genetic operators on the quality of solution and the runtime of the GA. The hybridization proposed was tested on instances from 29 to 246 cities. The obtained results are very satisfied regarding to the solution qualities and the execution time.

1 Introduction

The combinatorial optimization takes an important place in operations research, discrete mathematics and computer science. Indeed, it is to find the optimal solution of a combinatorial optimization problem characterized by the exponential number of combinations to explore. Moreover, the main objective of solving combinatorial problems methods is to find the optimal solution in a reasonable time in relation with the complexity of the problem.

The traveling salesman problem TSP is one of the classic combinatorial optimization problems belonging to the class "*NP-complete*"; it is widely studied for a long time by a considerable number of scientists and mathematicians because of

Sara Sabba
Mentouri University- Constantine
Computer Science Department, MISC Laboratory
sara.sabba@yahoo.fr

Salim Chikhi
Mentouri University- Constantine
Computer Science Department, MISC Laboratory
chikhi@umc.edu.dz

W. Zamojski et al. (Eds.): Complex Systems and Dependability, AISC 170, pp. 195–208.
springerlink.com

its importance in many fields of applications: problems of logistics, transport (commodities / people), and more generally all kinds of scheduling problems. Moreover, the problem is to find the shortest Hamiltonian path in a graph completely connected (N, A) [1] where N is the number of cities and while A represents the paths between these cities. It seems to be a simple problem but actually it is very difficult to be solved because the number of possibilities to find the right solution is very large: (N-1)! possible circuits.

Indeed, the genetic algorithm ***GA*** is a stochastic optimization algorithm [10], this algorithm is the very useful to solve a wide range of complex problems [14][15]. However, GA is often trapped in local minimum which means that the reproductive process of the genetic algorithm is stuck in negative solutions. To overcome this weakness, it is clear that the integration of a local search method with genetic operators is required In order to improve the quality of final solutions and diminish the execution time of the algorithm.

Consequently, several methods have been proposed in the literature to improve the performance of the genetic algorithm for solving the traveling salesman problem. These methods deal generally with crossover operators such as, Partially mapped crossover PMX [2], cycle crossover CX [3], order crossover OX [4], distance Preserving crossover DPX[5] and 2-exchange crossover heuristic ECH [11] or mutation operators such as, inversion, different swapping strategies, greedy optimization, k - opt steps (2-opt, 3-opt)[7][8].

In this paper we study the new heuristic local optimization Best 2-opt as a mutation operator able to guide the genetic algorithm to make better local search and to find best solutions in a reasonable time. For this purpose, we present at first the optimization strategy of this method, then, we demonstrate the impact of the integration of the Best 2-opt heuristic on the quality of solution and the runtime of the GA. Finally, we compare our results with already existed ones.

2 Genetic Algorithm

The genetic algorithm is a stochastic optimization algorithm based on the mechanisms of natural selection and genetics [10][17], it was designed by J. Holland by simulating evolution of the species, the theory developed by Charles Darwin. Therefore, the genetic algorithm is simple to be developed and it does not require detailed knowledge of the problem to be solved. GA Simply represents the solutions of the problem as vectors of integer, real or alphabet "individuals", all of these individuals "or population" represents a part of the search space, then it uses the genetic operators (selection, crossover, mutation) to produce other solutions which may be optimal.

Algorithm 1. Genetic algorithm

1. Initialize population randomly P

2. Evaluate initial population

3. WHILE stopping criterion is not satisfied do

3.1. Evaluate each individual's fitness

3.2. Select M% of population P

3.3. Produce a new population P ' by reproduction operators

3.3.1. Apply crossover operator

3.3.2. Apply Mutation operator

3.4. Replace a few individuals by the new population P '

3.5. End While

4Write the best solution

Fig. 1 The pseudo code of the Genetic Algorithm

2.1 *Encoding*

The traveling salesman problem is a scheduling problem; a classical representation as chains of binary is inadequate. Therefore, the solutions are better represented by permutations of the elements to be visited (each city is represented by an integer).

9	3	2	4	5	6	7	1	8

Fig. 2 Representation of a solution of 9 cities

2.2 *Selection Operator*

The selection consists to choose individuals who will be able to survive and reproduce to pass their characteristics to the next generation.

2.3 *Crossover Operator*

The crossover operator aims to create new solutions by combining information from two selected parents. However, several crossover operators have been created specifically to solve the TSP problem [2][3][4][5][11], because the classical crossing methods are not effective.

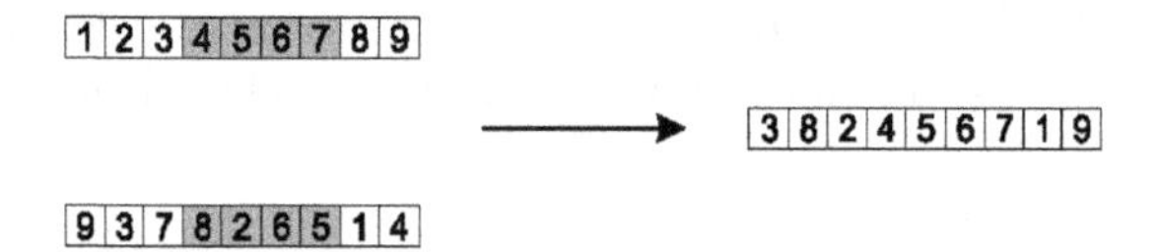

Fig. 3 Order Crossover OX

2.4 Mutation Operator

It is a very important operator; it allows making a modification of some produced individuals in the aim is to introduce diversity and to explore new regions of solution space. The mutation can be performed before or after crossing on the parents or on the new individuals.

Fig. 4 Reverse Mutation

The traditional genetic algorithm will often have a great tendency to converge to local minima and thus stuck in the wrong direction of the search space (adverse solutions).

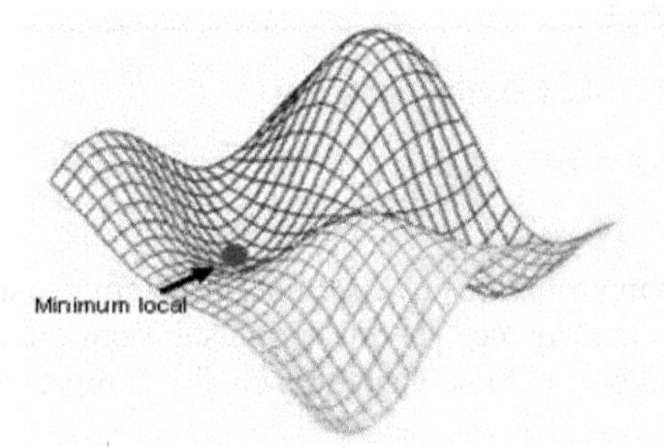

Fig. 5 GA is blocked in the bad space of solutions

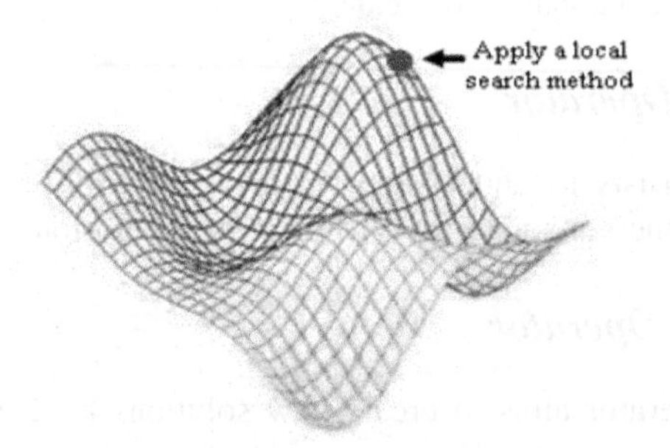

Fig. 6 A local search method helps GA to get out of local minimum

However, to avoid this problem, adding a method of local search (mutation) will change the research direction of the algorithm and ensure that it will be taken in several regions of space and so it is more likely to find optimal solutions.

3 2-Opt Local Research

Several local search methods can be applied as improvements heuristic to optimize the solutions found by the genetic operators, such as simulated annealing and tabu search ... etc. In fact, 2-opt , 3-opt[7], and Lin-Kernighan[8] are well-known improvement heuristics for solving TSP problem, they are iterative algorithms based on the exchange of 2,3 or a variable number of edges until the it reaches an improved solution comparing with the initial solution. The computational complexity of these methods is respectively O (n ^{2}), O (n3), O(nk).

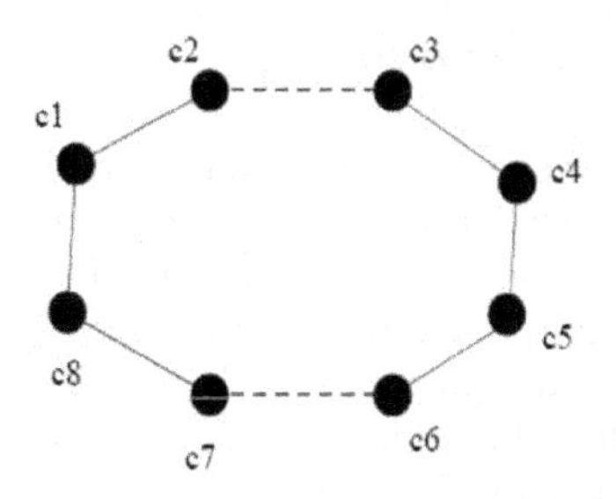

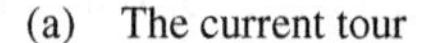
(a) The current tour

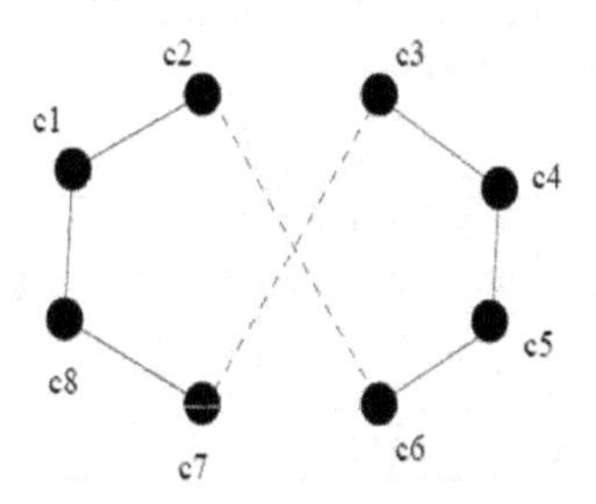

(b) The tour proposed by 2-opt

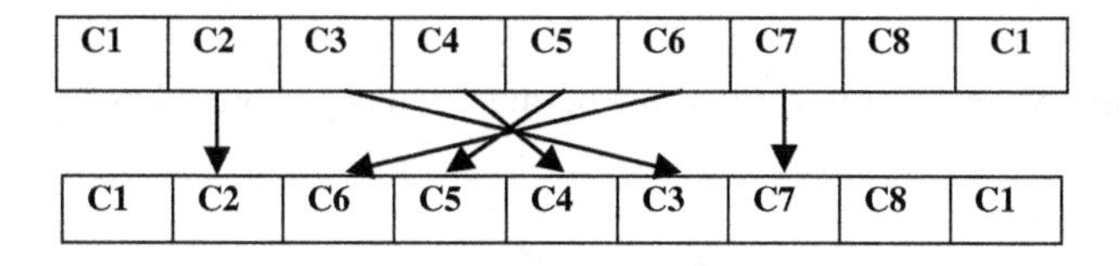

(c) The array structure

Fig. 7 2-opt swapping

The 2-opt heuristic is based on the conditional permutation of cities. In other words, select two segments in the current tour for example **c2c3** and **c6c7** then swap them (Fig. 7.c) if the equation (1) is satisfied:

Distance(c2,c3)+Distance(c6,c7) > Distance(c2,c6)+ Distance(c3,c7) (1)

Where Distance (a, b) represents the Euclidean distance between city A and B.

$$\text{Distance (a,b)} = \text{sqrt}((A_x\text{-}B_x)^2+(A_y\text{-}B_y)^2) \quad (2)$$

Thus, if the formula (1) is satisfied, remove the two segments **c2c3** and **c6c7** and replace them with two **c2c6** and **c3c7**, otherwise select two segments and repeat the test of the formula (1). This process can be repeated k times until the tour is optimized.

4 Best 2-Opt Local Research

In this paper we propose the local optimization heuristic Best 2-opt, as an effective method to improve the local search of the genetic algorithm. Indeed, the Best 2-opt method consists to select a position ***i*** in the current tour, then use the same formula (1) of the 2-opt heuristic to test whether the permutation of cities is authorized between the two segments « C_i C_{i+1} » and « C_n $C_{n+}1$», this verification will be performed with all segments which follow the segment « C_i C_{i+1} », n=n+2…L, where L is the Length of tour-2.

If the test of the formula (1) is satisfied, only the second city of the first segment and the first city of the second segment can be swapped, while the intermediate cities keep the same position.

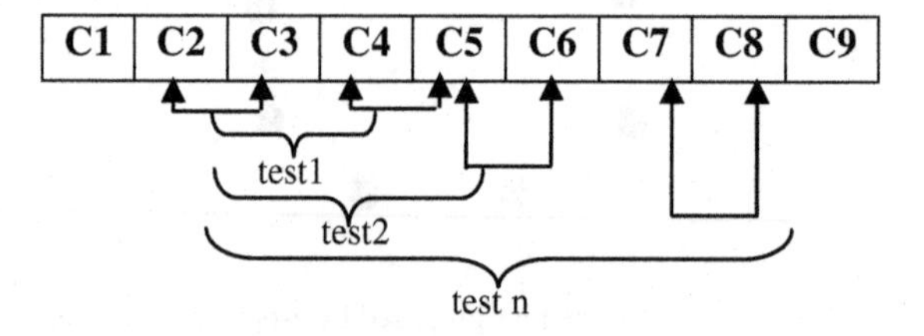

Fig. 8 Test phase

Each time when a test is verified the permutation is performed as follows:

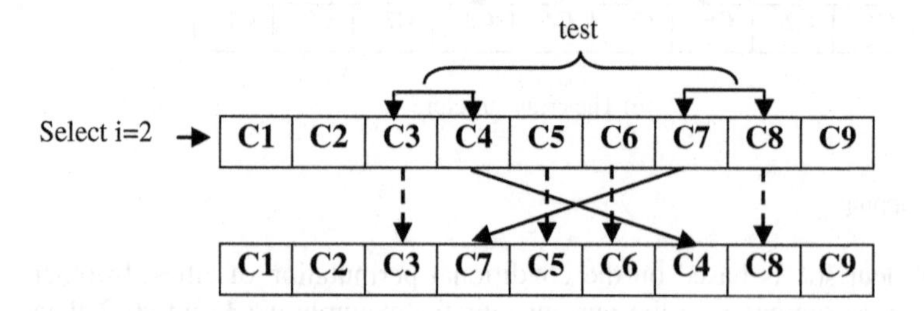

Fig. 9 The permutation of cities by the Best 2-opt method

Finally, when this process is complete, select the next position j = i +1 and repeat the same steps outlined for the position i. The Best 2-opt method can be repeated k times (in this paper we have fixed k = 5), to ensure that the tour is well optimized.

Algorithm 2. Best 2-opt local search.

```
1. Initial Tour : T
2. Improved Tour T'
3. k=0
4. Repeat
4.1. For i=0 to length(T)-2 do
4.1.1. For j=i+2 to length(T)-2 do
4.1.1.1.  If (1)=true then
4. 1.1.2.  T'= apply Best 2-opt permutation
4. 1.1.3.   endif
4.1.2.  endfor
4.2.  if ( T < T' ) then
4.2.1. T'= T ;
4.2.2.  Else
42.3. T= T';
4.3.  Endif
4.4. Endfor
4.5. k=k+1;
5.Until k>5
6.Write the best tour T'
```

Fig. 10 The pseudo code of the Best 2-opt local search

5 The Hybridization of Genetic Algorithm with Best 2-Opt

Genetic local search was first introduced by Moscato [16] as "Mimetic Algorithms"; the aim of this is to produce hopefully better individuals from existing ones. Indeed, in the last years several hybridizations of GA have been designed to solve the traveling salesman problem and the most local search methods proposed are generally integrated before or after the reproduction of new solutions. In this purpose, our tested algorithm is structured as follows:

- Every tour is coded as a vector of integers where each value represents a city.
- The initial population is randomly generated and only the best individuals are selected to produce the new population.
- The selected individuals (50% of the initial population) are crossed by the crossover operator PMX or OX, and the mutation operator is completely replaced by the Best 2-opt heuristic.
- The bad individuals are replaced by the individuals of the new population.
- The algorithm is repeated according to a fixed number of generations and in each generation the probability of applying crossover and mutation operator is 100%.

6 Experimental Results and Performance Comparisons

Based on the above discussion, this section is devoted to the representation of different hybridization results of the genetic algorithm with the optimization method Best 2-opt (solution quality and run time). The experimental study was tested on instances of symmetric TSP from 29to 264 cities.

6.1 Step 1

The aim of the first step is to show the strength of the Best 2-opt method over other local search methods. For this purpose, we applied the Best 2-opt method to the bad tours chosen randomly from different instances of the STSP problem (berlin52, eil51, st70, eil76, pr76, kroA100), to follow how this method improves bad solutions at each iteration k.

Each row in the table represents the result of permutation in the iteration k, excluding the last row that represents the difference between the length of the initial tour and the length of the tour improved.

Table 1 The results of optimization of the Best 2-opt method at each iteration k (test 1).

Instance	Berlin52	Eil51	St70	Eil76	Pr76	Kroa100
Initial tour	**9209**	**598**	**916**	**740**	**146129**	**33831**
k=1	9150	574	998	834	130376	30598
k=2	8920	535	893	681	125846	28552
k=3	8652	520	832	666	122495	27527
k=4	8620	517	772	657	120771	27570
k=5	**8552**	**445**	**770**	**636**	**117534**	**26075**
Initial Tour - Best Tour	**657**	**153**	**146**	**140**	**28595**	**7756**

Table 2 The results of optimization of the Best 2-opt method at each iteration k (test 2).

Instance	Berlin52	Eil51	St70	Eil76	Pr76	Kroa100
Initial tour	**9284**	**556**	**922**	**734**	**143288**	**29981**
k=1	9888	593	934	811	149914	33103
k=2	9510	543	894	732	140355	29300
k=3	8753	518	848	670	138623	27277
k=4	8591	503	793	664	130785	25251
k=5	**8324**	**496**	**765**	**664**	**129726**	**26607**
Initial Tour - Best Tour	**860**	**60**	**157**	**70**	**13562**	**3374**

Table 3 The results of optimization of the Best 2-opt method at each iteration k (test 3).

Instance	Berlin52	Eil51	St70	Eil76	Pr76	Kroa100
Initial tour	**9546**	**556**	**922**	**997**	**143288**	**28594**
k=1	8710	593	934	765	149914	24801
k=2	8634	543	894	635	140355	23955
k=3	8137	518	848	600	138623	22044
k=4	7958	503	793	589	130785	21360
k=5	**7958**	**496**	**765**	**581**	**129726**	**21360**
Initial Tour - Best Tour	**1588**	**60**	**157**	**416**	**20188**	**7234**

According to the values displayed in the three tables, we noticed that the tour length can be increased in the first iteration, these results are entirely normal because in the first permutation the order of cities will completely change which increases the length of the tour, while in the other iterations that order will get better step by step (the length of the turn improves iteration after iteration) until the obtaining of a well improved solution. Finally, we also noticed that the gain of optimization obtained by the method Best-2-opt is very high, which proves that this method is able to optimize the bad solutions in minimal time and that is what we will show in the second step.

Basing on the observations and previous results, we decided to eliminate the testing phase of the lap's length after each permutation (lines 4.2 - 4.3 of Algorithm 2). In fact the running time of the method and algorithm will be further reduced.

6.2 Step 2

After the elimination of the test phase of the Best 2-opt method, the second step is devoted to compare the error rate and the total running time of two hybridizations GA + standard 2-opt and GA + Best 2-Opt. (noted that the both algorithms are executed in the same machine).

The results are represented in Table 4: The first column shows the different instances of TSP used in the tests; the second column shows the results obtained by the GA + standard 2-opt algorithm and the third column shows the results obtained by the GA + Best 2-opt algorithm. The error rate is measured by the percentage to get the wrong solution and the run time required to obtain the optimal solution is measured by seconds.

Table 4 Comparison of the results obtained by GA+Best 2-opt and GA+ standard 2-opt algorithm.

TSP Instance	GA+Best 2-OPT		GA+ STANDARD 2-OPT		Best Results TSPLIB
	Time (s)	Error	Time (s)	Error	
Berlin52	0.1	0%	2.52	0%	7542
Bier127	15.41	0%	115.78	10%	118282
Eil101	2.88	0%	33.65	0%	629
Eil51	2.02	0%	4.37	0%	426
Eil76	3.86	0%	17.51	0%	538
KroA100	3.04	0%	8.46	0%	21282
Lin105	3.6	0%	11.55	0%	14379
Pr107	4.12	0%	17.83	0%	44303
Pr144	7.06	0%	52.63	0%	58537
Pr264	17.53	0%	75.39	0%	49135
Pr76	2.73	0%	9.10	0%	108159
Rat99	6.94	0%	31.2	0%	1211
St70	1.10	0%	7.28	0%	675
U159	18.57	0%	90	20%	42080
Wi29	0.01	0%	0.17	0%	27604

The experimental results show that the hybridization of the genetic algorithm with the optimizations methods Best-2-opt help it to get out from the local minima and find always the best solutions in the minimal time. Based on the previous analysis of step 1, this difference is very logical according to the optimization strategy used by this method.

6.3 Step 3

Indeed, to prove the performance and efficiency of the Best 2-opt method as a mutation operator we combined it with two different crossover operators OX and PMX and compared the obtained results with the results of the ECOGA algorithm [11] (which uses the ECH crossover operator and Improved 2-opt as a mutation operator) also with the results published in the TSPLIB library.

Table 5 Comparison of the results obtained by GAOX+ Best 2-opt, GAPMX+ Best 2-opt, ECOGA algorithms and the results published in the library TSPLIB.

Instances TSP	GAOX+ Best 2-opt	GAPMX+ Best 2-opt	ECOGA	TSPLIB
Eil51	426	426	426	426
Eil76	538	538	538	538
Rd100	**7905**	**7905**	7910	7910
KroA100	21282	21282	21282	21282
KroC100	20749	20749	20749	20749
KroD100	219294	219294	219294	219294
Pr107	44301	44303	44301	44303
Eil101	629	629	629	629
Ch130	**6105**	**6105**	6110	6110
Pr144	58537	58537	58535	58537
Ch150	**6526**	**6526**	/	6528
U159	42080	42080	42075	42080
kroA200	29369	29369	29369	29369
Tsp225	**3855**	**3876**	**3858**	3916
Pr264	49135	49135	49135	49135

The results shown in Table 5 by GAOX + Best 2-opt and GAPMX + Best 2-opt algorithms are obtained with an error rate of 0%. However, this results show the effectiveness of the Best 2-opt method as an operator of optimization, we obtained optimal solutions for most instances and for some others RD100, CH130, CH150 and tsp225. We have found better results than those found by the algorithm ECOG [11] and published in the TSPLIB library.

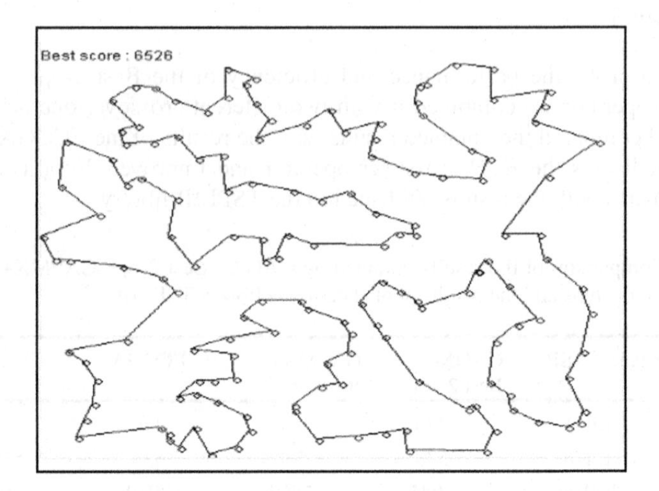

Fig. 11 The best tour of ch150.tsp

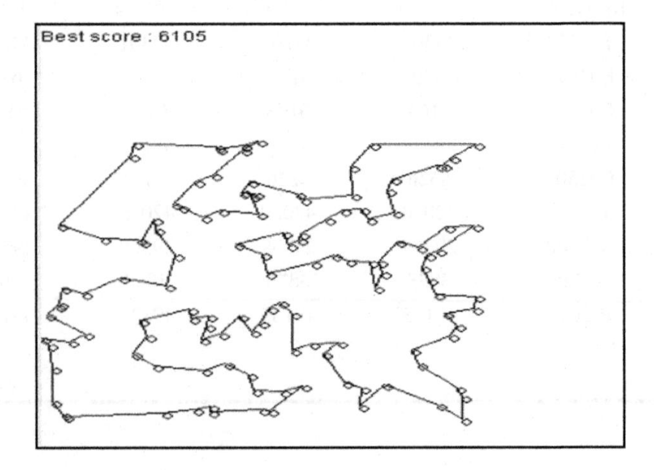

Fig. 12 The best tour of ch130.tsp

Finally, Compared with the others local search method the Best 2-opt method shows that it may be the best local optimization heuristic over other methods because it is able to avoid blocking on the local optima it find always the best solutions in a reasonable time, and it can ensure the performance of the final results and even find other optimal. The integration of this method to optimize the tours produced by the genetic algorithm is the right solution to ensure the performance of the results not only the quality but also the speed of execution.

7 Conclusion

In this paper we presented the genetic algorithm as a method of solving TSP problem, however GA has a high probability of being trapped in local minima and thus the evolution of the algorithm is stuck in poor solutions. To solve this problem, we proposed a new method for local optimization Best 2-opt. In fact, we studied the impact of integrating of this method with the genetic operators (selection, crossover), then we compared the results found by others existing.

At the end, through the obtained results, we proved that the integration of the improvement method Best 2-opt with the genetic operators for solving the TSP problem is the best solution to assure the quality of the solutions and the speed of execution.

References

[1] Dorigo, M., Gambardella, L.M.: Ant colony system: a cooperative learning approach to the traveling salesman problem. IEEE Transactions on Evolutionary Computation 1, 53–66 (1997)

[2] Goldberg, D.E., Lingle, R.J.: Alleles, Loci and the TSP. In: Proceedings of the First International Conference on Genetic Algorithms and Their Applications, pp. 154–159 (1985)

[3] Oliver, I.M., Smith, D.J., Holland, J.R.C.: A Study of Permutation Crossovers on the TS. In: Genetic Algorithm and Their Applications on Proceedings of the Second International Conference, pp. 224–230 (1987)

[4] Davis, L.: Applying Adaptive Algorithms to Epistatic Domains. In: Proceedings of the International Joint Conference on Artificial Intelligence, pp. 162–164 (1985)

[5] Freisleben, B., Merz, P.: New Genetic Local Search Operators for the Traveling Salesman Problem. In: Ebeling, W., Rechenberg, I., Voigt, H.-M., Schwefel, H.-P. (eds.) PPSN 1996. LNCS, vol. 1141, pp. 890–899. Springer, Heidelberg (1996)

[6] Lin, S.: Computer solutions of the traveling salesman problem. Bell Systems Journal 44, 2245–2269 (1965)

[7] Lin, S., Kernighan, B.W.: An effective heuristic algorithm for the traveling salesman problem. Operations Research 21, 498–516 (1973)

[8] Milan, D., Milan, T., Bojan, D.: Impact Of Graffing A 2-Opt Algorithm Based Local Searcher into the Genetic Algorithm. In: International Conference on Applied Informatics and Communications (AIC 2009), pp. 485–490 (2009) ISBN: 978-960-474-107-6

[9] Holland, J.H.: Adaptation in Natural and Artificial Systems. The University of Michigan Press, Ann Arbor (1975)

[10] Li, L., Zhang, Y.: An Improved Genetic Algorithm for The Traveling Salesman Problem. In: ICIC 2007. CCIS, vol. 2, pp. 208–216. Springer (2007)

[11] Hansen, P., Mladenović, N.: First Vs Best Improvement: An Empirical Stud. Science Direct. Discrete Applied Mathematics 154, 802–817 (2006)

[12] http://compt.ifi.uni-heidelberg.de/software/TSPLIB95/tsp/

[13] Jaszkiewicz, A.: On the performance of multiple-objective genetic local search on the 0/1 knapsack problem - a comparative experiment. Research report, Institute of Computing Science, Poznan University of Technology. RA-002 (2000)

[14] Wu, Y., Liu, M., Wu, C.: A Genetic Algorithm for Solving Flow Shop Scheduling Problems with Parallel Machine and Special Procedure Constraint. In: International Conference on Machine Learning and Cybernetics, Xi'an, China, vol. 3, pp. 1774–1779 (2003)
[15] Moscato, P.: On Evolution, Search, Optimization, Genetic Algorithms and Martial Arts: Towards Memetic Algorithms. Caltech Concurrent Computation Program (report 826) (1989)
[16] Koza, J.R.: Genetic Programming. Bradford /MIT Press (1992)

Representation of Objects in Agent-Based Lighting Design Problem

Adam Sędziwy

Abstract. Applying agent systems for solving large-scale design problems in particular in smart grid solutions, requires using proper representations at all levels of system description and specification. In the paper we introduce formally the hierarchical hypergraph representation of an urban space including both maps and physical objects like buildings. Such representation enables further decomposition of system model and performing parallel computations on it.

1 Introduction

Graph structures provide a flexible and widely used modeling framework for solving various types of problems in such areas as system specification, software generation, task allocation control [18, 8] or simulating a behavior and interactions in complex multiagent agent systems [17]. Moreover, graph transformations which properties were described in numerous works [6, 5, 4, 1] may be used to model dynamics of systems. The limitation for their applicability is time complexity of parsing or membership problems. That difficulty may be overcome however either by using distributed parallel computations paradigm [15] or by decreasing expressive power of a graph grammar if possible.

The example of the first method is GRADIS multiagent environment [9, 14] capable of performing such distributed graph transformations, representing both algorithmic and algebraic approaches [10, 11].

Multiagent systems are an effective way of solving various types of CAD related problems in such areas as the automotive industry [3], a support in constructional tasks [20], collaborative CAD systems [16] or an adaptive design [12]. A **La**rge-scale **i**ntelligent **l**ighting (shortly: LaSIL) is yet another example of such a problem.

Adam Sędziwy
AGH University of Science and Technology, Department of Automatics, al. Mickiewicza 30, 30-059 Kraków, Poland
e-mail: `sedziwy@agh.edu.pl`

W. Zamojski et al. (Eds.): Complex Systems and Dependability, AISC 170, pp. 209–223.
springerlink.com

The essence of a large-scale intelligent lighting problem in its basic form, is finding a distribution of lighting points (lamps) in an urban area and controlling their performances. This distribution has to satisfy given criteria [19] that include covering a considered area with a suitable luminosity and, on the other side, exploitation costs that are generated mainly by the power consumption. Note that criteria are local i.e. assigned to particular points, subareas or buildings. In the extended approach LaSIL includes also an intelligent, adaptive lighting control, supported by distributed sensors, providing information about an environment state. Since a LaSIL deliverable is reducing power consumption it may be seen as the important component of smart grid solutions.

The main difficulty related to a LaSIL is its high computational complexity. The following two examples illustrate the complexity issues. In the first one a city area composed of 1000 square blocks is given. Assuming that there are 5 lamps on each side of a block we obtain 20,000 lighting points to be distributed across a considered urban space. One has to add to this number the cooperating sensors and take into account a communication among lamps and a control center(s). The second example concerns a performance control complexity. Let us assume that we have 10 LED lamps, each working at 11 levels of luminosity (0,10%,... 100%). Thus we obtain 11^{10} possible states of a set of lamps. As a matter of fact the number of luminosity levels for a single LED luminary may be of order of 100. In summary, the first, static phase (i.e. while setting up an optimal distribution of lamps) of a LaSIL consists of multiple optimization subproblems in 3D space. In the second, dynamic phase (lighting control), optimizations following environment changes are performed. Note that the second phase is permanent i.e. optimizations triggered by environment's changes are made during all the lifetime of a system.

As the computations may be broken into separable tasks the first step to be done is finding an appropriate representation of a LaSIL enabling a problem decomposition. It should be remarked that using an unstructured description (e.g. pixel maps) for finding a LaSIL solution is ineffective and poorly convenient from the practical point of view. The sample structuring may be found in approaches applied to the similar problem, namely computer simulation of an adaptive illumination facility, where an environment descriptions are based on cellular structures (Situated Cellular Agents and Dissipative Multilayered Automata Network) [2].

The graph formalism seems to be the most convenient one for solving a LaSIL due to its correspondence with the problem structure and mentioned capability of modeling system architecture and dynamics (e.g. environment changes) that can be described by means of the graph grammar transformations. Using a graph description enables introducing hierarchical structures and thereby improving a tasks allocation. Note that all individual lamps and all sensors (some of them may be the mobile ones) may exchange messages. For that reason preparing a reliable communication layer gets the crucial design objective. The suitable graph-based model supports an architecture design and controlling an allocation of resources in this layer of an intelligent lighting control system. In the presented paper we introduce the hypergraph-based formalism used for modeling an urban area environment in a LaSIL. The proposed approach includes both street and building levels.

It is assumed that analytic (numeric) representations of city maps and buildings are given. In the case of maps it may have form of GIS (Geographic Information System) descriptions like GML (Geography Markup Language) or be expressed in any other geospatial numeric data format. For a building a list of vertices belonging to a bounding surface and analytic descriptions of curved edges/surfaces are sufficient.

The paper is organized as follows. In the Section 2 the hierarchical representation of a model is presented. In Sections 3 and 4 the hypergraph structures underlying a model (namely, a cartographic hypergraph and a face adjacency hypergraph) and a linkage between them are defined. The principles of a multi-agent system deployment in a graph environment and the future works are sketched in the Section 5. The paper conclusions are presented in the Section 6.

2 Hierarchy in LaSIL

The general idea of the multiagent graph-based approach to solving a LaSIL problem is its decomposition into a number of small subproblems expressed in a graph formalism, to process them in parallel by deploying and running multiagent system on such a distributed graph environment.

Any approach to solving a LaSIL has to reflect a hierarchical structure of a system (see Fig.1). In our problem we use a hierarchical, hypergraph description based on the two level hierarchy of hypergraphs. At the level 1 a city map is subdivided into an atlas of maps that cover the entire considered area. Both a whole city map and particular maps of an atlas are represented by *cartographic hypergraphs* defined in the Section 3. At the lower, second level we have architectonic objects which are specified by hypergraphs of another type, so called *face adjacency hypergraphs*. In the Section 4 the formal background of such specification based on the formalism presented in [7], will be introduced.

At a higher level (*Level 1* in Fig.1) we use coarse-grained description, transforming a considered area into a set of graph-based maps (an atlas). At a lower level

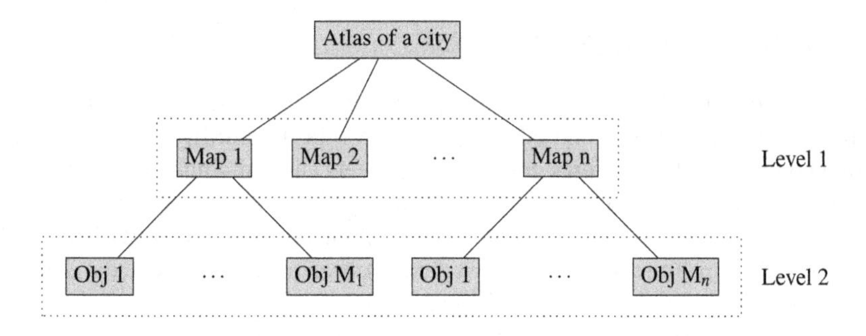

Fig. 1 Hierarchical description of an urban area: streets belong to the level 1, buildings to the level 2.

(*Level* 2) architectural details are introduced. Those data provide a detailed specification of buildings and other objects present in an urban area.

The linkage between hypergraphs of a first and a second level is dependent on a considered model. The detailed description of such an association lies beyond the scope of this paper.

2.1 Hypergraphs

The commonly used notion of a graph $G = (V, E)$ refers to a set of vertices V and a set of edges E (either directed or undirected) which may describe relations between vertices. It should be noted that expressive power of such a description is not sufficient in the case of k-ary relations (for $k > 2$) i.e. relations among more than two elements. An example is the relation $\mathscr{R}$ of having a common intersection, among subsets of S:

$$(S_1, S_2, \ldots S_k) \in \mathscr{R} \Leftrightarrow \bigcap_{i=1}^{k} S_i \neq \emptyset, \text{ where } S_i \subseteq S.$$

The structure of a normal graph may be enriched by introducing *hyperedges* which may be added either as a replacement of normal edges represented by E set (see Definition 2) or as a next element of a tuple defining G (see Definition 3) where the set of edges is divided explicitly into (hyper)edges connecting one or two vertices and all others. As a basis for our further considerations we formulate following definition.

Definition 1. *A **hypergraph** is a pair $H = (V, E)$ where V is nonempty set of vertices and $E = \{e : e \subset V\}$ is a set of hyperedges.*

Thanks to the ability of modeling the various types of spatial relations among objects the hypergraphs are an useful representation in an interior architecture design tasks. The example is a multi-agent system for a distributed design [13]. The underlying hypergraph structure possesses two kinds of hyperedges corresponding respectively to an object parts and relations.

3 Cartographic Hypergraph

In most cases an intuitive graph representation of maps is used: graph nodes correspond to junction points and edges model ways. In the LaSIL problem we deal with a composite 3D structures consisting of streets and buildings. To preserve the uniformity of a description at both levels we use the hypergraph map representation rather than the "intuitive" one (note that both approaches are equivalent). Thanks to this, a part of a given scene may be switched between an aggregated and a detailed description, dependently on an actual agent's need.

In a hypergraph-based map model approach, computations (e.g. street-level lighting optimizations) are enclosed within individual nodes.

In this section the formal hypergraph representation of maps is introduced.

Definition 2. *A* ***cartographic hypergraph*** *is a tuple of the form*

$$G = (N, H, att_N, att_H, lab_N, lab_H),$$

where N is a set of nodes, $H \subset \bigcup_{i>1} P_i(N)$ *is a set of hyperedges,* $att_N : N \longrightarrow \mathscr{A}_N$ *and* $att_H : H \longrightarrow \mathscr{A}_H$ *are node and hyperedge attributing functions respectively,* $lab_N : N \longrightarrow \mathscr{L}_N$ *and* $lab_H : H \longrightarrow \mathscr{L}_H$ *are node and hyperedge labeling functions.* $\mathscr{A}_N$ *and* $\mathscr{A}_H$ *denote sets of node and hyperedge attributes,* $\mathscr{L}_N$ *and* $\mathscr{L}_H$ *are sets of node and hyperedge labels. The family of cartographic hypergraphs is denoted as* $\mathscr{H}_{Cart}$.

Elements of the node set N correspond to such physical objects as streets, paths, squares and so on. Elements of H correspond to physical junction points of streets, paths and so on.

As it was mentioned above, representing a map as a hypergraph enables to enclose all actions related to finding an optimal lamp distribution for a single street or square, in an individual vertex of $G \in \mathscr{H}_{Cart}$. Thereby such a structure is well prepared for a hypergraph decomposition prior to computations parallelization.

Example 1.

In Figure 2 the sample map and the corresponding hypergraph are shown. The part of map consists of streets sections $s_1, \ldots, s_5$, and the square p which are represented by the cartographic hypergraph vertices. Their junction points marked with bolded points correspond to its hyperedges.

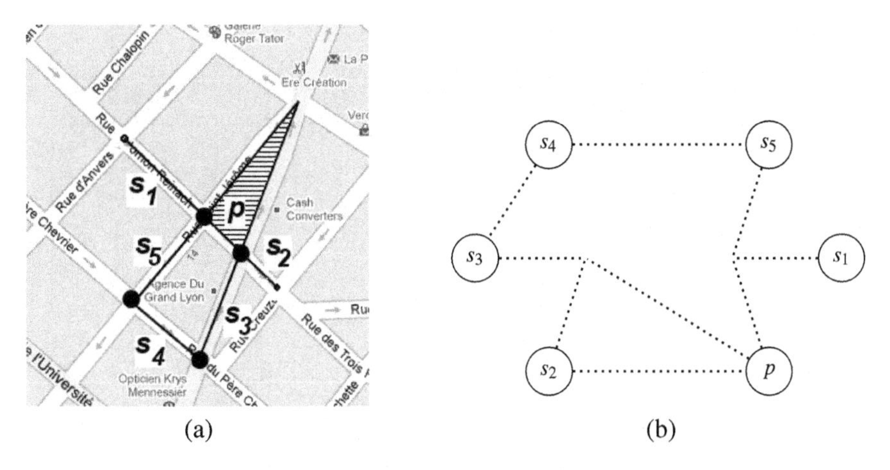

Fig. 2 (a) The sample map (b) Hypergraph representing the map

Example 2.

To demonstrate an order of magnitude of generated hypergraphs we transformed the rectangular areas (a square of the size 4 km^2) of the OSM city maps of four cities (Barcelona, Chicago, Rome, Tokyo). Fig. 3 presents selected areas.

The hypergraph associated with the map represents all streets, paths and so on. As the main features characterizing a hypergraphs we selected following descriptors:

- d_1 – number of nodes,
- d_2 – number of hyperedges,
- d_3 – number of *high* hyperedges i.e. hyperedges connecting at least three vertices,
- d_4 – average number of vertices incident with a *high* hyperedge.

The descriptor d_3 was calculated as an absolute value and as a percentage of all hyperedges in a given hypergraph. Table 1 demonstrates values obtained for areas shown in Figure 3

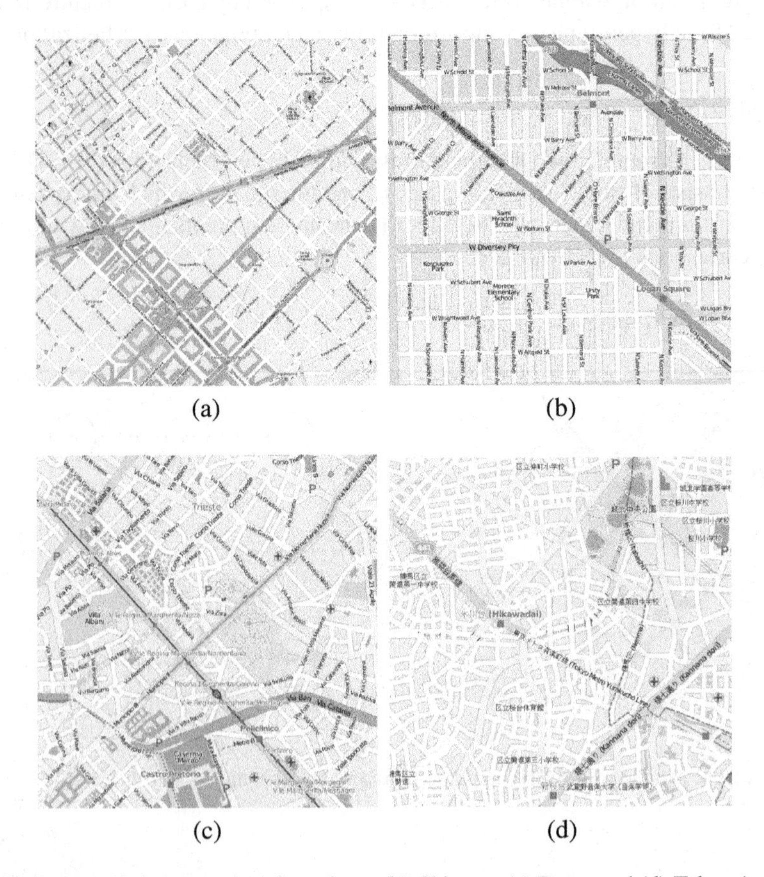

Fig. 3 The part of city map of (a) Barcelona, (b) Chicago, (c) Rome and (d) Tokyo (source: *www.openstreetmap.org*)

Table 1 Hypergraphs descriptors values

City	d_1	d_2	d_3	d_4
Barcelona	2918	2205	897 (40.7%)	3.71
Chicago	1989	1552	631 (40.7%)	3.49
Rome	2888	2361	790 (33.5%)	3.45
Tokyo	3259	2779	861 (31.0%)	3.33

Attributing functions.

Definition 2 restricted to nodes and hyperedges only gives a qualitative (topological) description of maps. The presence of two attributing functions, $att_N : N \longrightarrow \mathscr{A}_N$ and $att_H : H \longrightarrow \mathscr{A}_H$, enables including quantitative data like physical coordinates of junction points or information concerning adjacent buildings, necessary to model accurately given area. One can specify precisely structures of sets $\mathscr{A}_N$ and $\mathscr{A}_H$ relying on an actual problem specification. We assume that $att_N(v \in N)$ contains information concerning physical objects (buildings) adjacent to a given v. Note that $att_N(v) \in \mathscr{A}_N$ structure holding HFAHs/FAHs (defined below) of related buildings comprises a link between two levels of an urban area description.

4 Face Adjacency Hypergraph

The formal background for the level 2 description, related to physical objects (solids) like buildings is presented below. It consists of two notions. The first (FAH) corresponds to *simple* objects e.g. single buildings, the second one (HFAH) allows for both, modeling complex entities composed of a number of coupled solids and aggregating individual, adjacent objects into a complex one.

Let S be an object (solid) and ES, VS be sets of its edges and vertices respectively, that define two relations over the set FS of S faces:

- faces $f_1, f_2 \in FS$ are *edge adjacent* iff there exists an edge in ES common for f_1 and f_2,
- faces $f_1, f_2 \in FS$ are *vertex adjacent* iff there exists a vertex in VS belonging to f_1 and f_2.

Those relations underlay the graph representation of a solid named FAH (face adjacency hypergraph).

Definition 3 (FAH). *A **face adjacency hypergraph** (FAH) of an object S is a labeled hypergraph*

$$G = (N, A, H, lab_N, lab_A, lab_H, att_N, att_A, att_H),$$

where N is a nonempty set of nodes, $A \subset P_2(N)$ is a nonempty set of edges, $H \subset \bigcup_{i>2} P_i(N)$ is a nonempty set of hyperarcs, $lab_\mu : \mu \rightarrow \mathscr{L}_\mu$ for $\mu = N, A, H$ is a labeling function for vertices, arcs and hyperarcs respectively with corresponding set of labels $\mathscr{L}_\mu$; $att_\mu : \mu \rightarrow \mathscr{A}_\mu$ for $\mu = N, A, H$ is an attributing function for *vertices, arcs and hyperarcs respectively with corresponding set of attributes $\mathscr{A}_\mu$. Moreover following conditions are fulfilled:*

1. *For each face $f \in FS$, there exists a unique node in N corresponding to f and labeled f.*
2. *For every edge $e \in ES$ common for some faces $f_1, f_2 \in FS$, there exists a unique arc in A joining nodes labeled by f_1 and f_2 corresponding to faces f_1 and f_2, which is labeled by e.*
3. *Let $F_v \subset FS$ be the set of faces of S incident to the vertex $v \in VS$, and $N_v \subset N$ be the subset of nodes of G corresponding to the faces of F_v. Then for every $v \in VS$ there exists a unique hyperarc in H labeled by v and connecting the nodes of N_v.*
4. *Arcs in A incident to some node f of G are ordered according to the order of the edges of face f (note that edges of f in S form a loop).*
5. *Hyperarcs in H incident to some node f of G are ordered according to the order of vertices belonging to the face f of S.*
6. *Nodes of any hyperarc v in G are ordered according to the order of the corresponding faces incident to v in S.*

The family of FAHs will be denoted as $\mathscr{H}_{FA}$.

Attributing functions.

Similarly as in the Definition 2, the FAH object gets applicable for practical use if attributing functions att_N, att_A, att_H carry the complete information related to geometric features of a solid. That assumption, influencing a structure of $\mathscr{A}_N, \mathscr{A}_A$ and $\mathscr{A}_H$, is imposed.

Example.

Let us consider the cuboid S presented in Figure 4a with faces denoted by $f_1, \dots f_6$. For that solid we have following hypergraph representation (see Fig.4b) $H = (N, A, H, lab_N, lab_A, lab_H, att_N, att_A, att_H)$, where

- $N = \{f_1, \cdots f_6\}$,
- $A = \{\{f_1, f_3\}, \{f_1, f_4\}, \{f_1, f_5\}, \{f_1, f_6\}, \{f_2, f_3\}, \{f_2, f_4\}, \{f_2, f_5\}, \{f_2, f_6\}, \{f_3, f_5\}, \{f_3, f_6\}, \{f_4, f_5\}, \{f_4, f_6\}\}$,
- $H = \{\{f_1, f_3, f_5\}, \{f_1, f_3, f_6\}, \{f_2, f_3, f_5\}, \{f_2, f_3, f_6\}, \{f_1, f_4, f_5\}, \{f_1, f_4, f_6\}, \{f_2, f_4, f_5\}, \{f_2, f_4, f_6\}\}$,
- Attributing functions are defined using stub expressions:
  - att_N: $N \ni p \to$ Attributes of node p,
  - att_A: $A \ni \{p, q\} \to$ Attributes of the edge between p and q,
  - att_H: $H \ni \{p_1, p_2, \cdots p_k\} \to$ Attributes of the hyperedge connecting p_1, $p_2, \cdots p_k$.

No limitations are imposed on labeling functions lab_N, lab_A, lab_H in the considered example.

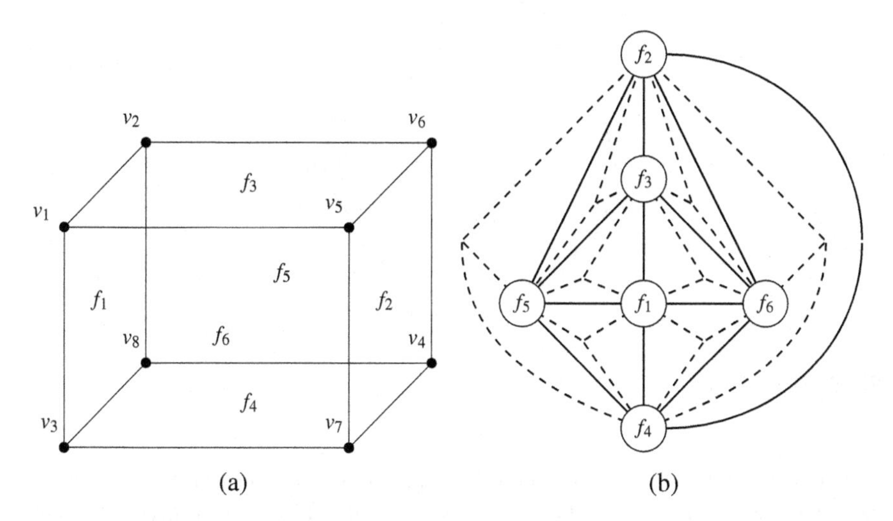

Fig. 4 (a) Cuboid – the sample solid (b) Hypergraph representation of cuboid shown in Fig.4a

It should be noted that each of these hyperedges corresponds to one vertex of the solid S according to the Definition 3. For the image clarity only hypergraph vertices are labeled in Figure 4b. Edges are drawn with a solid line and hyperedges with a dashed one.

Obtaining hypergraph representation of a solid

Before introducing the transformation of the analytic representation of a solid to the hypergraph one let us describe the solid specification being the input data format.

We assume that for a considered solid S a set of its vertices VS (and thus their coordinates) is known. Also the analytic form of all curved edges is given. Moreover the bounding surface of S is decomposable into a set of faces $FS = \{f_1, f_2, \ldots f_n\}$. On the other side a face f_i is described by an ordered set of bounding vertices

$$VS_i = (v_{i_1}, v_{i_2}, \ldots v_{i_k}), VS = \bigcup_{i=1}^{n} VS_i.$$

A set VS_i induces an ordered set of bounding edges $ES_i = (e_{i_1}, e_{i_2}, \ldots e_{i_k})$ where $e_{i_j} = (v_{i_j}, v_{i_{j+1}})$ and $v_{i_{k+1}} \equiv v_{i_1}$ (see Figure 5).

Note that geometric features of an edge are not explicitly determined in a graph model. The edge e_{i_j} may be parametrized by

$$v(t) = (x_0 + \delta_x(t), y_0 + \delta_y(t), z_0 + \delta_z(t))$$

for $t \in [t_1, t_2]$, where $\delta_k(t)\,(k = x, y, z)$ are either linear or nonlinear transformations with respect to t and $v(t_1) = v_{i_j}, v(t_2) = v_{i_{j+1}}$. A detailed (geometric) characteristic

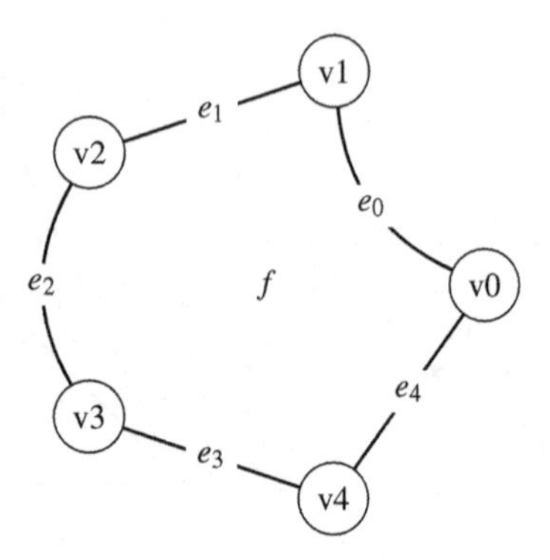

Fig. 5 The solid face

of e_{i_j} may be included in an attribute value returned by function att_A. Similar remark can be applied to a shapes of faces and node attributing function: $att_N(v \in N)$ may contain analytic surface specification of a face f represented by v.

For a cylindrical or ellipsoidal object that has no vertices and/or edges dummy nodes may be provided to stay in compliance with description used in the following algorithm.

Algorithm 1 produces hypergraph representation of given solid. It uses the queue Q of faces that are to be processed. Additionally we assume that each edge and vertex has a boolean flag named `processed` that is set initially to false; each face of S has the flag named `visited` that is initially set to false too. The algorithm objective is to find hypergraph representation of a solid S whose analytic description was discussed above. The procedure shown in Algorithm 1 produces also attributing functions (see lines 23,17,11) basing on coordinates and analytic description of objects (e.g. parametric characteristics of curves).

Time complexity of Algorithm 1 equals $\mathscr{O}(2NM)$. The `while` loop in line 6 is executed at most N times, where N is a number of faces in S; `foreach` loops in lines 12 and 19 are executed at most M times, where M denotes a maximal number of vertices of a face in S. Thus we obtain complexity $\mathscr{O}(2NM)$.

Definition 3 provides a description of 3D objects being the primitives from the perspective of an entire urban area. More complex systems being the sets of neighboring solids, may be described using the notion of hierarchical face adjacency hypergraph (HFAH).

Definition 4 (HFAH). *A* ***hierarchical face adjacency hypergraph*** *(HFAH) is a pair $g^* = (\mathscr{G}, T)$ where $T = (\mathscr{G}, E, att_E)$ is a tree representing a hierarchical structure of a solid, $\mathscr{G} \subset \mathscr{H}_{FA}$ is a set of FAHs, called components of g^*, which correspond to particular nodes of T, E denotes a set of edges of T and $att_E : E \longrightarrow A_E$ is an edge attributing function such that $att_E(e \in E)$ specifies the way of coupling of solids represented by nodes incident to e. $G_0 \in \mathscr{G}$ corresponding to the root of T is referred to as root component of g^*.*

Algorithm 1. GenerateHypergraph(S)

```
  input : S – a solid for which the hypergraph is to be generated
  output: G = (N,A,H,lab_N,lab_A,lab_H,att_N,att_A,att_H) – a hypergraph representation
          of S
1  begin
2     N ← ∅, A ← ∅, H ← ∅;
3     Mark all faces of S as unvisited;
4     Mark all vertices and edges of S as unprocessed;
5     Q ←≺ any face of S ;                          /* Enqueue any face of S */
6     while Q is nonempty do
7        f ←≺ Q ;                                   /* Dequeue f from Q */
8        if f is unvisited then
9           Mark f as visited;
10          N ← N ∪ {f};
11          Generate lab_N(f), att_N(f);
12          foreach bounding vertex v of face f do
13             if v is not processed then
14                Mark v as processed;
15                F_v ← set of vertex adjacent faces with respect to v, excluding f;
16                H ← H ∪ {F_v};
17                Generate lab_H(F_v), att_H(F_v);
18                Q ←≺ F_v;                     /* Enqueue all faces from F_v */
19          foreach bounding edge e of face f do
20             if e is not processed then
21                Mark e as processed;
22                A ← A ∪ {e};
23                Generate lab_A(e), att_A(e);
24    return G = (N,A,H,lab_N,lab_A,lab_H,att_N,att_A,att_H)
```

Component G_i is the *parent* of G_j if the node of T associated with G_i is the parent of the node of T associated with G_j. The family of all HFAHs will be denoted as $\mathscr{H}_T$.

Example.

In Fig.6a the example of a composite solid, named S, is shown. One can distinguish three component of S with corresponding FAHs denotes as G_0, G_1, G_2. The HFAH describing S has the form $g^* = (\{G_0, G_1, G_2\}, T)$ where the tree T is presented in Fig.6b. The cube associated with G_0 is the root component of T while child nodes G_1 and G_2 represent small cubes of S adjacent to the large one. Attributes a_1, a_2 specify the relationships between corresponding hypergraphs' nodes (i.e. faces of solids).

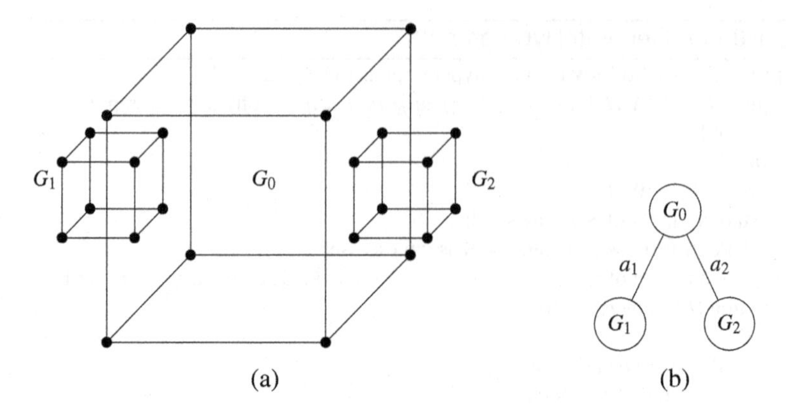

Fig. 6 (a) Complex object – S (b) HFAH tree of S

Although the above example presents the composite solid consisting of three adjacent primitives, a relation given by the edges of $T = (\mathcal{G}, E, att_E)$ can be extended over the pairs of objects which don't contact and their relationship may be described as *is located in the neighborhood parametrized by....* A parametrization of such a *weak* neighborhood will be also stored in values of att_E function. Thanks to this we can use HFAHs as a representation of arbitrary sets of neighboring buildings.

4.1 Linkage between Hierarchy Levels

Since we defined the notion of a hierarchical face adjacency hypergraph we can model composite systems consisting of numerous objects (buildings). In particular an urban neighborhood of a street or a place may be expressed in terms of multiple HFAHs.

Let $G = (N, H, att_N, att_H, lab_N, lab_H) \in \mathcal{H}_{Cart}$ and $v \in N$. Then $(g_1^*, g_2^*, \cdots g_k^*) \in att_N(v)$ represent sets of objects (buildings) in a neighborhood of the street v, where $g_1^*, g_2^*, \cdots g_k^* \in \mathcal{H}_T$. Additional attributes in $att_N(v)$ provide data concerning location details for $\{g_i^*\}$ HFAHs.

5 Multi-Agent System Deployment

As it was mentioned previously a main hypergraph (cartographic hypergraph) representing an entire system is decomposed into a set of sub-hypergraphs so called *partial hypergraphs*. A multiagent system performing computations on partial hypergraphs contains at least one type of agents denoted as CA (computational agent). Each partial hypergraph has a single CA ascribed to it. Each CA solves locally an optimization problem related either to static or to dynamic phase of LaSIL.

Initially, MAS deployed on a centralized hypergraph G_0 consists of a single, initiating CA, say A_0 ascribed to G_0. Next A_0 splits G_0 according to the

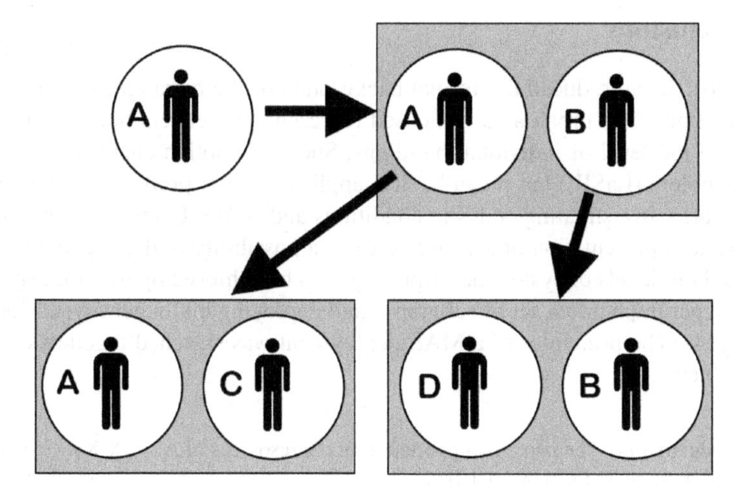

Fig. 7 Hypergraph decomposition performed by agents

decomposition rule $\mathscr{D}$: $G_0 \xrightarrow{\mathscr{D}} \{G'_0, G_1\}$. A new agent, A_1 is created and ascribed to G_1 by A_0. G'_0 remains maintained by A_0. At an arbitrary level of a decomposition we have: $G_i \xrightarrow{\mathscr{D}} \{G'_i, G_k\}$ and a new agent A_k is created by an agent A_i. The rule $\mathscr{D}$ specifies a stop criterion for the decomposition process performed by an agent A_i. Entire process runs recursively (Fig.7).

5.1 Future Works

Two following questions arise around a deployment process. The first problem to be investigated is identifying criteria and the methods of an optimal decomposition of a centralized hypergraph into a set of partial hypergraphs. That operation, prior to solving the static phase of LaSIL, influences the performance of further agent-based computations. In other words, a decomposition rule $\mathscr{D}$ has to be specified. Usually it depends on a G_i size: splitting process continues until a number of vertices hits a given range $[N-\varepsilon, N+\varepsilon]$. On the other side there also exist other factors which may be taken into account e.g. a number of connections with other partial hypergraphs. Since an initial decomposition doesn't produce an optimal set of hypergraphs a fine tuning of such a set has to be performed. For the normal graphs the problem was solved successfully [14]. For the hypergraphs the question is open.

The second issue concerns the specification of an interface type between partial hypergraphs. Interface specification is necessary to enable reassembling a centralized hypergraph from partial ones. For normal graphs *border nodes* play the role of such interfaces (see [14]). This step impacts communication among CAs and thus reassembling process complexity, but that problem is beyond the scope of that paper.

Both questions described above will be the subject of the further research.

6 Conclusions

In the paper we introduced the formal background of the hierarchical, hypergraph representation of an urban area. Proposed representation covers both, the level of streets and the level of individual buildings. Such description enables solving the design problem (LaSIL) but it can be also applied in other problems like planning escape routes, investigating acoustic conditions and so on. Using the hierarchical, graph based representation of a system creates a possibility of distributing computations at both levels of system description. It can be achieved by decomposition of a given hypergraph into a set of subgraphs and deploying multiagent system across that set [14]. The principles of a MAS deployment was sketched together with related problems.

Acknowledgements. The paper is supported from the resources of Alive & KIC-ing project and NCBiR grant no O ROB 002101/ID 21/2.

References

1. Baland, P., Corradini, A., Montanari, U., Rossi, F.: Concurrent sematics of algebraic graph transformations, pp. 107–187 (1999)
2. Bandini, S., Bonomi, A., Vizzari, G., Acconci, V.: Self-organization models for adaptive environments: Envisioning and evaluation of alternative approaches. Simulation Modeling Practice and Theory 18(10), 1483–1492 (2010)
3. Baumgart, S., Toledo, B., Spors, K., Schimmler, M.: PLUG: An Agent Based Prototype Validation of CAD-Constructions. In: The 2006 International Conference on Information and Knowledge Engineering (2006)
4. Corradini, A., Montanari, U., Rossi, F., Ehrig, H., Heckel, R., Löwe, M.: Algebraic approaches to graph transformation - part i: Basic concepts and double pushout approach. In: Handbook of Graph Grammars and Computing by Graph Transformations, Foundations, vol. 1, pp. 163–246. World Scientific (1997)
5. Ehrig, H., Heckel, R., Lowe, M., Ribeiro, L., Wagner, A.: Algebraic Approaches to Graph Transformation. In: Part II: Single Pushout and Comparison with Double Pushout Approach, pp. 247–312
6. Ehrig, H., Ehrig, K., Prange, U., Taentzer, G.: Fundamentals of Algebraic Graph Transformation. Monographs in Theoretical Computer Science. An EATCS Series. Springer-Verlag New York, Inc., Secaucus (2006)
7. De Floriani, L., Falcidieno, B.: A hierarchical boundary model for solid object representation. ACM Trans. Graph. 7(1), 42–60 (1988)
8. Ehrig, H., Engels, G., Kreowski, H.-J., Rozenberg, G.: Handbook of Graph Grammars and Computing By Graph Transformation: Applications, Languages, and Tools, vol. II. World Scientific Publishing Co., River Edge (1999)
9. Kotulski, L.: GRADIS – Multiagent Environment Supporting Distributed Graph Transformations. In: Bubak, M., van Albada, G.D., Dongarra, J., Sloot, P.M.A. (eds.) ICCS 2008, Part III. LNCS, vol. 5103, pp. 644–653. Springer, Heidelberg (2008)
10. Kotulski, L., Sędziwy, A.: On Complexity of Coordination of Parallel Graph Transformations in GRADIS Framework, DepCoS-Relcomex. In: 2009 Fourth International Conference on Dependability of Computer Systems, pp. 279–289 (2009)

11. Kotulski, L., Sędziwy, A.: Parallel Graph Transformations with Double Pushout Grammars. In: Rutkowski, L., Scherer, R., Tadeusiewicz, R., Zadeh, L.A., Zurada, J.M. (eds.) ICAISC 2010. LNCS, vol. 6114, pp. 280–288. Springer, Heidelberg (2010)
12. Kotulski, L., Strug, B.: Distributed Adaptive Design with Hierarchical Autonomous Graph Transformation Systems. In: Shi, Y., van Albada, G.D., Dongarra, J., Sloot, P.M.A. (eds.) ICCS 2007. LNCS, vol. 4488, pp. 880–887. Springer, Heidelberg (2007)
13. Kotulski, L., Strug, B.: Multi-agent System for Distributed Adaptive Design. Key Engineering Materials 486, 217–220 (2011)
14. Kotulski, L., Sędziwy, A.: GRADIS - the multiagent environment supported by graph transformations. Simulation Modeling Practice and Theory 18(10), 1515–1525 (2010)
15. Kreowski, H.J., Kluske, S.: Graph multiset transformation as a framework for massive parallel computation. In: Ehrig, H., Heckel, R., Rozenberg, G., Taentzer, G. (eds.) ICGT 2008. LNCS, vol. 5214, pp. 351–365. Springer, Heidelberg (2008)
16. Ligong, X., Zude, Z., Quan, L.: Multi-agent Architecture for Collaborative CAD System. In: 2008 International Conference on Computer Science and Information Technology, pp. 7–11 (2008)
17. Peng, W., Krueger, W., Grushin, A., Carlos, P., Manikonda, V., Santos, M.: Graph-based methods for the analysis of large-scale multiagent systems. In: Proceedings of The 8th International Conference on Autonomous Agents and Multiagent Systems 2009, pp. 545–552 (2009)
18. Rozenberg, G.: Handbook of Graph Grammars and Computing By Graph Transformation: Foundations, vol. I. World Scientific Publishing Co., River Edge (1997)
19. Sędziwy, A., Kotulski, L.: Solving Large-Scale Multipoint Lighting Design Problem Using Multi-agent Environment. In: Key Engineering Materials, Advanced Design and Manufacture IV, vol. 486, pp. 179–182 (2011)
20. Yabuki, N., Kotani, J., Shitani, T.: A Cooperative Design Environment Using Multi-Agents and Virtual Reality. In: Luo, Y. (ed.) CDVE 2004. LNCS, vol. 3190, pp. 96–103. Springer, Heidelberg (2004)

Formal Methods Supporting Agent Aided Smart Lighting Design

Adam Sędziwy, Leszek Kotulski, and Marcin Szpyrka

Abstract. In the paper we present the formal description of the agent-based light sensors (LSA - Light Sensors Agents) serving as data suppliers for a multiagent system controlling the distribution and work parameters of lighting points distributed across a given urban area. The cooperation and behavior of sensor agents are modeled and verified using Alvis modeling language.

1 Introduction

Multiagent systems support design process in various areas such as automotive industry [1] , support in constructional tasks [2, 3], collaborative CAD systems [4] or adaptive design [5]. The main benefit of their usage is possibility of computing parallelization in problems characterized by a high computational complexity. The problem being a background of the article concerns both finding an optimal distribution of light points in an urban area [6] and controlling their work parameters to minimize power consumption. Additionally the problem solution is constrained by local luminosity conditions which have to be fulfilled.

Designing a large scale system for planning distribution of luminaries in an urban environment is a complex task. First, one has to formalize description of an environment. Since it consists of streets, buildings and numerous infrastructure details, suitable structures matched to particular levels of a description are required. The graph representation seems to be the most relevant one for those purposes. On the other side, such a representation creates an environment of a deployment of an agent system. The graph structure data become a part of an agent's knowledge.

To achieve an optimal, adaptive lighting control we require gathering suitable physical data describing environment conditions like luminosity, traffic intensity and

Adam Sędziwy · Leszek Kotulski · Marcin Szpyrka
AGH University of Science and Technology, Department of Automatics, al. Mickiewicza 30, 30-059 Krakow, Poland
e-mail: {sedziwy,kotulski,mszpyrka}@agh.edu.pl

W. Zamojski et al. (Eds.): Complex Systems and Dependability, AISC 170, pp. 225–239.
springerlink.com

so on. We assume that this information may be captured automatically by means of auxiliary robot agents. There is no limitation for possible types of those entities thus they can model mobile robots responsible for probing luminosity levels, stationary induction loops or other, problem dependent devices.

To summarize the problem, we deal with a lighting system being designed for an urban area modeled by a distributed graph representation. Lighting system performance optimizations are made by multiagent system deployed on a distributed graph, on the basis of sensor data. In the article we focus on light sensors agents which will be described and analyzed in terms of the Alvis modeling language. The multiagent structure of a system will be sketched only to provide the problem context.

The process of designing an embedded system controlling a light sensor agent will be presented in the paper. We use Alvis toolkit here because it allows for parallel generation of an embedded system code and a corresponding labeled transition system (abbrev. LTS) description which enables formal verification of a system.

The structure of the paper is following. In the next section we present the related works. The basics of the Alvis modeling language are introduced in the Section 3. In the Section 4 the architecture of the considered system is presented. The structure of a mentioned robot agent and an embedded system design process are described in the Section 5. Formal verification of a model is presented in Section 6. Last section contains conclusions of the article.

2 Realated Works

Component models (CM) are widely used in the design and the development of the embedded software. An example is Koala model used in the software design for television products [7]. Using this model allows for independent development of configuration and components. Another sample of a component-based approach is SOFA model used in the communications middleware, component management, component design, electronic commerce, and security [8]. Component models may be used in a production of various types of software. Fractal CM for instance, may be used in the software design and development in such distant fields like operating systems and graphical user interfaces [9].

The important feature of a CM, absent in models presented above, is possibility of a formal verification of a designed system, in particular the absence of deadlocks. Such a verification may be made within ProCom model [10].

The Alvis formalism was selected for modeling embedded systems in the smart lighting problems due to the additional feature enriching the capability of the system formal verification. It offers the ability of an automated verification of composite embedded systems. Such a verification is performed on the basis of specifications of cooperating components forming entire system (see [11]). Thus a system may be easily verified at any phase of the design process, in particular after adding a new component to it. Besides the mentioned feature Alvis enables extensible system design. A new component may be added to the system or an existing one may be grained into several sub-components without the need of a model rebuilding.

3 Alvis Description

3.1 Model Layers

To prepare a system specification we define two layers of its Alvis model description: graphical and code ones. According to the Alvis convention a graphical layer is referred to as a communication diagram. It contains active agents drawn as rounded boxes and passive agents drawn as rectangles. Ports used for a communication are drawn as circles placed at edges of rounded boxes or rectangles. Alvis agents can communicate directly with each other using communication channels. A communication channel between two agents is defined explicitly and connects two ports. It is drawn as a line (or a broken line). An arrowhead points out an input port for a particular connection. Communication channels without arrowheads represent pairs of connections directed in opposite directions.

Table 1 Alvis statements in alphabetical order

Statement	Description
cli	Turns off the interrupts handlers.
critical {...}	Define a set of statements that must be executed as a single one.
delay ms	Delays an agent execution for a given number of miliseconds.
exec x = e	Evaluates the expression and assign the result to the parameter; the **exec** keyword can be omitted.
exit	Terminates the agent that performs the statement.
if (g1) {...} **elseif** (g2) {...} ... **else** {...}	Conditional statement.
in p **in** p x	Collects a signal via the port p. Collects a value via the port p and assigns it to the parameter x.
jump label	Transfers the control to the line of code identified with the label.
jump far A	Transfers the control to the agent A.
loop (g) {...} **loop** (**every** ms) {...} **loop** {...}	Repeats execution of the contents while the guard if satisfied.. Repeats execution every ms miliseconds. Infinite loop.
null	Empty statement.
out p **out** p x	Sends a signal via the port p. Sends a value of the parameter x via the port p; a literal value can be used instead of a parameter.
proc (g) p {...}	Defines the procedure for the port p of a passive agent. The guard is optional.
select { **alt** (g1) {...} **alt** (g2) {...} ... }	Selects one of the alternative choices.
start A	Starts the agent A if it is in the *Init* state, otherwise do nothing.
sti	Turns on the interrupts handlers.

The code layer is used to define data types used in a considered model, functions for data manipulation and a behavior of particular agents. The layer uses the Haskell functional language (e.g. the Haskell type system) and native Alvis statements. The set of Alvis statements is given in Table 1. To simplify the syntax, the following symbols have been used. `A` stands for an agent name, `p` stands for a port name, `x` stands for a parameter, `g`, `g1`, `g2`,... stand for guards (Boolean conditions), `e` stands for an expression and `ms` denotes milliseconds. Each agent[1] placed in a communication diagram must be defined in the code layer and vice-versa. For more details see [12] or [13].

3.2 Communication

Before discussing the communication between agents in the lighting control system environment let us recall some facts about the communication among agents in the Alvis model. This communication is the synchronous one i.e. sending a message by an agent blocks him until a message is received by a receiver agent. Alvis uses only two statements for a communication. The `in` statement for receiving data and `out` for sending. Each of those operations takes a port name as its first argument and, optionally, a parameter name as the second one. A communication between two active agents can be initialized by any of them. An initiating agent may perform either the `out` statement to provide some information and wait until a second agent receives it, or the `in` statement to express its readiness to receive some information and wait until a second agent provides it.

To describe a current state of an agent, we need a record consisting of four pieces of information:

- agent mode (*am*);
- program counter (*pc*);
- context information list (*ci*);
- parameters values tuple (*pv*).

The mode (*am*) is used to indicate whether an agent is running or waiting for an event. The program counter (*pc*) points out the current or the next step to be executed. The context information list (*ci*) contains additional information about the current agent's state e.g. a name of a port used in current communication. The parameters values (*pv*) tuple contains values of agent's parameters. For more details see [12].

3.3 Communication with Environment

Alvis agent may contain ports which are not used in any connection and specified inside the environment statement [14]. Such ports are called border ports and are used for a communication with the considered system environment (see Section 5). Border ports can be used for both collecting or sending some

[1] Precisely, non-hierarchical agent, specifically light sensor agent (LSA) introduced in the Section 4. Hierarchical agents that are also modeled by Alvis, are not considered here.

information to an external environment. Properties of border ports are specified in a code layer preamble with the use of the `environment` statement. Each border port used as an input one corresponds to at least one `in` clause. Similarly, each border port used as an output one is described by at least one `out` clause. Each clause inside the `environment` statement contains the following pieces of information:

- `in` or `out` keyword,
- the border port name,
- a type name or a list of permissible values to be sent through the port,
- a list of time points, when the port is accessible,
- optionally some modifiers: **`durable`**, **`queue`**, **`signal`**.

4 System Architecture

The graph formalism used in conjunction with distributed, parallel computations performed by multi-agent system is the approach allowing solving various types of problems which common property is high computational complexity. One of them is the large scale intelligent lighting (LaSIL) problem consisting of both an optimal distribution of luminaries and, in the sequel, the lighting control. The proposed approach is applied to the LaSIL in following phases: (i) transforming an initial problem to a centralized graph representation, (ii) decomposing this graph into a set of subgraphs according to given criteria [15, 16] and thereby dividing an initial task into a set of subtasks that may be solved in parallel (or almost in parallel), (iii) running *designer* agents deployed previously to particular subtasks. The primary goal of designer agents is optimizing a distribution of lamps in particular subareas. After this task is completed their objective is controlling lighting parameters to supply a proper lumonosity, basing on the environment's conditions.

In the LaSIL we enrich this generic schema by introducing two additional types of agents: the *light sensor agent* and the *broker agent*. Thus the architecture of the considered multi-agent system consists of three types of agents:

- *Designer agent* (DA) - which is the basic agent type, responsible for an allocation of lighting points for a given area and adjusting their work parameters according to given objective functions (including such parameters as luminosity levels or a power consumption). In some circumstances a designer agent may cooperate with other DAs, e.g. in border regions.
- *Light sensor agent* (LSA) - whose goal is to check luminosity levels in points specified in received requests. Thus LSA is assumed to control its physical movement and obtain data from sensors.
- *Broker agent* (BA) - which intermediates in communication between a designer agent and light sensor agents.

It should be remarked that for other kinds of sensors corresponding agents may be introduced to the system in a similar manner.

Figure 1 illustrates the architecture of relations among agents listed above. It is assumed that a light sensor agent is composed of two subsystems, denoted as S

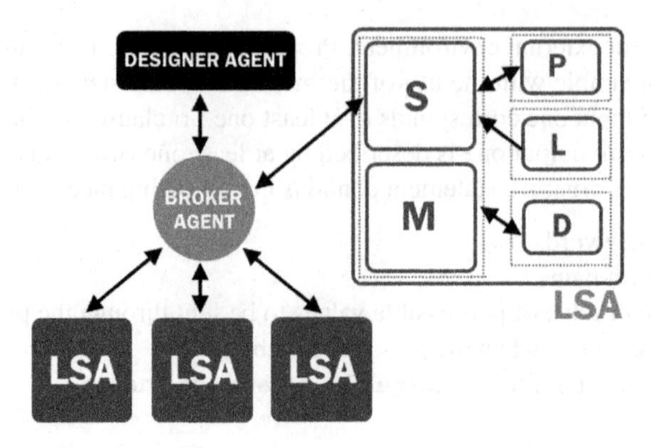

Fig. 1 Architecture of multiagent environment

(serving) and M (motion). S subsystem plays a key role in an LSA. It processes requests received from a broker agent and cooperates with M if needed. Additionally S may request data from light sensor device L (actual luminosity level) and location provider P (actual coordinates). If a current position cannot be determined then P returns an error message to S. M subsystem is responsible for a physical movement (performed by a driving unit D) of a light sensor agent. Note that dotted lines on Figure 1 denote logical borders of particular components of an LSA and its cooperating elements.

The LaSIL problem solution schema may be described as follows. Initially a map of an urban area is maintained by an initiating designer agent, say A. After transforming that map into a graph G, A begins a process of decomposition of G. Thus new designer agents are created and attached to newly obtained subgraphs (one designer agent is ascribed to one subgraph). When decomposition stops designer agents perform a distribution of lighting points on their subareas (subgraphs). When this phase is completed they enter the permanent phase of adjusting the work parameters of lamps, according to a given objective function. At the beginning of this phase a designer agent deploys light sensor agents to obtain luminosity levels in selected checkpoints. Those data are required to optimize an objective function. Described phase is performed iteratively.

Communication between a designer agent and its LSAs is accomplished with an intermediation of a BA. Thanks to that a workload related to communication with sensors is removed from a designer agent.

The requests sent by a designer agent may be of two types: in the first one (`READ`) a DA queries an LSA for an actual light level value in its current location. In a request of the second type (`MOVE`) it demands from an LSA changing its position (by sending a list of coordinates on a requested route, a list of subsequent transitions, or simply, a target coordinates). For `READ` request some additional parameters may be provided (e.g. a timestamp). For `MOVE`, new coordinates are specified.

Requests of both types are delivered to embedded subsystems of an LSA via a broker agent.

5 LSA Structure

In this section we present the process of defining the considered system, beginning from its preliminary specification. Alvis modeling language enables an incremental development of a system: having given system specification we may extend it by introducing additional embedded subsystems.

Before discussing the structure of an LSA we introduce the convention concerning port names. A name of a port has the form `A.p`, where the prefix `A` is an agent name and `p` is a local name of a port.

Initially we assume that an LSA consists of the single embedded system S only (see Figure 2). S contains seven border ports:

- the input port, `S.l_in`, collecting luminosity data consisting of a list of eight parameters (captured by eight analog light sensors), which is assumed to work in `durable` mode;
- the pair of ports, input and output ones, denoted as `S.b_in` (assumed to work in `signal` mode) and `S.b_out` respectively, for receiving/sending messages from/to a broker agent;
- the pair of ports, input and output ones, denoted as `S.p_in` (working in `signal` mode) and `S.p_out` respectively, for sending/receiving messages to/from a location provider;
- the pair of ports, `S.m_in` (working in `signal` mode)) and `S.m_out` to communicate with a device responsible for a physical movement of an LSA.

As it was mentioned above Alvis enables an extensible design, i.e. additional subcomponents of a system may be added at any level of design process. Exploiting this property of Alvis we introduce a subsystem M intermediating between S and a driving device, and supporting the logic of movement operations. Thus a light sensor agent consists of two embedded subsystems, denoted as S and M. After providing M subsystem to the system description, two border ports, `S.m_in` and `S.m_out`, get regular ones in the subsystem S. They are bound to corresponding ports of M, `M.m_out` and `M.m_in` respectively, by two communication channels. Besides that,

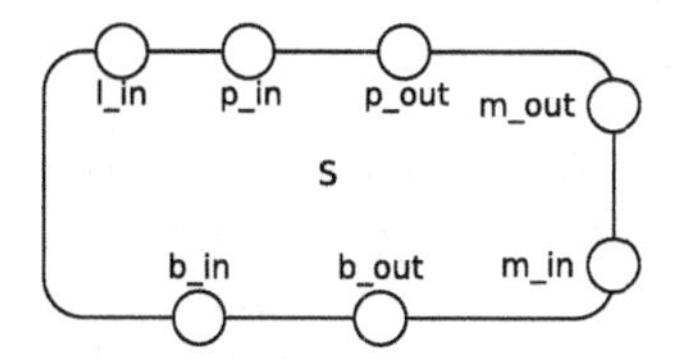

Fig. 2 Initial form of the LSA

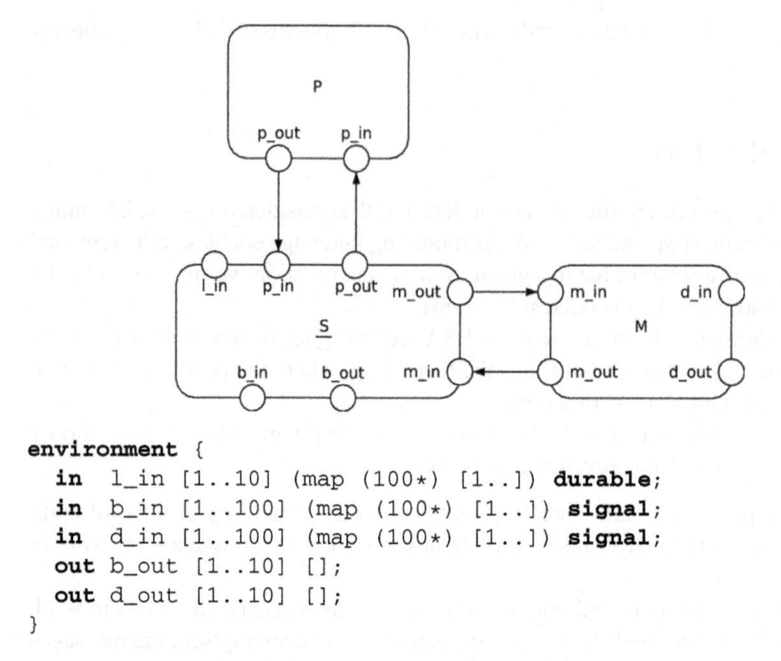


```
environment {
  in   l_in [1..10] (map (100*) [1..]) durable;
  in  b_in [1..100] (map (100*) [1..]) signal;
  in   d_in [1..100] (map (100*) [1..]) signal;
  out b_out [1..10] [];
  out d_out [1..10] [];
}
```

Fig. 3 Communication diagram for LSA subsystems and border ports specification

M exposes two border ports `M.d_in` (working in `signal` mode) and `M.d_out` to communicate with a driving unit. Additionally we introduce explicitly a location provider bound to `S.p_in` `S.p_out` ports. Servicing `MOVE` request requires co-operation with a location provider which supplies actual coordinates necessary to compute a route to a destination point. Next, M subsystem processes it and controls a driving unit accordingly on a basis of data received from a location provider. If any unexpected situation occurs (e.g. an obstacle in a path) a driving unit returns an appropriate error message which is forwarded to a designer agent via S subsystem and a corresponding BA. The communication diagram of an LSA and the code layer specifications of border ports are presented in Figure 3. Note that the incremental design presented above shifts a system logical boundaries.

6 Formal Verification

The next step following defining system components is a formal verification of a model and an analysis of its behavior (interactions between subsystems) by means of LTS graphs. The objective of the operation is discovering occurrences of undesired states in subsystems cooperation e.g. deadlocks. The verification phase consists of four steps:

```
data Task = READ | MOVE;

agent S {
  x :: Task = READ:
  resp :: Int = 0;
  dst :: Int = 0;

  loop {                        -- 1
    in b_in x;                  -- 2
    if(x == READ) {             -- 3
      in l_in y;                -- 4
      out b_out y; }            -- 5
    else {                      --    x == MOVE
      out p_out;                -- 6  query location provider
      in p_in resp;             -- 7  get location
    if(resp /= ErrMsg) {        -- 8
      dst = getShift x resp     -- 9
      out m_out dst             -- 10
      in m_in y; }              -- 11
    else {                      --     current location is n/a
      y = ErrMsg; }             -- 12
    out b_out y;                -- 13
  }
}
```

Listing 1 Code layer of agent S

```
agent M {
  errType :: Int = 0;
  req :: Int = 0;
  command :: Int = 0;

  loop {                        -- 1
    in m_in req;                -- 2
    command = decode req;       -- 3
    out d_out command;          -- 4
    in d_in resp;               -- 5
 -- set errType. 0->no error
    errType = decode resp;      -- 6
    out m_out errType;          -- 7
  }
}
```

Listing 2 Code layer of agent M

```
agent P {
  resp :: Int = 0;

  loop {                              -- 1
    in p_in;                          -- 2
    resp = getLocation;               -- 3
    out p_out resp;                   -- 4
  }
}
```

Listing 3 Code layer of agent P

1. Defining a code layer for agents S, M, P.
2. Generating transition diagrams LTS_S, LTS_M, LTS_P for respectively S, M, P.
3. Merging all LTS diagrams into a single composite LTS (abbrev. CLTS), starting from selected initial points of LTS_S, LTS_M, LTS_P (see [17])
4. Analyzing obtained CLTS against occurrences of deadlocks or other undesired properties.

Remark. We introduce the following indexing convention. A CLTS state obtained from LTS states numbered by i, j, k will be indexed by a triple (i, j, k).

Alvis codes of agents S, M and P are shown in Listings 1, 2 and 3; corresponding LTS graphs are presented in Figures 4, 5 and 6. To model a cooperation between M and S subsystems we merge both LTS graphs in the point where S, M and subsystems start to communicate with each other, namely in states 8 (for S), 1 (for M) and 1 (for P). The resultant diagram is shown in Figure 7.

The quantitative characteristics of the generated CLTS is presented in Table 2. The number of vertices in an LTS corresponds to a number of states of a given system/subsystem, the number of edges refers to a number of transitions between the states. It should be emphasized that a formal verification of more complex systems, i.e. consisting of more components described by graphs of higher orders, is hardly possible to accomplish without an automated method of generating composite LTS diagrams.

Table 2 Size characteristics of individual and composite LTS graphs

Graph	Number of vertices	Number of edges
LTS_S	21	30
LTS_M	11	15
LTS_P	6	18
CLTS	73	146

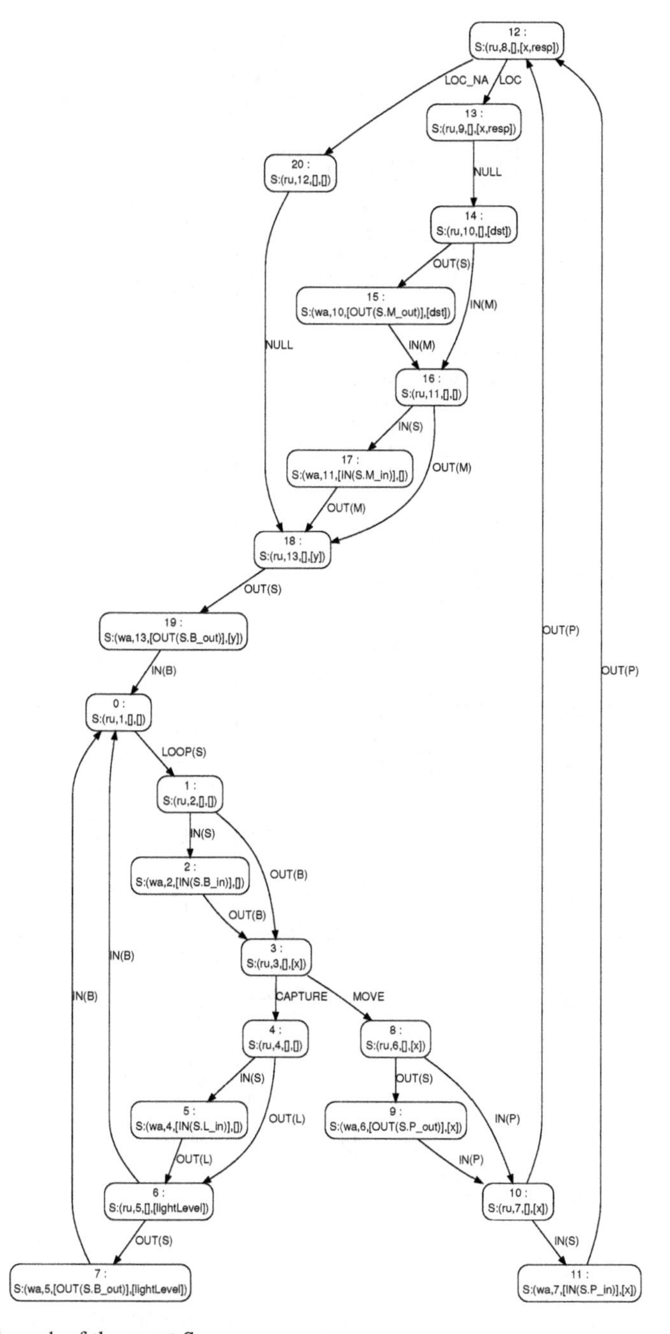

Fig. 4 LTS graph of the agent S

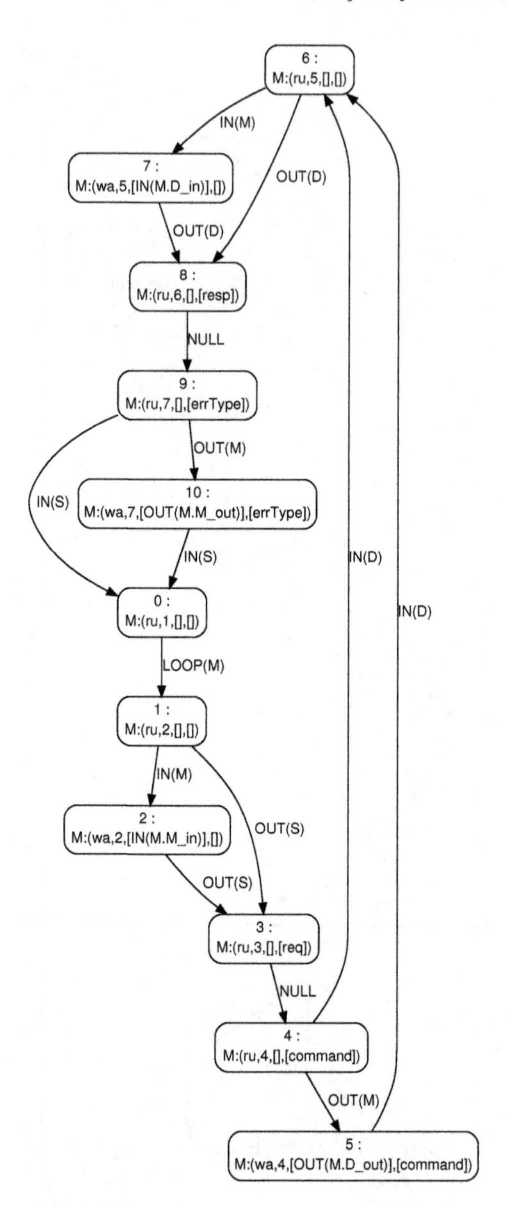

Fig. 5 LTS graph of the agent M

As it was stated previously we assume that a communication with the driving unit D is performed via `M.d_in` and `M.d_out` border ports of M. It implies that a signal from of a driving D unit is required to quit the state (17,5,2). Analogously, reading a message by a border agent B makes a system unsuspend from (19,2,2) (see the Table 3).

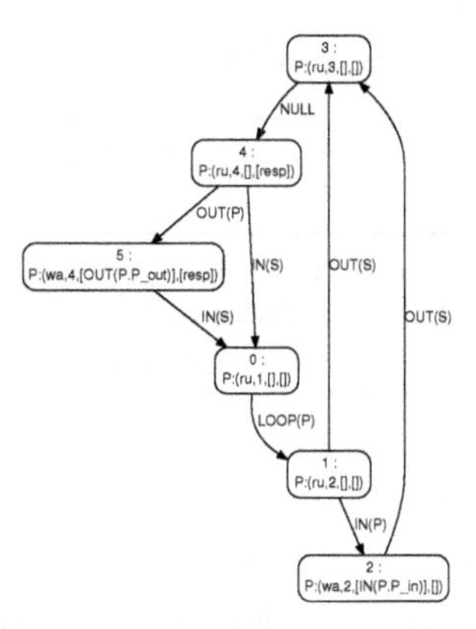

Fig. 6 LTS graph for the agent P

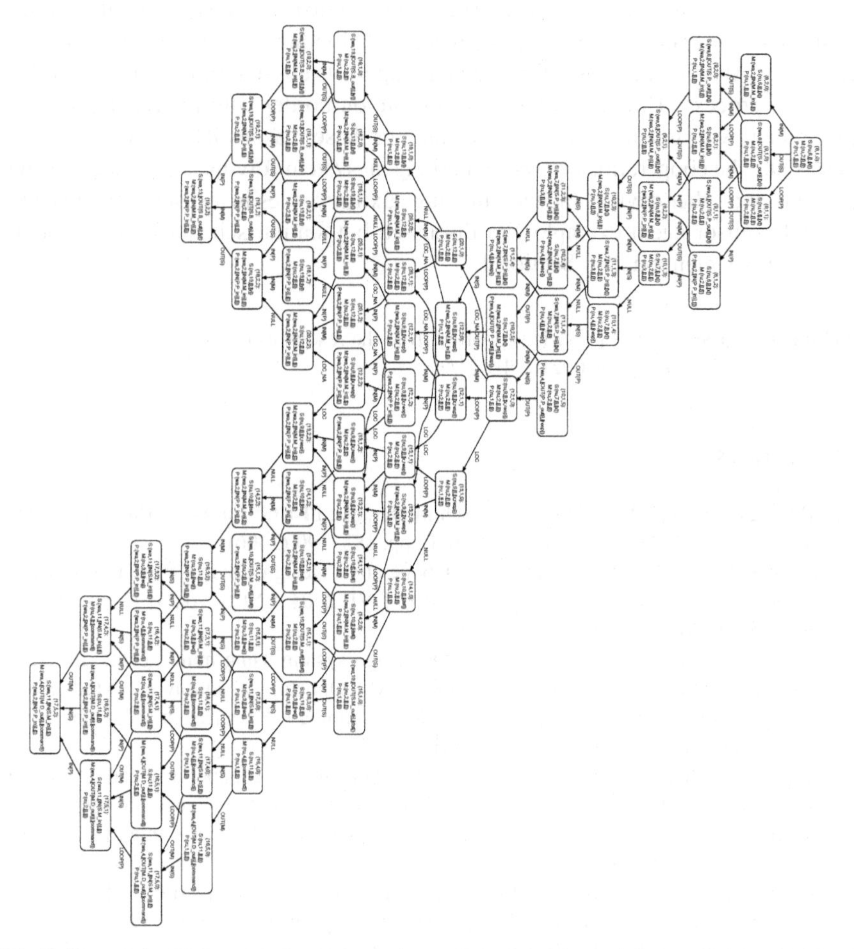

Fig. 7 Composite LTS graph for S, M and P. A state label is a triple consisting of ordinal numbers of states of S, M and P in their individual LTS graphs

Table 3 Details of "suspended" states of the CLTS

CLTS state	(17,5,2)	(19,2,2)
S state	(wa,11,[in(S.m_in)],[])	(wa,13,[out(S.b_out)],[y])
M state	(wa,4,[out(M.d_out)],[command])	(wa,2,[in(M.m_in)],[])
P state	(wa,2,[in(P.p_in)],[])	(wa,2,[in(P.p_in)],[])

7 Conclusions

Alvis language is the tool enabling flexible modeling of multiple cooperating embedded systems. It allows for creating a complete formal specification of such systems, containing a communication diagram, a code layer and individual and composite state diagrams (i.e. obtained by merging individual LTSs). Another important feature of Alvis is the possibility of extending a system description at any level of a design process. Thus we are able to operate on border (stub) ports and next replace them with given embedded systems. For these reasons Alvis was chosen for describing the cooperation between a multiagent system solving a large scale design problem and other devices (agents) playing auxiliary roles in a system. In the paper the light sensor agent example was discussed but similar schema is applicable to other entities mentioned in Section 1.

Acknowledgements. The paper is supported from the resources of Alive & KIC-ing project and NCBiR grant no O ROB 002101/ID 21/2.

References

1. Baumgart, S., Toledo, B., Spors, K., Schimmler, M.: PLUG: An Agent Based Prototype Validation of CAD-Constructions. In: IKE, pp. 183–190 (2006)
2. Yabuki, N., Kotani, J., Shitani, T.: A Cooperative Design Environment Using Multi-Agents and Virtual Reality. In: Luo, Y. (ed.) CDVE 2004. LNCS, vol. 3190, pp. 96–103. Springer, Heidelberg (2004)
3. Shitani, T., Yabuki, N.: A concrete bridge design system using multi-agents. In: Abraham, A., Dote, Y., Furuhashi, T., Köppen, M., Ohuchi, A., Ohsawa, Y. (eds.) Soft Computing as Transdisciplinary Science and Technology. Advances in Soft Computing, vol. 29, pp. 695–704. Springer, Heidelberg (2005)
4. Ligong, X., Zude, Z., Quan, L.: Multi-agent architecture for collaborative cad system. In: Proceedings of the 2008 International Conference on Computer Science and Information Technology. ICCSIT 2008, pp. 7–11. IEEE Computer Society, Washington, DC (2008)
5. Kotulski, L., Strug, B.: Distributed Adaptive Design with Hierarchical Autonomous Graph Transformation Systems. In: Shi, Y., van Albada, G.D., Dongarra, J., Sloot, P.M.A. (eds.) ICCS 2007. LNCS, vol. 4488, pp. 880–887. Springer, Heidelberg (2007)
6. Sędziwy, A., Kotulski, L.: Solving large-scale multipoint lighting design problem using multi-agent environment. Key Engineering Materials 486, 182–197 (2011)

7. van Ommering, R., van der Linden, F., Kramer, J., Magee, J.: The koala component model for consumer electronics software. Computer (33), 78–85 (2000)
8. Bures, T., Hnetynka, P., Plasil, F.: Sofa 2.0: Balancing advanced features in a hierarchical component model. In: Proceedings of the Fourth International Conference on Software Engineering Research, Management and Applications, pp. 40–48. IEEE Computer Society, Washington, DC (2006)
9. OW2_Consortium: The fractal project, `http://fractal.ow2.org`
10. Borde, E., Carlson, J.: Towards verified synthesis of procom, a component model for real-time embedded systems. In: Proceedings of the 14th International ACM Sigsoft Symposium on Component Based Software Engineering, CBSE 2011, pp. 129–138. ACM, New York (2011)
11. Szpyrka, M., Matyasik, P., Mrówka, R., Kotulski, L.: Formal modeling and verification of concurrent system with alvis. International Journal of Applied Mathematics and Computer Science (to appear, 2012)
12. Szpyrka, M., Matyasik, P., Mrówka, R.: Alvis – Modelling Language for Concurrent Systems. In: Bouvry, P., González-Vélez, H., Kołodziej, J. (eds.) Intelligent Decision Systems in Large-Scale Distributed Environments. SCI, vol. 362, pp. 315–341. Springer, Heidelberg (2011)
13. Szpyrka, M.: Alvis on-line manual (2011), `http://fm.ia.agh.edu.pl/alvis:manual`
14. Szpyrka, M., Matyasik, P., Mrówka, R., Kotulski, L.: Communication with environment in Alvis models. International Journal of Electronics and Telecommunications (to appear, 2012)
15. Kotulski, L., Sędziwy, A.: Gradis - the multiagent environment supported by graph transformations. Simulation Modelling Practice and Theory 18(10), 1515–1525 (2010); Simulation-based Design and Evaluation of Multi-Agent Systems
16. Kotulski, L., Sędziwy, A.: On the Effective Distribution of Knowledge Represented by Complementary Graphs. In: Jędrzejowicz, P., Nguyen, N.T., Howlet, R.J., Jain, L.C. (eds.) KES-AMSTA 2010. LNCS, vol. 6070, pp. 381–390. Springer, Heidelberg (2010)
17. Kotulski, L., Szpyrka, M., Sędziwy, A.: Labelled Transition System Generation from Alvis Language. In: König, A., Dengel, A., Hinkelmann, K., Kise, K., Howlett, R.J., Jain, L.C. (eds.) KES 2011, Part I. LNCS, vol. 6881, pp. 180–189. Springer, Heidelberg (2011)

Computational Support for Optimizing Street Lighting Design

Adam Sędziwy and Magdalena Kozień-Woźniak

Abstract. The design of an urban area lighting has to preserve compliance with existing standards and regulations but also satisfy non-formalized rules related to the functionality, reliability or energy efficiency. The next important step following the design process is ensuring the optimal performance of a lighting system. It may be accomplished by a suitable system control. Such a formulation of a problem implies the high computational complexity of the design tasks. For that reason it's necessary to develop an approach allowing to overcome the complexity problem. This article presents main factors determining the street lighting design and on the other side the formal methods providing an effective support in a design process.

1 Introduction

Intelligent street lighting systems make the important contribution to smart grid solutions. Thanks to them we are able to fit the actual end users demands and thereby minimize related energy consumption. Optimization of lighting performance may be made in two aspects. The first one is designing a distribution of lighting points, oriented for esthetic objectives and guaranteeing a suitable luminosity levels with minimal power supply. The second aspect concerns an intelligent control of a lighting system. Such a control is supported by the infrastructure layer (e.g. light sensors) but also computational methods capable of huge data processing, environment state predicting, decision making and so on.

Adam Sędziwy
AGH University of Science and Technology, Department of Automatics, al. Mickiewicza 30, 30-059 Kraków, Poland
e-mail: sedziwy@agh.edu.pl

Magdalena Kozień-Woźniak
Cracow University of Technology, Faculty of Architecture, ul.Warszawska 24, gmach WA, 31-155 Kraków, Poland
e-mail: magdalena.kozien@kozien.pl

W. Zamojski et al. (Eds.): Complex Systems and Dependability, AISC 170, pp. 241–255.
springerlink.com

It should be emphasized that designing a distribution of luminaries in an urban area can't be perceived as an optimization problem only. It's constrained by lighting design standards and good practices.

A typical process of the lighting design is aided by various tools (AutoCAD, 3ds Max, SketchUp, Dialux etc.) which are used dependently on the actual task such as design, visualization, making quantitative photometric characteristics of an area. The difficulty arises when an architect has to verify multiple variants of a design, corresponding for example to configuration of lamps or their parameters.

The idea of the support for creating lighting solutions may be found in [1]. It is the computer simulation environment of an adaptive illumination facility, where the considered area description is based on cellular structures, namely Situated Cellular Agents and Dissipative Multilayered Automata Network. Its functionality enables a simulation of an end product and some limited control capabilities rather than the support in the design phase.

The goal of this paper is twofold: 1) to present major issues related to a lighting design in the context of intelligent street lighting solutions 2) to define the formalism necessary for developing computational methods supporting a design process.

The article is organized as follows. In the next section we characterize main aspects of the lighting design taking into account a diversity of possible design solutions. In the Section 3 the comprehensive system model is provided. It contains the schema of computations and definitions of the underlying graph structures. The linkage between a designer and a supporting computational system is described in the Section 4. The final remarks are contained in the Section 5.

2 Architectural Background of Design Problem

Designing illumination in the city may be considered with reference to intelligent systems in two aspects: intelligent systems as designing tools and intelligent systems as the subject of a design.

2.1 *The Supported Design of Illumination*

The design of an illumination supported by intelligent information systems gives a chance of introducing solutions for problems of a high level of complexity or groups of variant solutions. Support can be related to an analysis of harmony between ruminations and the designing rules and assumptions, their effectiveness, economy, energy-saving.

Intelligent systems support a designers work and may serve:

- computer-aided design based upon the collection of initial data,
- verification of solutions,
- optimization of solutions,

- visualization and presentation of a design,
- simulation of the performance of an illumination.

Intelligent systems can be also introduced in activities including the choice of lamps and luminaries, their power, color, distribution angle etc. as well as the design of a layout of lamps and the distances between them, their composition, rhythm or control. Designing the illumination of roads, pavements and squares, based on the binding rules and individual needs, includes the distribution of luminous points, the delimitation of the height of street lamps and the spaces between them, the choice of the intensity and quality of lamps. Street luminaries, park luminaries on tall posts and short posts, wall lights, luminaries for earth constructions, luminaries for underwater installations etc. are designed.

Design aims at securing the illumination of public spaces in the city at night as well as preserving the energy efficiency and economy of solutions so that staying and moving in these spaces could be safe and comfortable. The basic assignments for artificial illumination in the city is to guarantee its safe usage and to support orientation in an urban space.

Securing the safety of moving in streets, squares and pedestrian crossings is related to two areas: the safety of movement and personal safety. The threshold parameters of illumination are strictly regulated [3, 12]. The kind and number of applied luminaries mostly depends on the adopted class of illumination. It is defined on the basis of an analysis of a lighting situation related to the kind and number of its users. The following elementary classes are distinguished: for traffic with medium and high speeds (ME), for traffic in conflict areas, such as intersections or roundabouts (CE), for pedestrian and bicycle sequences (S, A). Other classes of subordinate character are distinguished when it is necessary to identify people and objects, when there is a risk of violating the rules, in order to decrease the feeling of uncertainty (ES) or the need for seeing vertical surfaces (EU). Tables 01 and 02 present the list of ME and CE classes. The required parameters are defined depending on the class of illumination. These are the basic parameters:

- road surface luminance (for ME classes - see Tab.1),
- illuminance on a road area (for CE, S classes - see Tab.2),
- semi-cylindrical illuminance for the height of 1.5 m (for A, ES classes),
- vertical plane illuminance (for EU classes).

Classes of brightness and the illumination rate are also defined to reduce distracting illumination. The level of luminance depends on: the power of an installed lamp, the characteristics of light distribution, the location of luminaire. The more a lamp protrudes above a road area, the higher its luminance is; the lower a lamp is set, the higher its luminance gets. However, this causes more brightness and less evenness the design of illumination means a search for an optimal solution with the least possible costs of installation and maintenance as well as higher effectiveness of luminaries. Luminance changes depending on the atmospheric conditions. Computer-aided design in the field of controlling parameters required by the rules makes it possible to use some complex calculative methods without introducing any simplifications.

Table 1 ME lighting classes according to DIN EN 13201-2

Class	Road luminance in case of dry road surface			Threshold value	Ambient illuminance ratio
	Lw cd/m^2 [maintenance value]	U_o [minimum value]	U_l [minimum value]	T/w %[a] [maximum value]	SR_2 [b] [minimum value]
ME1	2.0	0.4	0.7	10	0.5
ME2	1.5	0.4	0.7	10	0.5
ME3a	1.0	0.4	0.7	15	0.5
ME3b	1.0	0.4	0.6	15	0.5
ME3c	1.0	0.4	0.5	15	0.5
ME4a	0.75	0.4	0.6	15	0.5
ME4b	0.75	0.4	0.5	15	0.5
ME5	0.5	0.35	0.4	15	0.5
ME6	0.3	0.35	0.4	15	-

[a] 5% higher admissible for lamps with low luminance

[b] This criterion is only to be used if no traffic surfaces with own photometric requirements are next to the road

Table 2 Table 02 CE lighting classes according to DIN EN 13201-2

Class	Horizontal illuminance	
	Ew lx [maintenance value]	Uo [minimum value]
CE0	50	0.4
CE1	30	0.4
CE2	20	0.4
CE3	15	0.4
CE4	10	0.4
CE5	7.5	0.4

This survey ascribes the source of the complexity of the problem of design to the multitude of parameters taken into account: classes of the illumination of spaces under consideration (Tab. 1, 2), kinds of luminaire (e.g. Trilux proposes 13 kinds of street luminaries which may be configured in 6-74 ways depending on the manner of installing lamps; Fig. 1 shows the photometric data of a sample luminaire), the manners of installation, work characteristics etc. Optimizing algorithms which act on the basis of some settled criteria, concerning both the technological and economic aspect (accordance with the norms, the costs of maintenance) and the esthetical and functional aspect, should make significant support in this process. Artificial light facilitates effective functioning after dusk. The perception of a public space in a city after dusk uses it as the main medium. Orientation in an urban space is related to various categories of spatial references, namely: transport sequences, variously shaped closed forms districts, outskirts, nodes, single architectural landmarks. Light is a medium which shapes a space it shows spaces and objects as well as creates their emotional climate. There is a tendency to reduce diffused light called night protection and preserve the impression of nighttime.

Apart from objective, quantitative and qualitative functional respects, illumination should also satisfy requirements concerning environmental, social and economic aspects. At the same time, illumination plays the following roles: strengthening the individuality and identity of places in the city, supporting its promotion by shaping moods, arousing esthetic sensations and the feeling of pleasure. They are related to a suitable choice of these and other parameters (e.g. light color, the appearance of luminaire) depending on the individual features of a place (related to restoration etc.). Light is an architectural material. The manner of perceiving a space changes together with the angle, intensity and color of light. Through the kind of applied light, its color, the degree of rendering colors, the manner and type of adopted luminaire, an urban interior assumes some unique features influencing a recipients mood and arousing the sensation of pleasure or comfort. The possibilities of acquiring the unique character of a given place is broadened by a combination of street illumination with object illumination which uses the decorative values of buildings and the richness of botanical motifs in parks.

Various lamps are used: bulbs, halogen lamps, tubular fluorescent lamps, mercury vapour lamps, metal halide lamps, sodium vapour lamps, optical fibres, light-emitting diodes (LEDs). They render different shades and levels of colors. They can be used in layouts with motion detectors (bulbs, halogen lamps, LEDs); they save energy (tubular fluorescent lamps, metal halide lamps, sodium vapour lamps,

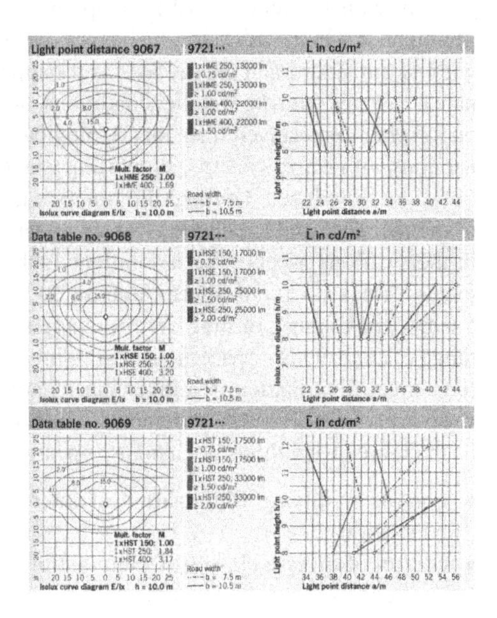

Fig. 1 Outdoor luminaire series Lumega by Trilux - selected data tables

LEDs). Control is applied to facilitate a dynamic change in the color of diodes (LEDs); lighting stages (LEDs) (see [17]) that change in time are used in RGB or AWB colors. The impact of the objective circumstances on a spectators subjective impressions (the proportions of a space, the basic dimensions of a space, the character of bodies and planes shaping a space, the participation of greenery in this impact, the character of perception cursory, contemplative) is taken into consideration, too. Computer renderings make it possible to verify and present the intended effects.

A designer ought to aim at minimizing electric energy consumption and preserving parameters required by the rules as well as optimizing solutions which realize other guidelines. Intelligent systems of design may serve to calculate predicted electric energy consumption, the possibility of savings, introducing interchangeable solutions, the costs of installation and maintenance.

2.2 *The Assignment of Optimal Control*

The issue of designing illumination in the city can be considered with reference to intelligent systems as the subject of a design. In this case, intelligent systems are introduced in a designing solution in order to manage illumination. These systems give a chance of introducing some advanced technical solutions, using dynamic illumination, optimizing the costs of maintenance, increasing energy efficiency and limiting an unfavourable impact on the environment (reduced CO_2 emission).

In illumination management, the work of an entire layout, some of its parts or each lamp can be controlled. It is possible to control the functioning of each lamp through smooth regulation of its luminous flux depending on road users needs, especially with reference to LED lamps which practically offer a smooth scale of illuminance. On the other hand, changes of this illuminance take place immediately without any additional energy loss as it happens in the case of CFL lamps. Illuminance changes with the weather conditions, the intensity of traffic, time modes based on information received from meteorological stations or illuminative cameras. Connections with alarm systems, supporting a layout in changeable and complex conditions, are introduced. Optimal illumination management aims at reducing the costs of maintenance and installation as well as satisfying the inhabitants and users needs in their various aspects. Such solutions include interactive darkening which minimizes the costs without deteriorating safety. Control can comprise the brightness of a lamp (darkening), turning on/off, changing luminous layouts or sequences. The range of control may concern one lamp (inbuilt luminaire control), a group of lamps (e.g. street illumination in a given class), complex layouts of lighting groups with various parameters and requirements. The broader the range of a controlling system is, the more possibilities of changes appear:

- control of diverse illumination of interconnected streets (on account of sight adaptation)
- control of diverse classes of illumination between neighboring areas (the difference cannot exceed two classes),

Factors able to change the required and expected parameters of illumination (e.g. its intensity) are as follows:

- changes of traffic intensity considering the time of the day and night (motion detectors, time switches),
- changes of the weather (illuminative cameras),
- time limits for illumination depending on the season (drivers with an inbuilt astronomical calendar, light detectors),
- changes in the manner of maintenance related to some events in the city (e.g. an urban square/a car park, a road/a pedestrian sequence).

One should note that the number of working modes for a single lamp (LED) generates a large space for potential designing solutions. For instance, we receive 10^5 possible modes for a street which includes a group of five lamps in ten modes. This example neglects dynamics related to traffic, changeable weather conditions, the time of the day etc.

3 Formal Model of Computations

In this section we introduce the formal model of an urban space components being under consideration. It includes street layout (i.e. streets, squares, pavement etc.) and adjacent buildings. As the most convenient formalism to be laid under the developed model the graph formalism was selected. Such the choice is implied by computational features of such structures. Their properties enable using graphs in parallel, distributed computing ([6, 7, 8]) which is the key issue of the approach.

The general idea of a lighting design support is dividing a problem into a set of subproblems that may be solved independently or almost independently (if some border areas overlap). Next, an agent system is deployed across the obtained set. Particular subtasks are performed by assigned agents. The idea of employment of agent systems to solving CAD related problems was already investigated e.g. in constructional tasks [18], the automotive industry [2], collaborative CAD systems [9] or the adaptive design [11].

The detailed schema of the approach is following. A city map M is transformed into the graph $G^{(M)}$ which is decomposed (reversibly) into the set of subgraphs $\{G_i^{(M)}\}_{i=1,...N}$. Each subgraph $G_i^{(M)}$ has an agent A_i ascribed which is referred to as a designer agent. A_i may create additional, auxiliary agents $\{A_i^k\}_{k=1,...M}$ performing computations at the level of particular objects (buildings). Figure 2 illustrates the deployment schema. Dotted lines represent borders between particular subgraphs at the map level. Black figures denote designer agents ($\{A_i\}_{i=1,...N}$), dark gray ones auxiliary agents ($\{A_i^k\}_{k=1,...M}$) ascribed to particular subsets of buildings.

Particular designer and/or auxiliary agents are capable of solving optimization tasks implied by an actual design problem (see the Section 2), taking into account all imposed constraints (e.g. those included in Tables 1 and 2) . They may provide some suggestions concerning the distribution of luminaries, an optimal configuration of system components etc.

This approach introduces the parallelism and thereby allows to overcome problems related to computational complexity.

To enable a problem decomposition we need to have a suitable formal representation of a system. The most convenient one is a graph one.

A structure of an urban space area has two levels. At the first level we have a street layout, at the second one particular buildings and other objects. Since the decomposition of computations is made at the both levels we need to ensure a formalism which ensure good performance of an agent system, especially with respect to the inter-agent communication. For the street layout we use *slashed form* of graph representation (see Def. 2) At the building level the hierarchical hypergraph representation for modeling the objects is used (see Subsection 3.2). The detailed description of both graph models is presented below.

3.1 Street Layout Representation

The concept of the slashed form of a centralized graph aims reducing coupling among subgraphs (generated by border nodes) in a distributed representation and thereby simplify operations performed by maintaining agents in a distributed environment. The basic idea of that approach is splitting edges rather than the multiple replication of existing nodes of a centralized graph as it was made in RCG environment [10].

Definition 1 *(Σ^v, Σ^e, A)-graph is a triple $G = (V, E, \varphi)$ where V is nonempty set of nodes, $E \subseteq V \times (\Sigma^e \times A) \times V$ is a set of directed edges, $\varphi : V \longrightarrow \Sigma^v$ is a labeling function, Σ^v and Σ^e denote sets of node and edge labels respectively and A is a set of edge attributes. We denote the family of (Σ^v, Σ^e, A)-graphs as $\mathscr{G}$.*

Definition 1 modifies the (Σ^v, Σ^e)-graph notion (see [10]). The edge structure is changed from $V \times \Sigma^e \times V$ to $V \times (\Sigma^e \times A) \times V$ to enable encapsulating all required data in edge attributes. These data include slashing details (e.g. geometric coordinates) but also problem specific information (e.g. architectural details of adjacent buildings). (Σ^v, Σ^e)-graph definition can be also extended e.g. by introducing an attributing function for nodes, but such extensions will not be considered here.

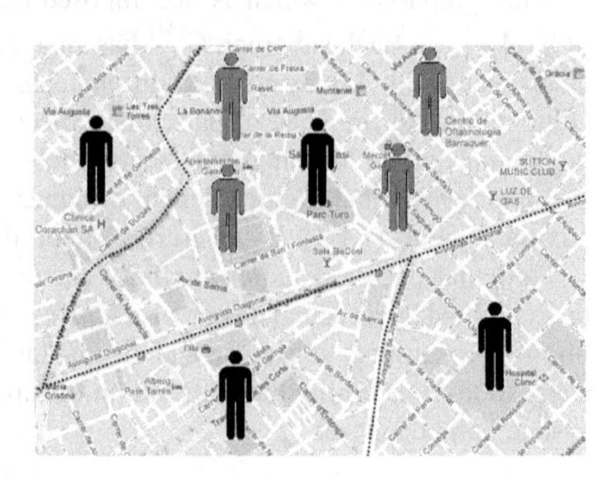

Fig. 2 Agent system deployment

Definition 2 (Slashed form of G**)** *Let* $G = (V, E, \varphi) \in \mathcal{G}$. *A set* $\{G_i\}$ *of graphs is defined as follows.*

- $G_i = (V_i, E_i, \varphi_i) \in \mathcal{G}$ *and* $V_i = C_i \cup D_i, C_i \cap D_i = \emptyset$, *where* C_i *is a set of* ***core nodes***, D_i *denotes a set of* ***dummy nodes*** *and* $\varphi_i \equiv \varphi|_{V_i}$,
- $\bigcup_i C_i = V$ *where* C_i, C_j *are mutually disjoint for* $i \neq j$,
- $\forall v \in D_i \exists ! v' \in D_j (i \neq j)$ *such that* v' *is the replica of* v; $\forall v \in D_i \deg(v) = 1$,
- $\forall e \in E_i :$ *e is incident to at last one dummy node.*

An edge incident to a dummy node is called a ***border edge***. *The set of all border edges in* G_i *is denoted as* E_i^b. *A set* $E_i^c = E_i - E_i^b$ *is referred to as a set of* ***core edges*** *of* G_i.

Let $M = \Sigma^e \times A$, *then a set* $\{G_i\}$ *as defined above is referred to as a* ***slashed form*** *of G, and denoted* $\not{G}$, *iff following conditions are satisfied:*

1. $\forall G_i^c = (C_i, E_i^c, \varphi_i|_{C_i}), \exists H_i \subset G : H_i \overset{\alpha}{\simeq} G_i^c$ *(*α *denotes an isomorphic mapping between graphs) and* H_i, H_j *are disjoint for* $i \neq j$.
2. $\exists f : M^2 \to M$ *– a bijective mapping* $\forall (e, e') \in E_i^b \times E_j^b (i \neq j)$ *such that (i)* $e = (x_c, m, v) \in C_i \times M \times D_i$, $e' = (v', m', y_c) \in D_j \times M \times C_j$, *(ii)* v' *is a replica of* $v : \exists ! e_{ij} = (x, m_e, y) \in E$ *such that* $x_c = \alpha(x), y_c = \alpha(y)$ *and* $f(m, m') = m_e$. e_{ij} *is called a* ***slashed edge*** *associated with replicated dummy nodes* v, v'.
3. $\forall e = (x, m, y) \in E : (i) \exists ! e_c \in E_i^c$ *for some i, such that* $e_c = \alpha(e)$ *or (ii)* $\exists ! (v, v') \in D_i \times D_j$ *for some* i, j, *such that e is a slashed edge associated with v and* v'.

$G_i \in \not{G}$ *is called a* ***slashed component*** *of G.*

f mapping recovers labeling/attributing data of a slashed edge basing on a labeling/attributing of given border edges.

In Figure 3 the centralized and the slashed form of the given graph G are shown. To preserve the clarity of images we neglect the attributing/labeling of graph edges. Core nodes are marked as circles and dummy ones as squares. The following indexing convention is used for slashed components (see Fig. 3b). A core node index has the form (i, k) where i is an unique, within $\not{G}$, identifier of a slashed component G_i and k is an unique, within G_i, index of this node. A dummy node index has the form $(-1, k)_r$ where k is a globally unique identifier of a node. Additionally, a subscript r denotes a reference to a slashed component (or its maintaining agent) hosting a replica of a given dummy node. Using such a subscript allows for immediate localization of a replica. To simplify the notation subscripts will be neglected within the text, unless needed. Note that a dummy vertex and its replica share a common index and differ in reference subscripts only: $(-1, k)_{r_1}, (-1, k)_{r_2}$.

Example

The following example depicts a decomposition of the graph G representing street layout of the square area selected from the Tokyo *OpenStreetMap* (OSM) map [13] (see Fig.4a). The selection size is $4km^2$. For the clarity of the Figure 4b we zoomed only the fragment of the graph G. This graph represents layer consisting of `highway`-tagged objects only. The number of vertices of G is $|V| = 2779$, the

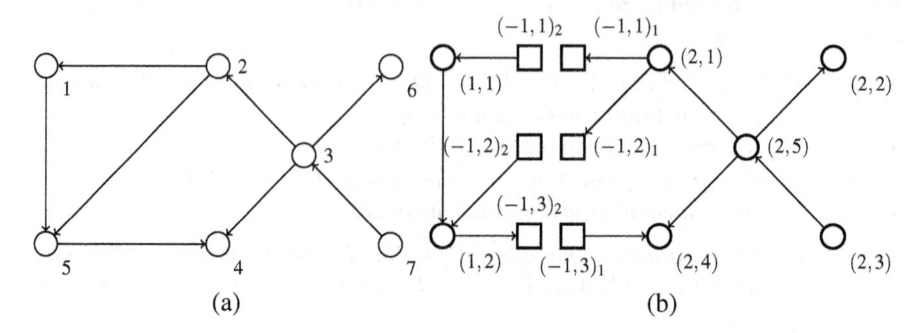

Fig. 3 (a) Graph G (b) $\not{\mathcal{G}}$ representation

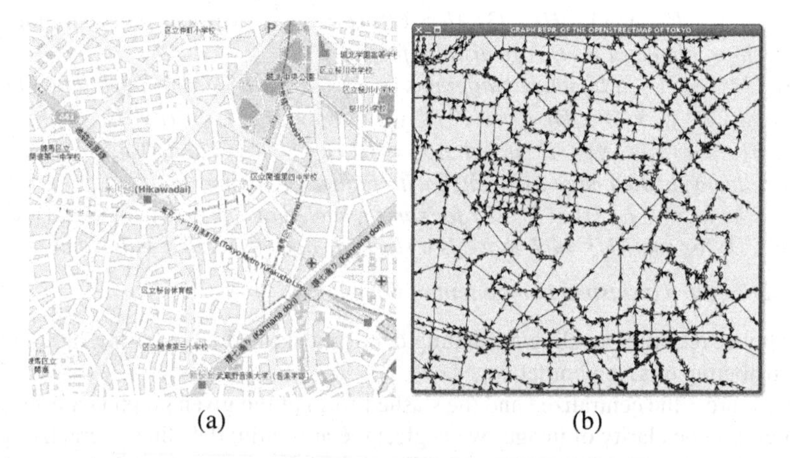

Fig. 4 (a) Selection from Tokyo city plan (source: *www.osm.org*) (b) Distributed graph representation of the street layer of the map 4a

number of edges $|E| = 3252$. After obtaining the graph from the OSM map we made the rough decomposition (based on BFS algorithm) of G into 137 slashed components having at most 50 vertices. The total number of interconnections between the slashed components was 460 i.e. approximately 3.4 per a slashed component.

3.2 Representation of Objects

The formal background for the level 2 description, related to physical objects (solids) like buildings consists of two notions: the face adjacency hypergraph (FAH) and the hierarchical face adjacency hypergraph (HFAH) [4]. Formal definitions of both structures were introduced in [16]. Here only the short outline will be given.

Hypergraphs belonging to the FAH family denoted as $\mathcal{H}_{FA}$, model simple objects like single buildings. A face adjacency hypergraph is a tuple of the form

$$G = (N, A, H, lab_N, lab_A, lab_H, att_N, att_A, att_H) \in \mathcal{H}_{FA},$$

where N is a nonempty set of nodes (solid faces), $A \subset P_2(N)$ is a nonempty set of edges (solid edges), $H \subset \bigcup_{i>2} P_i(N)$ is a nonempty set of hyperarcs (solid vertices), $lab_\mu : \mu \to \mathcal{L}_\mu$ for $\mu = N, A, H$ is a labeling function for vertices, arcs and hyperarcs respectively with corresponding set of labels $\mathcal{L}_\mu$; $att_\mu : \mu \to \mathcal{A}_\mu$ for $\mu = N, A, H$ is an attributing function for vertices, arcs and hyperarcs respectively with corresponding set of attributes $\mathcal{A}_\mu$. Attributing functions att_N, att_A, att_H contain complete data related to geometric features of a solid. This assumption, influencing structures of $\mathcal{A}_N, \mathcal{A}_A$ and $\mathcal{A}_H$, is imposed.

Example

Let us consider the cuboid S presented in Figure 5a with faces denoted by $u_1, \ldots u_6$. For that solid we have following hypergraph representation (see Fig.5b) $H = (N, A, H, lab_N, lab_A, lab_H, att_N, att_A, att_H)$, where:

- $N = \{u_1, \ldots u_6\}$,
- the set A consists of pairs of nodes which correspond to faces which form particular edges of S: $A = \{\{u_1,u_3\}, \{u_1,u_4\}, \{u_1,u_5\}, \{u_1,u_6\}, \{u_2,u_3\}, \{u_2,u_4\}, \{u_2,u_5\}, \{u_2,u_6\}, \{u_3,u_5\}, \{u_3,u_6\}, \{u_4,u_5\}, \{u_4,u_6\}\}$,
- the set H consists of tuples of nodes which correspond to faces meeting in a given vertex of S: $H = \{\{u_1,u_3,u_5\}, \{u_1,u_3,u_6\}, \{u_2,u_3,u_5\}, \{u_2,u_3,u_6\}, \{\{u_1,u_4,u_5\}, \{u_1,u_4,u_6\}, \{u_2,u_4,u_5\}, \{u_2,u_4,u_6\}\}$,
- attributing functions carry geometric characteristics of entities:
  - $att_N : N \ni p \to$ *Coordinates of vertices of the face represented by p*,
  - $att_A : A \ni \{p,q\} \to$ *Coordinates of endpoints of the physical edge common for p and q*,

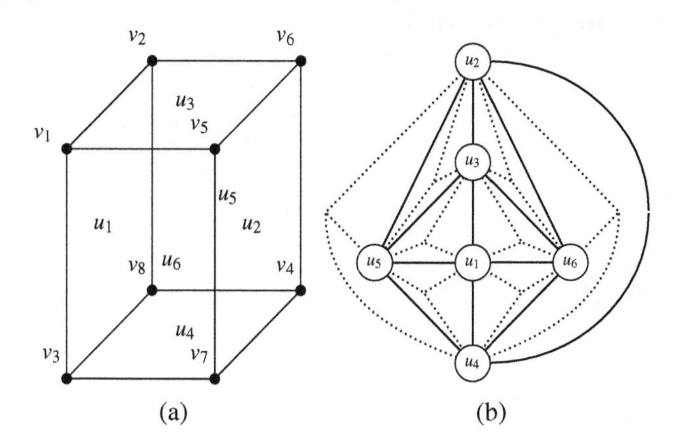

Fig. 5 (a) Cuboid – the sample solid (b) Hypergraph representation of cuboid shown in Fig.5a

- $att_H : H \ni \{p_1, p_2, \dots p_k\} \rightarrow$ *Coordinates of the physical vertex common for* $p_1, p_2, \dots p_k$.

For the clarity only hypergraph vertices are labeled in Figure 5b. Edges are drawn with the solid line and hyperedges with the dotted one.

To extend the applicability of the hypergraph notation to complex objects we use the notion of a HFAH which allows for modeling complex objects consisting of multiple coupled solids. The kind of coupling is not strictly determined: simple objects may either physically contact or remain in some distance.

A graph $g^* = (\mathscr{H}, T)$ belonging to the family of HFAHs denoted as $\mathscr{H}_T$, is a tree T which vertices are from the set $\mathscr{H} \subset \mathscr{H}_{FA}$. Additionally we introduce the edge attributing function att_E such that $att_E(e \in E(T))$ gives a full specification of coupling between solids represented by nodes incident to e.

Example

In Fig.6a the example of a composite solid, named S, is shown. One can distinguish three component of S with corresponding FAHs denotes as G_0, G_1, G_2. The HFAH describing S has the form $g^* = (\{G_0, G_1, G_2\}, T)$ where the tree T is presented in Fig.6b. The cuboid associated with G_0 is the root component of T while child nodes G_1 and G_2 represent small cubes of S adjacent to the large one. Attributes a_1, a_2 specify the relationships between corresponding hypergraphs' nodes (i.e. faces of solids).

Let $g^* = (\mathscr{H}, T) \in \mathscr{H}_T$, where $T = (\mathscr{H}, E, att_E)$. Removing any edge $e \in E$ form T causes that it splits into two disjoint subtrees, say T_1 and T_2, such that $g_i^* = (\mathscr{H}, T_i) \in \mathscr{H}_T$ for $i = 1, 2$. The process may be continued recursively until we obtain a set of individual nodes representing particular FAHs.

Subtrees of a tree representing a composite solid, determine natural borders for the problem decomposition being prior to an agent system deployment. For example in the Figure 6b the dotted lines suggest the decomposition for three subproblems related to respectively G_0, G_1 and G_2.

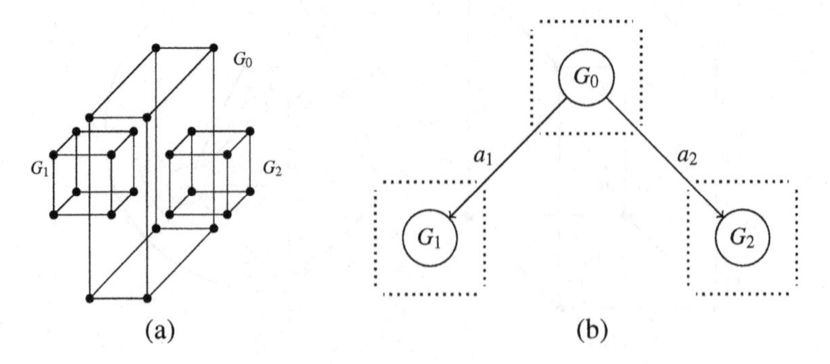

Fig. 6 (a) Complex object – S (b) HFAH tree of S

3.3 Linkage between Model Levels

The graph and hypergraph representations introduced for streets and buildings in previous sections should be linked formally to obtain the comprehensive system model. Let $G = (V, E, \varphi) \in \mathscr{G}$ be the (Σ^v, Σ^e, A)-graph. The key issue is defining the structure of the set of attributes A. We assume that

$$A \ni a \stackrel{def}{=} (C^{(a)}, \{(c_i^{(a)}, g_i^{*(a)})\}_{i=1,\ldots,N_a}),$$

where $C^{(a)}$ is the set of coordinates which describe completely geometric features of a relevant edge (the structure of $C^{(a)}$ may be neglected now) and $(c_i^{(a)}, g_i^{*(a)})$ is a pair consisting of the HFAH $g_i^{*(a)}$ representing a subset of buildings adjacent to a relevant street and the parameter $c_i^{(a)}$ specifying the physical location of a root component of $g_i^{*(a)}$.

4 Moving Formal Model to Application Level

From the designer's perspective our model is intended to play two roles. First is exploring a space of solutions. As it was explained before its out of the human reach to find and investigate all acceptable scenarios in a reasonable time. The second role is providing some architectural suggestions especially in the initial phase of a design process. Note that such suggestions may be generated on the basis of prior (archival) solutions. Using artificial intelligence methods enables preparing such the samples-based suggestions, profiled for a given problem context.

Since a human makes an ultimate decision which variant of a suggested solution to choose (or how to combine or modify them), we can say about the Human Aided Design (HAD).

To enable transforming expectations and demands of professionals into computer (algorithmic) actions one has to structure the domain of a problem. As it was remarked in the Section 3, the graph model was selected for this purpose. Besides the mentioned advantages it provides the scalability. Also the specification of architect's requirements imposes defining the language which allows to express those requirements in terms applicable in used algorithms.

The generic schema (Fig. 7) of HAD illustrates interaction between an architect creating a lighting design and the system supporting the design process, presented in the paper (note that Fig. 7 doesn't contain neither in-process iterative interactions nor the scenario of suggesting a *start point* solution for a design).

The basic role of a human in this schema is preparing an architectural model of a considered area (e.g. in AutoCAD) and define objectives for a design. Those objectives are determined by at least tree factors: lighting standards (see Tables 1, 2 or [12]) – luminosity levels have to satisfy norms, energy saving constraints – minimizing power consumption and other non normative requirements, in particular esthetical ones. Basing on a CAD model the relevant graph representation is

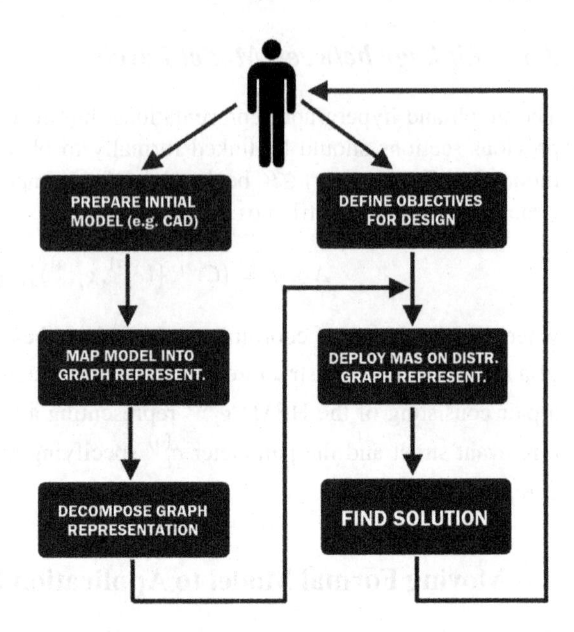

Fig. 7 The role of supporting system in the design process

generated and decomposed. On the other side design objectives are supplied to an agent system which is deployed on distributed graph to find an optimal solution. Finally, such a solution is returned to an architect.

5 Conclusions

In the paper we outlined main factors determining the design of lighting solutions in urban spaces. It is constrained by normative requirements but also functional and esthetic demands. The additional question related to the economic aspect of a solution is designing the optimal system control. In this case one has take into account except existing lighting standards but also the energy saving performance. For supporting a process of selecting an optimal solution from a numerous set of variants we propose distributed, agent-based computations and the formal model of underlying structures describing an urban area. The schema of such a supported design task was discussed.

Acknowledgements. The paper is supported from the resources of Alive & KIC-ing project and NCBiR grant no O ROB 002101/ID 21/2.

References

1. Bandini, S., Bonomi, A., Vizzari, G., Acconci, V.: Self-organization models for adaptive environments: Envisioning and evaluation of alternative approaches. Simulation Modeling Practice and Theory 18(10), 1483–1492 (2010)

2. Baumgart, S., Toledo, B., Spors, K., Schimmler, M.: PLUG: An Agent Based Prototype Validation of CAD-Constructions. In: The 2006 International Conference on Information and Knowledge Engineering (2006)
3. Czyżewski, D., Żagan, W.: Kompleksowe ujcie problematyki owietlenia miast. In: Konferencja Naukowo-Techniczna z cyklu "Energooszczędność w oświetleniu DIODY LED, Pozna (2011)
4. De Floriani, L., Falcidieno, B.: A hierarchical boundary model for solid object representation. ACM Trans. Graph. 7(1), 42–60 (1988)
5. Ehrig, H., Engels, G., Kreowski, H.-J., Rozenberg, G.: Handbook of Graph Grammars and Computing By Graph Transformation: Applications, Languages, and Tools, vol. II. World Scientific Publishing Co., River Edge (1999)
6. Kotulski, L.: GRADIS – Multiagent Environment Supporting Distributed Graph Transformations. In: Bubak, M., van Albada, G.D., Dongarra, J., Sloot, P.M.A. (eds.) ICCS 2008, Part III. LNCS, vol. 5103, pp. 644–653. Springer, Heidelberg (2008)
7. Kotulski, L., Sędziwy, A.: On Complexity of Coordination of Parallel Graph Transformations in GRADIS Framework, DepCoS-Relcomex. In: 2009 Fourth International Conference on Dependability of Computer Systems, pp. 279–289 (2009)
8. Kotulski, L., Sędziwy, A.: Parallel Graph Transformations with Double Pushout Grammars. In: Rutkowski, L., Scherer, R., Tadeusiewicz, R., Zadeh, L.A., Zurada, J.M. (eds.) ICAISC 2010. LNCS, vol. 6114, pp. 280–288. Springer, Heidelberg (2010)
9. Ligong, X., Zude, Z., Quan, L.: Multi-agent Architecture for Collaborative CAD System. In: 2008 International Conference on Computer Science and Information Technology, pp. 7–11 (2008)
10. Kotulski, L.: On the control complementary graph replication. In: Mazurkiewicz, J., et al. (eds.) Models and Methodology of System Dependability, Monographs of System Dependability, vol. 1, pp. 83–95. Oficyna Wydawnicza PW, Wrocław (2010)
11. Kotulski, L., Strug, B.: Distributed Adaptive Design with Hierarchical Autonomous Graph Transformation Systems. In: Shi, Y., van Albada, G.D., Dongarra, J., Sloot, P.M.A. (eds.) ICCS 2007. LNCS, vol. 4488, pp. 880–887. Springer, Heidelberg (2007)
12. Polska Norma PN-EN 13201 Oświetlenie dróg (Polish lighting standards – in Polish). PKN, Warszawa (2005)
13. OpenStreetMap, `http://www.osm.org`
14. Rozenberg, G.: Handbook of Graph Grammars and Computing By Graph Transformation: Foundations, vol. I. World Scientific Publishing Co., River Edge (1997)
15. Sędziwy, A., Kotulski, L.: Solving Large-Scale Multipoint Lighting Design Problem Using Multi-agent Environment. In: Key Engineering Materials. Advanced Design and Manufacture IV, vol. 486, pp. 179–182 (2011)
16. Sędziwy, A.: Representation of Objects in Agent-based Lighting Design Problem, Submitted to DepCoS-RELCOMEX 2012 Conference (2012)
17. Wiórek, A.: Światło dla miasta. In: Zawód-Architekt, 05/2010, pp. 72–74 (2010) (in Polish), `http://www.zawod-architekt.pl/download/ZA_2010_05.pdf`
18. Yabuki, N., Kotani, J., Shitani, T.: A Cooperative Design Environment Using Multi-Agents and Virtual Reality. In: Luo, Y. (ed.) CDVE 2004. LNCS, vol. 3190, pp. 96–103. Springer, Heidelberg (2004)

Monitoring Event Logs within a Cluster System

Janusz Sosnowski, Marcin Kubacki, and Henryk Krawczyk

Abstract. Resolving complex problems on cluster systems we have to take into account threats related to system dependability. We faced this problem in relevance to the developed project of the KASKADA platform targeted at managing heavy multimedia processing in a supercomputer environment (Galera cluster in Intel technology). Having analyzed the experience of other authors in the area of cluster dependability analysis and anomaly predictions we have found the need of developing appropriate strategy of monitoring the KASKADA platform within the Galera cluster. In the chapter we deal with the problem of exploring the event logs generated within the nodes of the platform. The main goal was to study the specificity of the platform operational image in the event space, explore the morphology and informative contents of logs and develop methods of anomaly detection. For this purpose special tools have been used.

1 Introduction

Recently large cluster systems are used to resolve complex problems. Hence dependability of such systems is of great importance. This issue is mainly the domain of the cluster owners and administrators. In the literature there are many papers devoted to this issue [1,13,16,21]. A special attention is paid to various monitoring techniques targeted at detection of errors or anomalies as well as prediction of possible problems [2,4,10,14,15]. The published results usually relate to large systems monitored for long time periods. Hence they provide quite interesting reliability data and prove the practical usefulness of the monitoring techniques to predict problems or optimize system usage, etc. Such research needs a full access to the monitored system. In many cases we use a cluster system with a limited

Janusz Sosnowski · Marcin Kubacki
Institute of Computer Science, Warsaw University of Technology, ul. Nowowiejska 15/19, Warsaw 00-665, Poland
e-mail: jss@ii.pw.edu.pl, M.Kubacki@ii.pw.edu.pl

Henryk Krawczyk
Faculty of Electronics, Telecommunications and Informatics, ul. Gabriela Narutowicza 11/12, Gdańsk-Wrzeszcz 80-233, Poland
e-mail: hkrawk@eti.pg.gda.pl

W. Zamojski et al. (Eds.): Complex Systems and Dependability, AISC 170, pp. 257–271.
springerlink.com

access to its resources (e.g. contracted nodes, disc area, interconnection ports, etc.). The system provider assures some level of dependability. Nevertheless it is reasonable to check dependability and related issues by the users. We faced this problem within a project of the KASKADA platform developed in Technical University of Gdansk on the base of a large cluster system Galera [7] with 1300 Intel Xeon 4-core processors. This platform is used for developing and running data stream processing applications.

In order to evaluate dependability of the developed KASKADA platform and related applications several monitoring mechanisms have been integrated with this platform. They are targeted at 3 aspects (application level, platform task management, resource utilization) and provide appropriated data logs. Application level logs are specified by designers and relate practically to application quality issues. Task monitoring gives some coarse grained view of the platform operation, start/termination times of tasks, task errors, etc. Fine grained (low level) image of the platform operation is created by system event logs within nodes and performance logs related to preprogramed variables. This work concentrates on this low level monitoring in particular strategies of monitoring and analysing event logs. The main contribution was to find normal operational profiles, morphology of logged events and their significance. We have also outlined possible correlations with higher level monitoring. To deal with these problems we have adapted previously developed tool QLogAnalyser [9].

Section 2 is a survey of research on monitoring cluster systems. It shows the benefits of monitoring and proves some uniqueness of cluster systems. Section 3 describes the specificity of the KASKADA platform and the assumed strategy of monitoring using appropriate tools. Section 4 presents some practical results of our approach to monitoring. Final conclusions and further research are summarized in section 5.

2 Monitoring Cluster Systems

Due to hardware, software and interconnection (network) complexity in cluster systems it is considered that they are susceptible to various faults (in particular transient faults, software failures, etc. In [20] the authors report high percentage (about 75%) of hardware faults (CPU- 79%, disc – 6%, memory 4%). Software faults contributed 10-15%, environmental, human and network errors contributed a few percent, similarly undetermined errors. In [16] the distribution of hardware, program, environment and network faults was: 60%, 20%, 1%, and 18%, respectively (averaged over 22 systems). However depending upon the system these figures fluctuated e.g. 27-70% and 5-25% for hardware and program faults. These statistics confirm dependability problems in cluster systems, hence arises the need of monitoring and reporting such cases. Failures can be reported automatically by the system as well as by the administrators and users. We must be conscious that not all failures are detected automatically (moreover they can be mitigated or tolerated by the system). Moreover it is reasonable to observe the system operation and predict the possible appearance of a failure; this may even result in failure avoidance. This concept attracted many researchers and the proposed solutions based on system monitoring. Most publications concentrate on event and performance logs collected and analysed for specific cluster systems e.g. [5,6,13,21] and references therein.

Event logs hold various information which can help identifying problems or confirm normal operation [17,18]. Typically event log specifies sequential number of the recorded event, event type (e.g. mechanism generating the event), timestamp (e.g.: date, hour, minute, second), identifier of the job which activated the event, localization (cluster node, device), event description, service or module touched by the event, severity level, etc. The formats of event records may differ upon the system so some fields can be easily recognized others may create some problems, especially event descriptions. Quite often a single problem may result in generation of many events. Hence it is reasonable to introduce time and spatial filtering of event logs. Typically time filtering combines events related to the same localization and the same job ID if their timestamps are within a specified time window (e.g. 10 s). Spatial filtering may combine events related to the same job ID, they are within the specified window time, but may differ in localization. In event log analysis we must be conscious of some format fluctuations (especially in Unix and Linux systems) and interpretation ambiguities. Special filtering and data exploration techniques have to be developed (compare section 3) to derive system operation features and peculiarities.

In tab. 1 we give some event log statistics related to 5 cluster systems [13] (from the Top500 ranking list of 2006): S1 – Blue/GeneL (P – 131072, M - 32768 GB), S2 – Thunderbird (P – 9024, M – 27072GB), S3 - Red Storm (P – 10880, M – 32640GB), S4 - Spirit (ICC2) (P – 1028, M – 1024GB), S5 - Liberty (P – 512, M – 944GB). In the brackets we specify the number of processors (P) and memory capacity (M).

Table 1 Log statistics for cluster systems S1-S5

System	Period [days]	Capacity [B]	Message length [B]	Byte rate [B/s]	Message rate [M/s]	Alarms [%] (x)
S1	215	1.2	254.6	65.0	0.26	7.3 (0.34)
S2	244	27.4	129.9	1298.1	9.99	1.5 (0.006)
S3	104	30.0	136.9	3337.6	24.34	0.7 (0.086)
S4	558	30.3	111.0	628.3	5.64	63.5 (0.003)
S5	315	22.8	86.1	835.8	9.71	0.001 (0.428)

The collected logs covered different time periods (100-500 days) and their capacity is different. Normalizing this capacity to 100 days we still get relatively high dispersion 0.56 – 28.8 [GB]. Scaling this per a single processor (proc) we get 4.27 KB/proc for S1; 2.65 MB/proc for S3; 14.1 MB/proc for S5. Such big dispersion range results from different operational profiles and monitoring schemes. It has also a direct impact on the number of generated message bytes [B/s] and messages [M/s] per second. Average message length is in the range 100-250 bytes (depending upon the system). It is worth noting that event logs can be efficiently compressed – gzip provides 0.1; 0.21; 0.04; 0.06 and 0.03 compression ratio for S1-S5 systems, respectively. In our Galera system this was 0.05.

Alarm messages contribute 0.001% - 63.5% of log entries. Systems S4 and S5 significantly differ from the remaining ones. These alarms were attributed to hardware (98.04%), software (0.08%) and other (1.88%) problems (in total 170 million of alarm messages in 5 systems). Many alarms may relate to the same problem. Hence using appropriate filters we can reduce the presented percentage by the factor (x) given in the bracket in the last column of tab. 1. Taking into account filtered alarms we get different distribution (averaged over all systems): 18.78%, 64.01% and 17.21% of hardware, software and other problems. It is worth noting that this distribution is different for different systems: S1 (3.7%, 43.3%, 53.0%), S2 (43%, 36.5%, 20.5%), S4 (2.8%, 97.2, 9.0%), S5 (1.9%, 98.1%, 0%). The system administrators introduced various alarm categories: S1 – 41, S2 – 10, S3 – 12, S4 – 8, S5 – 6, attributed to specific software (e.g. kernel) or hardware (e.g. SCSI, memory ECC circuitry, LAN transmission) modules.

Alarm qualification needs more comments. Standard system log qualifications introduce different message severity levels. System administrators may have different opinion on this qualification. For BL/G cluster [13] the automatic qualification produced the following distribution: 18.02%, 0.03%, 0.41%, 2.37%, 0.49% and 78.68% related to the message categories *fatal, failure, severe, error, warning, info*. This shows dominating information messages. The categorization defined by the system administrators resulted in 99.8% fatal and 0.02% failure messages. In the case of Red Storm system (S3) automatic original qualification resulted in the following distribution 0%, 0%, 6.09%, 7.95%, 8.45%, 14.74%, 61.63% and 1.14% within message categories *emerg, alert, crit, err, warning, notice, info, debug*. The administrators categorization produced the following distribution 98.69%, 0.75%, 0.02% and 0.54% for *crit, err, warning* and *info* categories. This confirms the need of individual approach to log qualification. In practice this process should be correlated with operational profiles and administrator activities. For example many log reports of high severity during some maintenance or system upgrade actions are not critical. On the other hand some problems have no impact in the event logs, so other symptoms has to be found (we faced this problem in practice).

The presented statistics we can confront with those provided in [10] they relate to two configurations of the system Blue Gene L: SDSC – 3072 computer nodes with two-core CPUs and 384 I/O nodes, 1.5 TB RAM memory; ANL – 1024 computer nodes, 32 I/O nodes, 500 GB RAM memory. The number of collected events in logs was: ANL - 5 887 771 and SDSC - 517 247. This corresponds to 2.27GB and 463MB of data collected within 112 and 132 weeks, respectively. The average message length was for ANL – 391B, for SDSC – 895B, so this is much higher than in systems from tab. 1 (86-250B). The number of registered events is not correlated with the number of CPUs. Moreover the system with the weaker configuration (ANL) produced more logs – this resulted from an intensive system tests performed during one week (they generated 1.5 million of events), these tests sensitised hardware errors not visible during normal workload. Similarly as in [13] the collected raw events were filtered in time and spatially (to eliminate some redundancy). Moreover some non-important attributes have been skipped; some others not precised were corrected. The time stamp of events was with the granularity of seconds. The event filtering process resulted in significant reduction coefficients of event logs: for ANL – (0.0098, 0.0046, 0.0040), for SDSC – (0.099, 0.0084, 0.0079) where subsequent coefficients in the brackets relate to time window of 10, 300 and 400s, respectively. Higher reduction in the case of ANL system resulted from a large number of events (in fact redundant) generated during the erroneous situation. Here arises the problem of selecting appropriate time window (trade-off between high reduction coefficient and a danger of losing some important information).

An important issue is event categorization. In [13] this has been done by the authors and system administrators. They assumed two-level hierarchical model. In the higher level they defined 10 categories (hardware, kernel, application, etc.). Within each of these categories (lower level) they specified 2 – 46 detailed subcategories (the biggest number for the kernel category). Hence in total they received 219 subcategories. System Blue Gene in an automatic way attributes event severity levels e.g. *fatal, failure, warning*, etc. Unfortunately many events specified as *fatal* or *failure* were not critical as well as some without this specification in fact were critical. Hence event categorization has been refined using the collected experience and system administrators' knowledge (they defined 69 types of fatal events).

The presented statistics prove the need of individual approach to each considered systems, collecting events for a long time and exploring the log morphology (syntax and semantics), accumulating users or administrators' remarks, etc. We have initiated such program for the KASKADA platform.

3 Monitoring Strategy of the KASKADA Platform

Developing the monitoring schemes for the KASKADA platform we had to take into account its specificity and available logs (section 3.1). In this research we concentrate on system event logs. The most important issue was the analysis of log morphology and appropriate processing procedures (section 3.2).

3.1 Outline of the KASKADA Platform

Computer centres provide a proper hardware core for multimedia processing which includes: well developed and high bandwidth network infrastructure, high performance supercomputers (especially computational clusters with hundreds of computational nodes), a huge and efficient data storages. However, a typical supercomputer environment is usually focused on the execution of computational tasks, collected as batches within long queue systems. This solution is well suited for scientific computations, usually performed offline without user interactions. On the other hand, the processing of surveillance data, for currently observed activities, usually needs to be performed online with nearly real-time performance and close user cooperation, so that critical situations can be immediately recognized and handled.

KASKADA (Context Analysis of Camera Data Streams for Alert Defining Applications platform) is a special middleware supporting developing multimedia applications and their efficient executions. It is one of the layers of a parallel environment for processing multimedia data streams coming from cameras located in various geographic areas: houses, streets, stadiums, railway stations, airports. The architecture of the environment is shown in Fig. 1. It facilitates heavy multimedia processing in a supercomputer environment (Galera cluster in Intel technology). The main goal of the KASKADA platform is to join multimedia data streams with tasks representing suitable functionality of the user application. Then such pairs (data, task) are located on cluster nodes in order to minimalize execution time of the whole applications. Different allocation strategies are considered and chosen according to type of processing tasks. The platform offers a set of services (in the sense of the SOA standard) to facilitate developing different applications. These services can be directly taken into account in construction of service oriented scenarios applications. In such cases the platform is able to map a service into sequence of suitable tasks, and execute them in parallel taking into account current state of cluster node loads.

User application level Development, testing, execution
KASKADA platform level task and resource management, managing service repositories, monitoring mechanisms
Hardware level Super computer Galera with application servers, data stream archives
Network infrastructure Provides source data (e.g. from cameras) by optical fibber connections

Fig. 1 Parallel processing environment for data stream analysis

We distinguish three areas of platform management related to multimedia streams, services, and events (summarized in tab. 2). “Multimedia stream” delivers data flowing continuously from their producers (e.g. video cameras) to the consumers – analysis tasks on the platform. A part of the platform functionality provided to the tasks (implemented algorithms) or directly to the external applications or users is called a “service”. An “event” is a detection of a specific situation (state of the platform) or object properties occurring in computations performed by the task implementing specific detection algorithms, for example an indication of a shape suspected to be a gun in the video incoming from a monitoring camera. Here arises the need of processing multimedia data streams coming from cameras located in various geographic areas: houses, streets, stadiums, railway stations, airport, etc. It is worth noting that classical well known management strategies (e.g. MPI platform) usually require implementation of management mechanisms in each application which is quite expensive solution, this problem is avoided in KASKADA [7].

Depending upon the complexity of the implemented application an appropriate number of computation nodes is allocated to KASKADA. The operation of the platform as well as the application can be monitored at different levels: system level (Linux event and performance logs), platform management level (e.g. service/task execution times, delays, errors), and application specific logs (introduced by the developers). The KASKADA monitor controls and manages tasks/services and in particular it assures:

Table 2 Management strategies in KASKADA

Management strategy	*Functionality*
Stream level maintain massive load of multimedia data	play, stop, archive, replay, distribute, load balance, multiply
Service level process the user/application requests	invoke, finish, monitor, kill, assign tasks
Event level provide means to communicate the processing state to the user/application	generate, store, distribute, filter, relay

- Managing task and services (initialization, termination, cancelling, periodical checking of states and correctness of execution)
- Monitoring allocated and used processor, memory and network resources
- Reporting detected errors
- Creating and archiving trace logs
- Monitoring correct operation of processing nodes

This is some kind of coarse grained monitoring. In the subsequent section we concentrate on fine grained monitoring related to system event logs. Unix systems provide various standard logs holding information on the system operation. Typically in event logs the system registers the start and end of appearing events, actions within software, detected errors, abnormal states, etc. In the case of clusters we have separate logs for each computing node. The structure and the information contents of system logs may be more or less regular with some specified data fields (e.g. date timestamp, source of information, severity level). However the stored messages are usually in a loosely textual form.

3.2 Event Log Processing

Depending upon the goal of the log analysis (e.g. error detection and diagnosis, finding operational profiles or trends, identifying anomalies) we are looking for different events or their sequences. In many cases this is reduced to a search of a specific event category e.g. reboots, errors, warnings. We must be conscious that beyond these well-known events there are many other interesting events worth tracing. They can be selected manually, however in the case of very large logs this must be supported with some automatization (e.g. based on data exploration or data mining). In this process it is important to identify static and variable (parameter) fields within log lines. This identification is called log file abstraction [6,12]. Having found abstracted log files we can perform log analysis more efficiently. In fact log abstracting results in finding log classes specified with the use of regular expressions [9].

In the literature event abstracting problem has been outlined at some general level and is based on checking the frequency and position of words in the event records. Unfortunately details of algorithms are not shown; moreover results do not refer to real event logs. Dealing with this problem in the considered cluster subsystem we have started with getting some knowledge on event morphology performing some searches (with various keywords, or specified regular expressions). Following this preliminary analysis we have developed multilevel abstracting scheme which combines regular expressions with searching constant and variable word phrases.

Looking for the representative event classes we have started with log preprocessing. In particular we skip the time stamp field and replace it with the type of log e.g. syslog), node and PID fields are replaced with symbol (*). For example a raw syslog entry:

Jul 1 12:01:44 g108 ntpd[2659]: synchronized to 192.168.19.201, stratum 2 is subsequently transformed with identified fields placed in square brackets (node name g108, process name ntpd, its PID and the message field):

<Syslog>[g108][ntpd][2659][synchronized to 192.168.19.201, stratum 2]
<Syslog>(*)[ntpd](*)[synchronized to 192.168.19.201, stratum 2]

All entries of this type can be represented by the following log class:

<Syslog>(*)[ntpd](*)[synchronized to (...), stratum 2]

where symbol "(...)" replaces the variable field (IPs of different clock synchronization servers).

To identify log classes we have developed the following algorithm:

<u>Log transformation algorithm (variant A//**B**)</u>

1. **for** every line (record) in the log entry LE **do**
   {create a list L of subsequent words in LE
   for every word W on position P in L **do**
   {count[W][P] := count[W][P] + 1}
   }
2. **for** every line (record) in the log LE **do**
   {create a list L of subsequent words in LE
   create a list (LC) of the number of word occurrences in LE line; *elements of list LC are defined by appropriate counters: count[W][P] - taking into account the word and its position*
   sort LC in an ascending way
   median = middle element of LC; related to the position defined by *length(LC)/2 rounded to lower value*
   for every word W on position P in L **do**
   {**if** count[W][P] >= median // **or '> median**"
   then W is a fixed element in the line
   else W is parameter (variable)
   if the previous word was not parameter **then** replace W with (...)
   add W to R *(result record)*}
   }

This algorithm can be implemented in version A or B (small difference noted in bold after //). Having transformed log entries we initiate a procedure to combine them into equivalent classes. Implementing the algorithm we use QMap list (taken from Qt library - http://doc.qt.nokia.com/qq/qq19-containers.html) for handling the count[W][P] structure. This list stores pairs (key, value) ordered by key. It is organized as skip-list where each node has several forward pointers (to speed up searching – O(log n)) and one backward pointer. The key is the concatenated W and P.

For an illustration we present some results of using A or B algorithm targeted at the message field. The syslog file related to the period 1.06.11-30.09.11 resulted in total in 248368 events (excluding cron events). Abstracting the timestamp and PID (replaced by wildcard (*)) we have got 117294 different events. Applying to this reduced set algorithm A we have got 12624 event classes, abstracting nodes we have reduced this set to 5197 classes. For algorithm B these figures are 5589 and 383, respectively. The analysed log used 2502073 words (separated by spaces), within this set we have detected 116213 unique words, among them 981 words which do not include digit characters. The number of word positions in the message field was up to 25. This statistics confirms the need of automatic log abstracting.

Most events were generated by kernel, excluding these events we have got 31 event classes. Within this reduced set we have identified an interesting event class (the wildcards (*) replace node and PID numbers, respectively):

<Syslog>(*)[udevd-event](*)[unlink_secure:
chmod(/dev/bus/usb/004/027, 0000) failed: No such file or directory]

It is related to nonstandard disconnection of USB device (e.g. pen drive) – the file of such device disappears so a failure situation is reported. Finding such event in a small set is simple, but it would be difficult to find it within hundreds of thousands raw events.

Having selected warning events algorithm A generated 3 classes:

<Syslog>(*)[postfix/qmgr](*)[warning: (...) clock]
<Syslog>(*)[postfix/qmgr](*)[warning: (...) to (...)]
<Syslog>(*)[postfix/sendmail](*)[warning: (...)]

The result of algorithm B was a little bit different (more detailed):

<Syslog>(*)[postfix/qmgr](*)[warning: backward time jump (...) -- (...) clock]
<Syslog>(*)[postfix/qmgr](*)[warning: backward time jump (...) -- (...) to (...)]
<Syslog>(*)[postfix/sendmail](*)[warning: fork: Resource temporarily unavailable]

These classes correspond to 34, 17, and 24736 raw events, respectively. The first two classes relate to clock synchronization problems, the third one signals (with sendmail) various system limitations encountered during process execution (e.g. due to too many processes).

Having selected events with errors in syslog file (including kernel events) algorithm A reduced it to 66 classes, most of them related to LustreError (distributed file system generated errors caused by temporary disc array problems), moreover 5 classes related to segfaults (memory protection problems with specified localizations and error types):

<Syslog>(*)[kernel](*)[(...) segfault at (...) rip (...) rsp (...) error 14]
<Syslog>(*)[kernel](*)[(...) segfault at (...) rip (...) rsp (...) error 15]
<Syslog>(*)[kernel](*)[(...) segfault at (...) rip (...) rsp (...) error 4]
<Syslog>(*)[kernel](*)[(...) segfault at (...) rip (...) rsp (...) error 6]
<Syslog>(*)[kernel](*)[(...) segfault at 0000000000000000 rip (...) rsp (...) error 4]

Distribution of events in these classes was 4, 3, 1, 5 and 1, respectively (14 events in total). Algorithm B generated only two classes (with distribution 9 and 5):

<Syslog>(*)[kernel](*)[(...) segfault at (...) rip (...) rsp (...) error (...)]
<Syslog>(*)[kernel](*)[(...) error 6]

Having selected events generated by NTP demon (handling clock synchronization in a node) algorithm A generated 15 classes, among them the following are the most interesting (with distribution 2, 9, 53, 361, 81232, and 22), comments added in italic:

<Syslog>(*)[ntpd](*)[frequency error (...) PPM exceeds tolerance 500 PPM] *- problem with the system clock*
<Syslog>(*)[ntpd](*)[no servers reachable] *- unavailable servers NTP (switched off)*
<Syslog>(*)[ntpd](*)[ntpd exiting on signal 15] *- proces killing with SIGTERM (comand kill, or restart)*
<Syslog>(*)[ntpd](*)[synchronized to (...) stratum (...)]
<Syslog>(*)[ntpd](*)[synchronized to (...) stratum 2] *- synchronization problems*
<Syslog>(*)[ntpd](*)[unable to bind to wildcard socket address 0.0.0.0 - another process may be running - EXITING] *– NTP erroneously connected 2 times*

In syslog there are a lot of events generated by DHCP client. DHCP is a protocol to dynamically assign IP configuration for workstations - in our case this configuration was assigned to the cluster nodes. On each node there is a client daemon that requests periodically an IP configuration. Generally after operating system boot DHCP client tries to discover DHCP servers by sending broadcast messages (DHCPDISCOVER). If an offer is send from any of DHCP servers that detected DHCPDISCOVER packet, client requests a configuration via DHCPREQUEST packet. IP configuration is assigned for a specified amount of time, called "lease time". After its expiration, client needs to renew configuration with another

DHCPREQUEST packet (DHCPDISCOVER is not needed, because client already knows the IP of the DHCP server). These periodic renewals are logged to the system log. In general lease time should be constant. Sometimes problems appeared with repeated requests due to the lack of responses; another anomaly was lack of requests for a longer time.

Algorithm A identified 3 classes of events related to DHCP (with distribution of 6030, 5865, 5865 events, respectively):

```
<Syslog>(*)[dhclient](*)[(...) on eth0 to (...) port (...)]
<Syslog>(*)[dhclient](*)[bound to (...) -- renewal in (...) seconds]
<Syslog>(*)[dhclient](*)[DHCPACK from 192.168.19.201]
```

The first class corresponds to DHCP request events (generated via eth0 network card to various DHCP servers IPs and ports). The second class relates to IP and lease time (renewal) assignments by DHCP servers. The third class corresponds to confirmations from DHCP server (with specified IP).

The cron file for the first two weeks of September contained 236071 events, applying algorithm A we have got 234 event classes, which has been further reduced to 22 classes by abstracting nodes. Similarly algorithm B produced 189 and 15 classes, respectively. Within these classes we have found one with "popenfail" related to 100-264 erroneous situation in one node during 4 days (exhaustion of the process number limit).

It is difficult to trace logs manually. On one hand we have different types of events, sometimes with some parameters of more or less important value, on the other hand it is important to know the source of event, its location in time and space (e.g. node number). So in the analysis we are interested in various statistics global or filtered (according to various attributes), raw or aggregated events, abstracted (classified) events, etc. Moreover it is reasonable to identify frequent (dominating events), critical events, event profiles, etc. The scope of possible questions is large and open. So it is reasonable to create appropriate data base (in fact data warehouse) with some analytical, data exploration and visualization instruments. For this purpose we have adapted two tools developed in Institute of Computer science WUT: QLogAnalyzer and QperfAnalyser. They were primarily dedicated for collecting and analyzing event and performance logs in servers and workstations via LAN. They cooperate with special agents installed in the monitored system which fill in the associated data base. Hence we have the possibility of monitoring and analyzing many computers. Adapting this approach to the considered cluster we load cluster logs into the data base using a special program (adapted to the log formats and their sources). Moreover we have extended the statistical and visualization capabilities with a special attention to cluster node multiplicity (node correlations).

4 Visualization of Event Distributions

Having analysed the morphology of event logs (including event classification - abstraction) we can perform more detailed studies to identify critical, anomalous states or to find the characteristics of the operational profile. This process can be supported with appropriate visualization tools. In the developed QLogAnalyser we have the possibility of visualizing events in various time perspectives, filtered (e.g. based on regular expressions) or aggregated. For example hour, daily or monthly perspectives show event distribution over subsequent hours, days or months (events are summarized over hours, days or months). Aggregated hour views show distribution of events within 24 hours, for each hour we sum all events of this hour for each day of the considered period. Similarly we can generate aggregated daily distribution over all week days (summarized events for each week day over the considered period). Aggregated distributions usually relate to system activity during day and night hours or week days (e.g. low activity on weekends).

The generated distributions can be limited to specified event classes (e.g. errors, warnings) or unique events. Moreover they can be presented for individual nodes (on a single or independent plots) or summarized over selected group of nodes, etc. In fig. 2-6 we give some plots from KASKADA with comments. The time scale in fig. 2-4 covers 4 months with single day resolution. Explanation of some anomalies needed cooperation with the administrators. This also showed the need of storing administrator or users reports for future log analysis.

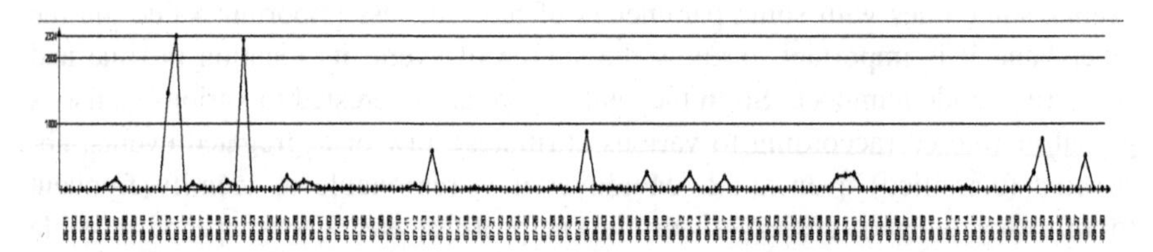

Fig. 2 Daily distribution of error events (the maximal number of errors per day 2324): all events summarized over all nodes (excluding cron popenfail) – about 89% related to LustreError (section 3), most caused by two disc array failures (two high pulses within the first month)

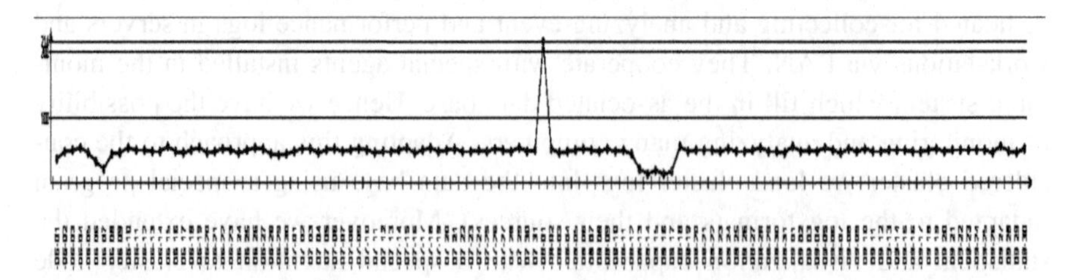

Fig. 3 Daily distribution of DHCP requests (section 3) summarized over nodes (maximal value 214 events): the positive peak relates to the network problem in two nodes, moreover two negative pulses relate to other problems caused by the file system

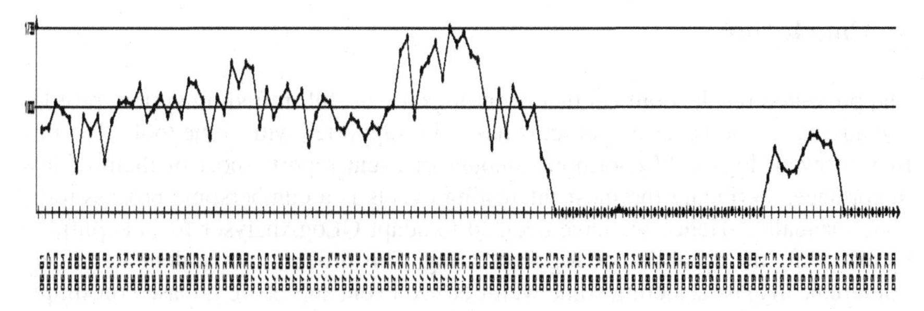

Fig. 4 Daily distribution of *NTP synchronized* events (section 3) summarized over all nodes (maximal value 1739 events): fluctuations relate to some synchronization problems, low activity periods related to switched off NTP server

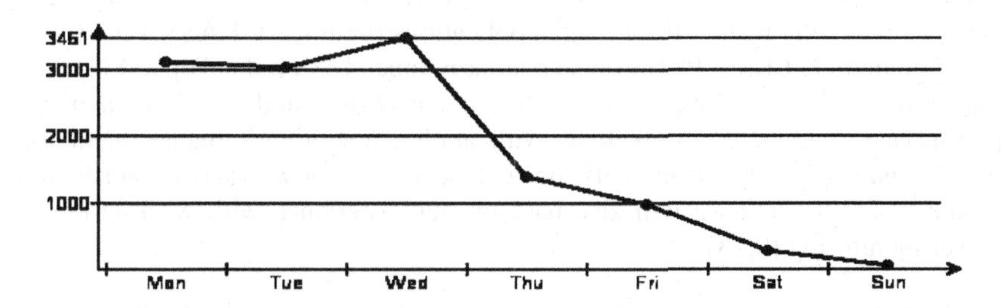

Fig. 5 Aggregated weekly distribution of errors (summarized over all nodes): descending trend relates to lower users activity during weekends

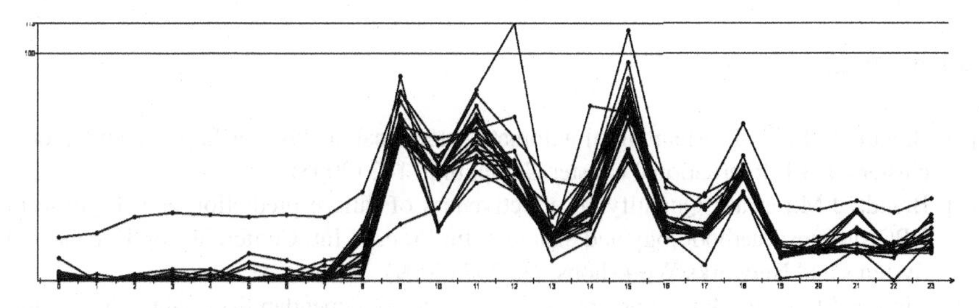

Fig. 6 Aggregated hour distribution of errors (without cron) for 22 nodes (maximal value 113 errors, time axis markers correspond to subsequent hours starting from 0 to 23): dominating errors in node g019, fluctuations result from daily workload profile composed of service development hours and its testing hours (typical cycles of 1-1.5 hours)

Tracing shapes of the plots we can identify some anomalies (e.g. spikes, drop-outs) and refer them to logged events. Deeper analysis may involve correlating them with other log records (e.g. application or task logs) as well as with performance logs.

5 Conclusion

The presented results confirm that event logs are useful in finding system profiles and anomalies, however this process has to be supported with some tools. The raw files of event logs hold enormous amount of event reports most of them of low importance, so tracing the most interesting events is a cumbersome process to be done manually. Hence we have decided to adapt QLogAnalyser for this purpose by extending the capability of log abstracting. This process can be done iteratively (hierarchically) by selecting some well-defined event subsets (e.g. based on simple regular expressions [9]) and applying the proposed algorithms. Another issue is the problem of visualizing the selected events to find other correlations or identify anomalous spikes, trends etc. This has been also assured in QLogAnalyser.

The usefulness of the proposed approach has been illustrated in relevance to real logs collected in a cluster subsystem (KASKADA). The future research is targeted at collecting results from many applications running on KASKADA (within the upgraded GALERA PLUS cluster- rank number 163 on the Top500 list) for a longer time. Moreover we plan to refine event classes and correlate them with performance parameters as well as with application or task management logs. Log visualization function will be extended to show various correlations (compare [11]). Here we will also base on our experience with workstation and server monitoring [8,18].

Acknowledgement. This work has been realized within the project MAYDAY EURO 2012, Operational Innovative Economy Program 2007-2013, Priority 2, "Infrastructure area B + R".

References

[1] Brandt, J.M., et al.: Meaningful automated statistical analysis of large computational clusters. IEEE International Cluster Computing, 1–2 (2005)

[2] Brandt, J.M., et al.: Quantifying effectiveness of failure prediction and response in HPC systems: methodology and example. In: Proc. of Int. Conference on Dependable Systems and Networks Workshops, pp. 2–7 (2009)

[3] Cinque, M., et al.: A logging approach for effective dependability evaluation of computer systems. In: Proc. of 2nd IEEE Int. Conf. on Dependability, pp. 105–110 (2009)

[4] Gmach, D., et al.: Workload analysis and demand prediction of enterprise data center applications. In: 10th IEEE Int. Symposium on Workload Characterization, pp. 171–180 (2007)

[5] Hassan, A.E., et al.: An industrial case study of customizing operational profiles using log compression. In: Proc. of ACM ICSE, pp. 713–721 (2008)

[6] Huang, L., et al.: Symptom based problem determination using log data abstraction. In: Proceeding of ACM CASCON 2010 Conference of the Center for Advanced Studies on Collaborative Research, pp. 313–326 (2010)

[7] Krawczyk, H., Proficz, J.: KASKADA – multimedia processing platform architecture. In: Proc. of the International Conference on Signal Processing and Multimedia Applications. SIGMAP 2010, pp. 26–31 (2010)

[8] Król, M., Sosnowski, J.: Multidimensional monitoring of computer systems. In: Proc. of IEEE Symp. and Workshops on Ubiquitous, Autonomic and Trusted Computing, pp. 68–74 (2009)
[9] Kubacki, M., Sosnowski, J.: Analysing event log profiles in Linux systems. In: Borzemski, L., et al. (eds.) Information system Architecture and Technology, Web Information Systems Engineering, Ofic. Wyd. Polit. Wroc., pp. 136–144 (2011) ISBN 978-83-7493-630-9
[10] Lan, Z., et al.: A study of dynamic meta-learning for failure prediction in large scale systems. J. Parllel Distribiuted Systems 70, 630–643 (2010)
[11] Makanju, A., et al.: Log View: Visulaizing event log clusters. In: 6th IEEE Conf. on Privacy, Security and Trust, pp. 99–108 (2008)
[12] Naggapan, M., Vouk, M.A.: Abstracting Log Lines to Log Event Types for Mining Software System Logs. In: Proceedings of Mining Software Repositories (co-located with ICSE 2010), pp. 114–117 (2010)
[13] Oliner, A.J., Stearley, J.: What Supercomputers Say: A Study of Five System Logs. In: Proc. of the IEEE/IFIP Int. Conference on Dependable Systems and Networks, pp. 576–584 (2007)
[14] Oliner, A.J., Aiken, B.: Online detection of multi component interactions in production systems. In: Int. Conf. on Dependable Systems and Networks, DSN 2011, pp. 49–60 (2011)
[15] Salfiner, F., Lenk, M., Malek, M.: A survey of online failure prediction methods. ACM Comput. Survey 42, 10:1–10:42 (2010)
[16] Shroeder, B., Gibson, G.A.: A large scale study of failures in high performance computing systems. IEEE Trans. on Dependable and Secure Computing 7(4), 337–350 (2010)
[17] Sosnowski, J., Poleszak, M.: On-line monitoring of computer systems. In: Proc. of IEEE DELTA Workshop, pp. 327–331 (2006)
[18] Sosnowski, J., Król, M., Machnicki, J.: Techniques and goals of monitoring computer systems. In: Górski, A. (ed.) Information Systems Architecture and Technology, Advances in Web-Age Information Systems, pp. 235–246. Oficyna Wydawnicza Politechniki Wro-cławskiej (2009) ISBN 978-83-7493-479-4
[19] Sosnowski, J., Król, M.: Dependability evaluation based on system monitoring. In: Ali, A.-D. (ed.) Computational Intelligence and Modern Heuristics, pp. 331–348. Intech (2010) ISBN 978-953-7619-28-2
[20] Xue, Z., et al.: A survey of failure prediction of large scale server cluster. In: 8th ACIS Int. Conference on Sofr\t. Eng., Artificial Intelligence, Networking and Parallel/distributed Computing, pp. 733–738 (2007)
[21] Li, Y., Zheng, Z., Lan, L.: Practical online failure prediction for Blue Gene/P: Period-based vs. Event-driven. In: Proceedings of the IEEE/IFIP International Conference on De-pendable Systems and Networks Workshops, PFARM Workshop, pp. 259–264 (2011)

Implementing AES and Serpent Ciphers in New Generation of Low-Cost FPGA Devices

Jarosław Sugier

Abstract. New generations of FPGA devices that are being continuously developed provide the designers with extended capabilities and create new options for implementation of contemporary ciphers. This work presents implementations of the two best algorithms of the AES contest – Rijndael and Serpent – in Spartan-6 devices from Xilinx and compares them with equivalent effects that were obtained in architectures of the previous generation. The included results allow for evaluation of implementation cost vs. efficiency in contemporary FPGA chips for these two cryptographic algorithms and also provide some conclusions about how the situation changes with development of new, more powerful programmable architectures.

1 Introduction

Dependable operation of numerous contemporary computer systems rely on data protection and this is assured with appropriate encryption methods. Among symmetric ciphers with secret key the AES algorithm is used as a standard solution in most of the applications with Serpent cipher being the main comparable alternative.

In this work we investigate various options for low-cost hardware implementations of the two ciphers and especially look at the changes that were caused in this area by new generation of Spartan-6 family of FPGA devices from Xilinx. The text is organized as follows: after presenting the two algorithms in chapter 2, the various hardware organizations of the cipher unit are introduced in chapter 3. Finally, chapter 4 discusses size and performance parameters that were obtained after implementation of all the variants in Spartan-6 and, for comparison, in Spartan-3 chips.

Jarosław Sugier
Wrocław University of Technology
Institute of Computer Engineering, Control and Robotics
ul. Janiszewskiego 11/17, 50-372 Wrocław, Poland
e-mail: `jaroslaw.sugier@pwr.wroc.pl`

W. Zamojski et al. (Eds.): Complex Systems and Dependability, AISC 170, pp. 273–287.
springerlink.com

2 The AES Contest: Rijndael vs. Serpent

The first widely used encryption algorithm, the Data Encryption Standard (DES), was developed by IBM and standardized by US National Institute of Standards and Technology (NIST) in 1977. In mid-90s its strength was seriously questioned by successful attacks ([13]) and in January 1997 NIST issued a first call for a successor algorithm, to be called an Advanced Encryption Standard or AES. In response 15 new cipher proposals were submitted from several countries. After two conferences organized to promote public examination of the methods (AES1, August 1998, and AES2, March 1999) the five finalists were announced in August 1999. Their scores in a voting which was organized during the AES2 conference were as follows:

- Rijndael: 86 positive votes, 10 negative;
- Serpent: 59 positive, 7 negative;
- Twofish: 31 positive, 21 negative;
- RC6: 23 positive, 37 negative;
- MARS: 13 positive, 83 negative.

After the last AES3 conference in April 2000, the final decision was announced which was consistent with the AES2 poll: the Rijndael was chosen as the winner. Under the new name of AES it was announced the U.S. Federal Information Processing Standard 197 (FIPS 197) in November 2001 ([10]).

Serpent and Rijndael belong to the same class of round-based cipher algorithms and bear significant resemblance. Both algorithms are symmetric block ciphers that are examples of substitution-permutation networks (SPN). Their processing consists in a set of *rounds*, with every round being a specific set of *elementary operations* executed repeatedly over a given *block of data*. Independently from cipher (data) path there is a separate processing path whose task is to provide every round with its individual *key*, generated form user-supplied secret *external key*.

To summarize the distinction between the two ciphers shortly, it is often said that Rijndael is faster (having fewer rounds) but Serpent is more secure. After the NIST final decision most of the attention concentrated on Rijndael for obvious reasons, but second-to-the-winner Serpent still deserves some consideration because of its advantages that won significant appreciation during the AES contest. It is worth noting that in the AES2 ballot it was the Serpent that received the least number of negative votes.

2.1 The AES (Rijndael) Algorithm

The Rijndael cipher ([10]) was initially developed by two Belgian cryptographers, Joan Daemen and Vincent Rijmen, and the finally approved AES standard, strictly speaking, is its subset with fixed block size of 128b[it] and allowed key sizes of 128, 192 or 256b. To focus the discussion in this paper we consider exclusively the AES-128 version, i.e. we assume size of the key to be 128b.

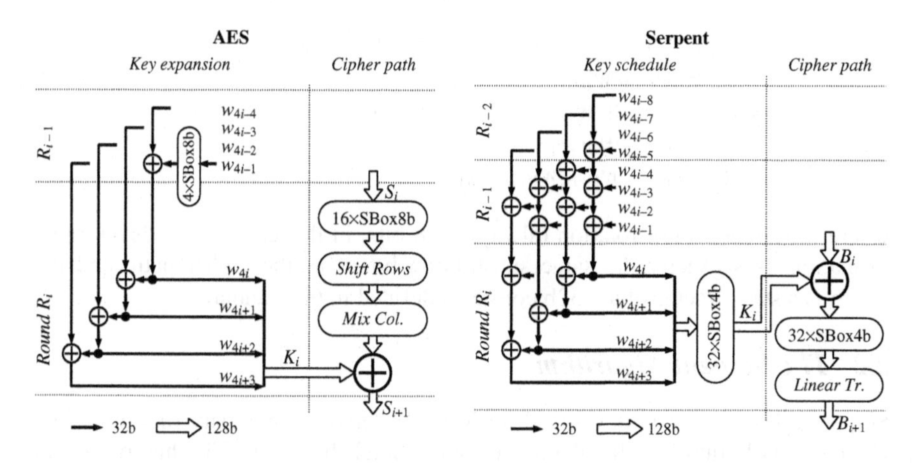

Fig. 1 Data flow in a single round of the AES (left) and Serpent (right) ciphers

Since AES allows only one block size of 128b, it always operates on 16B[yte] chunks of data that form a 4x4B array, termed *the State*. For 128b key, processing of the State during the encryption is divided into exactly 10 rounds plus one auxiliary executed at the beginning of the process.

Let P be a 128b plaintext, S_i – a state block that enters the i-th round R_i, K – external (user) key, K_i – round key, C – encoded ciphertext. The complete data path of the AES can be expressed with the following equations:

$$\begin{aligned} S_1 &:= P \oplus K \\ S_{i+1} &:= MC(\ SR(\ SBox\ (\ S_i\ )\ )\ ) \oplus K_i \qquad i = 1 \ldots 9 \\ C &:= SR(\ SBox\ (\ S_{10}\ )\ ) \oplus K_{10} \end{aligned}$$

That is, the initial round (numbered as 0) consists only of addition of the external (user) key while every regular round number 1 to 9 contains four elementary state transformations executed in specific order: *byte substitution SBox*, *row shifting SR*, *column mixing MC* and *addition* (XOR) *of the round key*. The last round (number 10) does not include column mixing but the other three operations remain unchanged. Additionally, rounds 1÷10 use *extended keys* that need to be generated from the user key by a separate key *expansion* routine. Execution of a regular round (1÷9), along with generation of its key, is shown in the left part of Fig. 1.

The key expansion routine, in turn, operates on 32b words w_i, $i = 0..43$, which, after computation, are directly copied to the round keys K_i. The first four words are initialized with bits from the user key:

$$\{w_0, w_1, w_2, w_3\} := K$$

and then every group of four words that creates one round key is computed as follows for $i = 1..10$:

$$w_{4i} := SBox(\, w_{4i-1} <<< 8\,) \oplus \mathtt{Rcon}[\, i\,] \oplus w_{4i-4}$$
$$w_{4i+1} := w_{4i} \oplus w_{4i-3}$$
$$w_{4i+2} := w_{4i+1} \oplus w_{4i-2}$$
$$w_{4i+3} := w_{4i+2} \oplus w_{4i-1}$$
$$K_i := \{w_{4i}, w_{4i+1}, w_{4i+2}, w_{4i+3}\}$$

where <<< denotes left rotation (always by 8 bits, in this case), the *SBox*() transformation uses exactly the same substitution boxes as the cipher path, and the `Rcon` is a static vector of ten 32b constants defined in the standard.

2.2 The Serpent Algorithm

Serpent ([1] – [3]) was developed by Ross Anderson (University of Cambridge Computer Laboratory), Eli Biham (Technion Israeli Institute of Technology), and Lars Knudsen (University of Bergen, Norway). In the version that was submitted for the contest the method operates on 128b blocks of data with 256b external key. If the user supplied key is shorter (call for the standard allowed also key lengths of 128 and 192b) simple expansion procedure is applied which ensures that the method always starts with the full 256b key. The transformation flow is divided into 32 almost identical rounds with every round using its own 128-bit round key generated by the *key schedule*; since the last round needs two keys, total of 33 different round keys are required.

In addition to the symbols defined above, now let the data block that enters the i-th round is denoted as B_i. Before the plaintext block enters the procedure a special bit reordering – so called Initial Permutation *IP* – is performed (this reordering has no cryptographic significance and was introduced only for bit-sliced implementations). The plaintext P after permutation gives block B_0, which is the input to the first round number 0. The output of the last round, R_{31}, after application of the Final Permutation *FP* (which is an inverse of IP) gives the ciphertext C.

The complete data path from the plaintext P to the ciphertext C can be formally represented by a sequence of the following equations:

$$B_0 := IP(\,P\,)$$
$$B_{i+1} := LT(\,SBox_{i \bmod 8}(\,B_i \oplus K_i\,)\,), \quad i = 0 \ldots 30$$
$$B_{32} := SBox_7\,(\,B_{31} \oplus K_{31}\,) \oplus K_{32}$$
$$C := FP(\,B_{32}\,)$$

Operation of a single round, together with generation of its key, is shown in the right part of Fig. 1. As the first transformation, the block B_i is XOR-ed with the round key K_i that is supplied by the key schedule, and then the resulting vector is passed through substitution boxes. The specification defines 8 different S-Boxes numbered 0 … 7 with each round R_i using S-Box number i mod 8. The vector created by S-Boxes finally undergoes *linear transformation LT*, giving block B_{i+1} that is the input to the next round. In the last round R_{31} the linear transformation is replaced with XOR operation with the extra last key K_{32} and therefore two keys are required in this round, to the total of 33 keys in the whole process.

The key generation in Serpent is no less involved. The schedule generates first a set of 32-bit *prekeys* w_i which are later used for computation of round keys. The starting 8 prekeys numbered from –1 to –8 are filled with bits of the external (user) key K (after its expansion to 256b, if necessary):

$$\{w_{-1}, w_{-2}, \dots w_{-8}\} := K$$

and then 132 prekeys $w_0 \dots w_{131}$ are generated by the following affine recurrence:

$$w_i := (\, w_{i-1} \oplus w_{i-3} \oplus w_{i-5} \oplus w_{i-8} \oplus \phi \oplus i \,) <<< 11$$

where ϕ is the fractional part of the golden ratio $\left(\sqrt{5}+1\right)/2$ represented as 32-bit vector (0x9E3779B9 in hexadecimal notation).

The final round keys are calculated from the prekeys using the same set of 8 substitution boxes that are defined for the cipher path. The general rule is that the key K_i is computed from a group of four prekeys w_{4i}, w_{4i+1}, w_{4i+2} and w_{4i+3} that undergoes bit substitution and reordering:

$$K_0 := IP(\, SBox_3(\, w_0, w_1, w_2, w_3 \,)\,)$$
$$K_1 := IP(\, SBox_2(\, w_4, w_5, w_6, w_7 \,)\,)$$
$$\dots$$
$$K_{31} := IP(\, SBox_4(\, w_{124}, w_{125}, w_{126}, w_{127} \,)$$
$$K_{32} := IP(\, SBox_3(\, w_{128}, w_{129}, w_{130}, w_{131} \,)$$

To avoid repetitive use of the same substitution as later in the round, during computation of K_i the schedule uses S-boxes number $(3 - i) \bmod 8$.

3 Implemented Architectures

Apart from relative simplicity of elementary operations at the binary level, ease of hardware implementation of the AES and Serpent algorithms comes from the fact that their processing flow is composed of (almost) identical rounds that are repeatedly executed over a given block of data. This leads to many potential processing schemes that blend different flavours of combinational, pipelined and iterative architectures ([4] – [9], [11] – [12], [14] – [16]).

In this study efficiency of hardware implementation of both ciphers will be tested using four essential types of processing: combinational, half (cipher-only) pipelined, fully (both cipher and key) pipelined, and iterative. For brevity, every implementation will be given a name starting with a letter A (for AES) or S (for Serpent) with indication of its type that follows: C (combinational), HP (half, i.e. cipher-only pipelined), FP (fully, i.e. both cipher & key, pipelined) and I (iterative). Since the AES pipelined architectures (AHP and AFP) can be optionally implemented with or without utilization of block-RAM resources in the FPGA chip, this leads to the total of 10 different implementations which will be investigated.

3.1 Combinational Dataflow

In this organization hardware structure closely follows flow of the data that is being encoded. All rounds of the cipher (11 for AES and 32 for Serpent) are implemented as separate hardware modules that create a continuous combinational path from the input registers (plain text *P*) to the output registers (cipher text *C*). In-between, the module operates as a combinational function that maps 128 + 128 = 256 input bits (data + key) into 128 output bits (cipher). The only registers used in this design are located in the *P* and *C* ports. The *K* input is not registered thus only the $P \rightarrow C$ path is taken into account by the implementation tool during optimization of the propagation speed.

In both cases (AC and SC) the design was specified by porting the specification to the VHDL language using strict RTL style: there were no instances of library elements, no sequential (procedural) descriptions were inserted and the code was free from references to any specific hardware attributes. After definition of all internal signals as `std_logic_vector` type, particular elementary operations were defined as separate entities with exception of key mixing, which was implemented simply with built-in `xor` operator at the place of their occurrence. Substitution boxes, both 8b (AES) and 4b (Serpent), were defined according to general templates recommended for ROM specification. AES row shifting and rotations required in key expansion or key schedule were treated as simple bit reordering in `std_logic_vector` signals and expressed with concurrent signal assignments (in hardware implementations, as opposed to software realizations, these transformations are done exclusively in routing and actually do not require any logic). The other operations: column mixing *MC* and linear transformation *LT* at the binary level end up as pure XOR networks and were represented with due number of concurrent assignments. The cascade of the modules that implement individual cipher rounds was easily constructed with a single `for...generate` statement which improved greatly conciseness and clarity of description.

A diagram describing structure of these architectures would mostly reproduce Fig. 1 hardly introducing any new information and, for brevity of this work, it is not included.

3.2 Cipher-Only Pipelining

The general idea of pipelining is to introduce evenly spaced registers along the combinational path so that in its synchronized operation several blocks of data can be processed at the same time during every clock cycle. In the combinational architectures of both ciphers the natural points of placing the pipeline registers are the signals S_i / B_i that cross boundaries of cipher rounds; this transforms each round into one pipeline stage. In technical terms such organization can be interpreted as a complete outer loop pipelining ([5]) and yields 11 pipeline stages for AES vs. 32 for Serpent. This means that valid output appears 11 or 32 clock cycles after input and although it does not improves the latency (which is actually worse than in the case of combinational propagation due to non-zero flip-flop switching time and non-ideal pipelining) the throughput (amount of data processed in unit time) rises enormously thanks to the parallel processing of multiple data in pipeline stages.

In this version of the architecture the key generation path remains combinational and this fact slows down changes of the external key during operation of the unit: loading a new key input invalidates the pipeline contents for 11 or 32 clock ticks until new data fill all the cipher stages. This may exclude this architecture from environments with frequent key changes but if the key can remain constant most of the time it is the optimal organization in terms of both speed and size.

Adding large amount of registers (128b × number of pipeline stages) may seem to be a substantial increase in resource usage but in case of FPGA architectures this increase is easily absorbed by the array. In these devices a flip-flop is included in every logic cell right at the output of combinational configurable element (Look-Up Table, LUT) so the only actual difference is that now some of them are used for registering the LUT signal while in combinational organization they were left unused. This usually does not affect the total number of occupied logic cells but just increases their utilization.

3.3 Full Pipelinaing

The drawback of the half-pipelined architecture – incompatibility with applications that require frequent changes of the key – can be a significant weakness in many applications. In general it is not recommended to encode large amounts of data with the same key because the attacker could get some information about it without breaking the cipher, namely by statistical analysis of the encoded stream.

To prepare the encryption unit for loading a new key with every block of data the key generation path should be pipelined in an equivalent way as the cipher path. More precisely, the pipelined key generator should provide the cipher stage with relevant key together with data which leads to conclusion that the key must be computed one clock cycle *before* the data is processed.

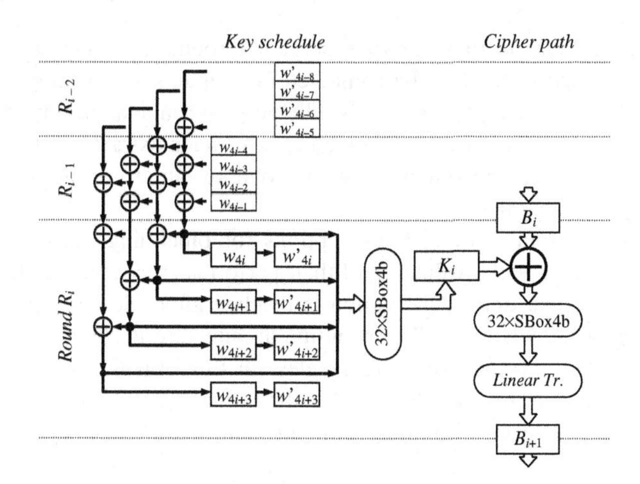

Fig. 2. Single round in fully pipelined implementation of the Serpent algorithm (SFP)

There is no problem with such organisation of the AES cipher: since in the first pipeline stage the round 0 uses external (user) key, its special preparation is not required. Instead, during the first clock cycle when the S_1 vector is computed, simultaneously the K_1 key can be prepared form K so that it is ready for round R_1 in the next cycle. The consecutive rounds work in the same way: R_i (i.e. S_{i+1}) is computed in parallel, simultaneously with preparation of the K_{i+1}.

Looking at the diagram in Fig. 1, the registers would be added right in the places of signals w_i and S_i. Thus, in case of the AES, the workflow of cipher and key paths was mapped in a natural way onto operation of the two pipelines in hardware.

In Serpent, in turn, situation at first looks similar: since computation of the round keys depends on prekeys w_i, these signals must be stored in pipeline registers. But the first problem is that, due to more complex key data dependency, computation of K_i in stage i depends on prekeys from not only stage $i - 1$ but also i – 2, so additional registers – denoted as w' – are required for storing previous values of w and feeding them *two* stages down the pipeline. This factor alone doubles the number of the key schedule registers. Moreover, w' registers are not located at LUT outputs – they are loaded with data from another registers – which is not an advantageous configuration for FPGA implementation.

Secondly, the last cipher round – R_{31} – needs two keys, so it must be split into two stages: the first one contains key mixing with bit substitution and the second one performs only final key mixing. An alternative solution – computation of two keys K_{31} and K_{32} in one clock cycle – is not a good option: the key schedule is relatively complex and a combinational path generating two keys would introduce unacceptable long delay holding back performance of the whole unit. Splitting the last round into two stages increases the total latency to 33 clock cycles but, compared to solution with 32 stages but with computation of K_{31} and K_{32} in one clock cycle, the shorter clock period compensates this more than adequately.

Another problem is that the first Serpent's round does not use unmodified external key; instead, K_0 must be computed in a regular way as any other key and during that the data in cipher path must idle going through a dummy (empty) stage added right at the beginning of the pipeline. This adds extra 128 flip-flops (which is a negligible increase compared to the total resource consumption) but also extends pipeline length to 34.

Detailed descriptions of different options for pipelining Serpent unit can be found in [14] along with evaluation of their performance vs. size trade-offs. It was shown that the final optimum solution is reached after adding registers not only for w_i but also for K_i signals. The resulting architecture is shown in Fig. 2 where pipeline registers are marked as rectangles. In this organization new computed values of prekeys w are not only stored in the flip-flops, but in the same cycle they go through the SBoxes evaluating new K_i value which is latched in the extra registers. As a result the longest combinational path (which decides about maximum frequency of operation of the whole unit) now runs from registers K_i to B_{i+1} and does not contain any elements belonging to key computation. Within the key schedule, on the other hand, there are two paths, both originating from *w/w'* flip-flops: the first one computes next values of *w* and the second one extends additionally through S-boxes to the registers K_i. Such a distribution of the elements in the combinational paths turned out to be the most balanced configuration for optimal (highest) performance.

The increase in speed that results from this amendment is accomplished at the cost of 33×128 = 4224 flip-flops but it was shown that this did not incurred any increase in total number of occupied logic cells – all the new registers were located at the outputs of the LUT elements used for implementation of the SBoxes and were absorbed in cells already occupied ([10]).

3.4 Iterative Loop

The two iterative architectures proposed in this study – AI and SI – are based on the structure of one round taken from the fully pipelined architectures (AFP and SFP). Such a single round was supplemented with necessary multiplexing logic (loading the data in – looping back – loading the data out) and a simple controller responsible for counting the repetitions of the loop (round numbers) and supervising the multiplexers. The controller, in its minimal form, comprises a single "idle/busy" register and a round counter. In both architectures number of clock cycles required for encoding one block of data was identical to the number of pipeline stages in AFP / SFP implementations. Every clock cycle completes processing which corresponds to one stage of the pipeline (usually equivalent to one cipher round, apart from the above discussed exceptions for the Serpent case).

One issue needs to be pointed out here, though. While in the AES there is just one SBox transformation used in all rounds in both data and key processing, the Serpent defines 8 different SBoxes, each one being applied in exactly four rounds in the cipher path and in another four rounds in the key path. In iterative organization where just one "universal" round is realized in hardware this means that the "universal" SBox must be created which includes the contents of all 8 regular substitution tables and additionally provides extra 3b input for selection signal. Such a solution is not elegant because, effectively, the SBox becomes a 7-input function (4b data + 3b selection) in place of a 4-input one, which makes its implementation with FPGA resources notably more complicated. For this reason one-round iterative implementation is usually not recommended for Serpent; instead it is proposed to implement 8 rounds in hardware with the data block looped back 4 times during the encoding (8 × 4 instead of 32 × 1). Nevertheless, such organization was not implemented in this study for consistency of examined solutions.

4 Implementation Results

All the 4 above architectures were implemented in Spartan-6 and, for comparison, in the previous family of Spartan-3 devices from Xilinx. There was 8 designs in total (4 for each cipher) and *the same* code was implemented twice in Xilinx ISE Design Suite version 13.4, for the two different target devices selected. Implementation was fully automatic, without any hand-made fine tuning neither in placement nor in layout. Since it turned out that AES pipelined architectures (AHP and AFP) can be optionally implemented with or without utilization of block RAM resources available in the FPGA chip, this gave the total of 10 different cases. The AHP and AFP architectures implemented with block RAM are marked with "_B' suffix.

From Spartan-6 family a middle-sized chip XC6SLX75 was selected as a representative test platform and it served this role very well but selection of Spartan-3 device was more difficult. The initial plan was to use Spartan-3E sub family intended for general, logic-optimized projects. As it soon turned out, even the largest 3E chip – XC3S1600E – was too small for combinational and pipelined AES designs. In other contemporary Spartan-3 families: I/O optimized Spartan-3A, flash-memory based Spartan-3AN and DSP oriented Spartan-3A DSP, only the largest Spartan-3A DSP chips were large enough but this family is optimized for different type of processing. Therefore it was decided to revert to, nowadays somewhat obsolete, initial Spartan-3 family, and to select the XC3S2000 device.

The results are presented in Tables 1 and 2. In general, different types of architectures behave as expected: the combinational organizations give the shortest latency, pipelining is the only way to maximize throughput, and the iterative units

Table 1 Implementation results for the Spartan-6 device (XC6SLX75-3)

	Available	AC	AHP	AHP_B	AFP	AFP_B	AI	SC	SHP	SFP	SI
Slice registers	93296	256	1536	256	2944	1664	817	256	4224	16768	806
Slice LUTs	46648	8997	9087	3946	8884	3376	1367	16888	15523	22029	1566
Slices	11662	2680	2529	1324	2352	1216	493	5243	4590	6629	536
RAMB8s	344			80		86					
F_{max}[MHz]		24.4	195	154	215	168	160	7.95	196	169	180
Latency [Tclk]		1	11	11	11	11	11	1	32	34	34
Latency [ns]		41.0	56.4	71.2	51.2	65.6	68.9	126	163	202	189
Throughput [Gbps]		3.05	24.4	19.3	26.8	20.9	1.81	0.99	24.5	21.1	0.66
Mbps / Slice		1.17	9.87	14.9	11.7	17.6	3.77	0.19	5.46	3.26	1.26
Max path: logic/routing [%]		21 79	28 72	38 62	30 70	38 62	21 79	15 85	25 75	18 82	32 68

Table 2 Implementation results for the Spartan-3 device (XC3S2000-5)

	Available	AC	AHP	AHP_B	AFP	AFP_B	AI	SC	SHP	SFP	SI
Slice registers	40960	271	5061	2771	3913	3913	781	256	4224	16768	783
Slice LUTs	40960	34566	30426	25328	29976	24583	7986	18939	22708	26876	3995
Slices	20480	17428	18799	14274	16103	13220	5948	9900	11793	18377	2145
RAMB16s	40			20		20					
F_{max}[MHz]		11.8	83.5	77.0	106	101	77.0	6.35	143	125	96.2
Latency [Tclk]		1	11	11	11	11	11	1	32	34	34
Latency [ns]		84.8	132	143	104	109	143	158	224	272	353
Throughput [Gbps]		1.47	10.4	9.62	13.2	12.6	0.88	0.79	17.9	15.6	0.35
Mbps / Slice		0.09	0.57	0.69	0.84	0.98	0.15	0.08	1.55	0.87	0.17
Max path: logic/routing [%]		27 73	24 76	30 70	19 81	34 66	28 72	30 70	33 67	28 73	31 69

are unsurpassed if smallest possible resource utilization, at the cost of low performance, is needed. It is worth noting, however, that from all the 4 architectures applied to the two ciphers only the two pipelined AES organizations were implemented with the use of block RAM resources in both Spartan-6 and Spartan-3 chips. In Spartan-6 this resulted in remarkable savings in other resources (slices, registers and LUTs) which utilization dropped roughly by half, but the performance was also affected although not so evidently (approx. 20% drop in the throughput). On Spartan-3 platform, on the other hand, the difference was not so apparent. In cases of other architecture / cipher combinations the implementation tool did not choose to use block RAM units, although the VHDL code did include templates of ROM definitions (for SBox specification) and they were properly detected in reports of the synthesis tool.

Looking at Figs. 3 and 4 we can better evaluate the results and see some remarkable relations. Comparing the effectiveness of AES and Serpent implementations in Spartan-6 it is seen that the AES is able to achieve notable better performance with significantly lower resource utilization: in combinational organization the AES reaches 307% of the Serpent's throughput with 51% of its slice size, while for the fully pipelined and iterative architectures these numbers are, respectively, 127% - 35%, and 275% - 92%. In Spartan-3 family, on the other hand, the relation is different: although generally the AES is able to reach higher levels of throughput (with one exception of the fully pipelined designs), its size is also bigger (again with exception of the xFP units). For combinational, fully pipelined and

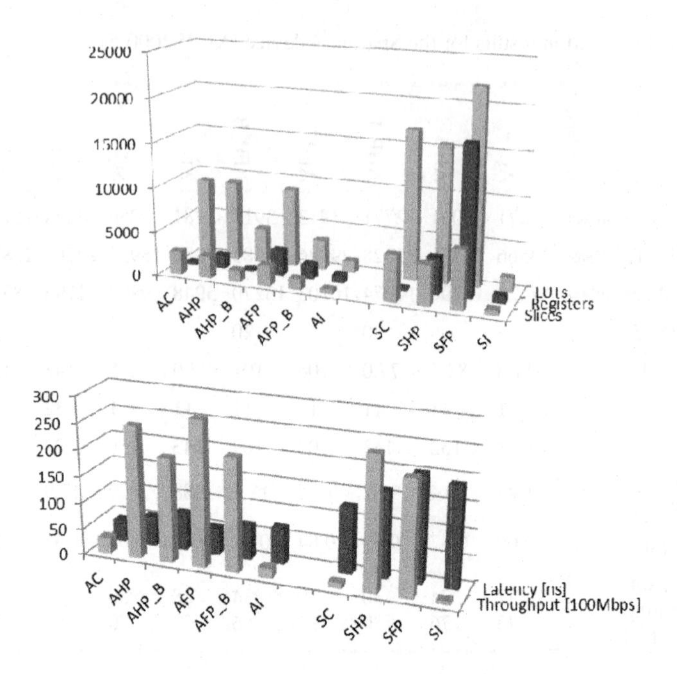

Fig. 3 Size and performance of AES and Serpent implementations in a Spartan-6 device

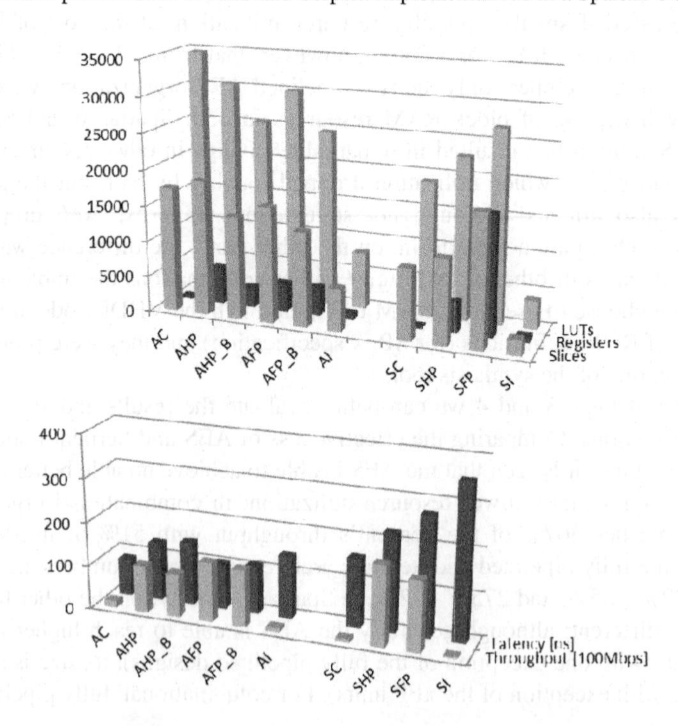

Fig. 4 Size and performance of AES and Serpent implementations in a Spartan-3 device

iterative architectures throughput and slice size ratios are, respectively, 186% - 176%, 84% - 88%, and 247% - 277%. This lead to conclusion that the new architecture of Spartan-6 family is better suited for implementation of the AES operations than the previous one.

This observation is confirmed by Fig. 5. In this graph we visualize size and performance not across different architectures implemented on the same platform but between the two platforms. What is instantly seen from the bars in the last row (slice ratio) is that while for AES the size ratios are in the range from 6.5 to 12 (meaning that the slice size is bigger by this factor in Spartan-3 than in Spartan-6), for Serpent these ratios are from 1.9 to 6 – so switching form Spartan-3 to Spartan-6 is much more beneficial for the AES implementations. Also the throughput ratio is in the range of 0.4 ÷ 0.6 for AES and 0.5 ÷ 08 for Serpent, indicating that the new family brings more progress for the Rijndael cipher.

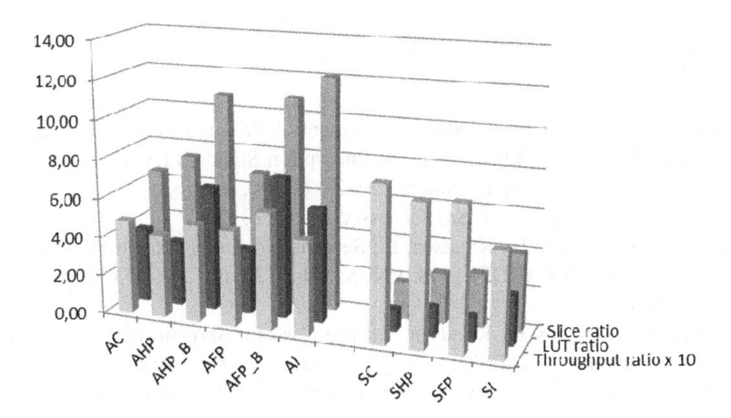

Fig. 5 Spartan-3 vs. Spartan-6 – ratio of performance and size parameters

This difference between the two methods can be explained looking at the most resource-hungry elementary operation in FPGA: the substitution function. In Spartan-3 every output bit of the AES SBox, being an 8-input function, requires 256 / 16 = 16 4-input LUTs for storing the substitution table plus some additional LUTs for multiplexing their outputs (in terms of ROM organization: for address decoding). Even not counting the extra multiplexing logic this needs 128 x 16 = 2048 LUTs in each round and 20480 LUTs in the entire 10-round cipher (vs., for example, total of 40960 LUTs in the whole mighty XC3S2000 chip). On the other hand, LUT elements in Spartan-6 are truly 6-input tables (but can also be configured as two 5-input LUTs what provide some amount of flexibility) so every SBox output is generated by 256 / 64 = 4 LUTs with much simpler multiplexing (which can be done in dedicated fast multiplexers and not in LUTs) and the whole ciphers needs 4 x 128b x 10 rounds = 5120 LUTs. In Serpent, in contrast, the SBoxes are 4-input functions so they perfectly fit already in Spartan-3 LUTs and moving to Spartan-6 does not bring any improvements in this aspect – in fact, Spartan-6 LUTs generating Serpent SBox functions are utilized not to their full potential.

5 Conclusions

We take for granted that new generations of FPGA chips bring larger sizes and faster operation but new architectural developments can sometimes change more than mere design performance and utilization parameters. In case of AES and Serpent ciphers new organization of array resources that was introduced with Spartan-6 family, especially larger Look-Up Tables used for generation of combinational functions in the design, substantially changed feasibility of various implementation options of the ciphers.

In the old Spartan-3 devices combinational and pipelined organizations of the AES units were unacceptable resource hungry and Serpent, despite much higher number of rounds, was a better option for these kinds of processing. The new Spartan-6 chips changed this situation and, effectively, advantage of the Serpent algorithm is again mainly in its better cryptographic strength.

References

[1] Anderson, R., Biham, E., Knudsen, L.: Serpent: A Proposal for the Advanced Encryption Standard. In: The First Advanced Encryption Standard (AES) Candidate Conference, Ventura, California, August 20–22 (1998), http://www.cl.cam.ac.uk/~rja14/serpent.html (accessed March 2012)

[2] Anderson, R., Biham, E., Knudsen, L.: Serpent and Smartcards. In: Quisquater, J.-J., Schneier, B. (eds.) CARDIS 1998. LNCS, vol. 1820, pp. 246–253. Springer, Heidelberg (2000)

[3] Anderson, R., Biham, E., Knudsen, L.: The Case for Serpent. In: Third AES Candidate Conference (AES3), New York, USA, April 13–14 (2000), http://csrc.nist.gov/archive/aes/index.html (accessed March 2012)

[4] Chu, P.P.: RTL Hardware Design Using VHDL. John Wiley & Sons, New Jersey (2006)

[5] Gaj, K., Chodowiec, P.: Comparison of the hardware performance of the AES candidates using reconfigurable hardware. In: Third AES Candidate Conference (AES3), New York, USA, April 13–14 (2000), http://csrc.nist.gov/archive/aes/index.html (accessed March 2012)

[6] Krukowski, Ł., Sugier, J.: Designing AES cryptographic unit for automatic implementation in low-cost FPGA devices. Int. J. Critical Computer Based Systems 3(1/2/3), 104–116 (2010)

[7] Lázaro, J., Astarloa, A., Arias, J.R., Bidarte, U., Cuadrado, C.: High Throughput Serpent Encryption Implementation. In: Becker, J., Platzner, M., Vernalde, S. (eds.) FPL 2004. LNCS, vol. 3203, pp. 996–1000. Springer, Heidelberg (2004)

[8] Liberatori, M., Otero, F., Bonadero, J.C., Castineira, J.: AES-128 Cipher. High Speed, Low Cost FPGA Implementation. In: Proc. Third Southern Conference on Programmable Logic, Mar del Plata. IEEE Comp. Soc. Press, Argentina (2007)

[9] Mroczkowski, P.: Implementation of the block cipher Rijndael using Altera FPGA. Military University of Technology, Warsaw (2000)

[10] National Institute of Standards and Technology, Specification for the Advanced Encryption Standard (AES). Federal Information Processing Standards Publication 197 (2001), `http://csrc.nist.gov/publications/PubsFIPS.html` (accessed March 2012)
[11] Osvik, D.A.: Speeding up Serpent. In: Third AES Candidate Conference (AES3), New York, USA, April 13–14 (2000), `http://csrc.nist.gov/archive/aes/index.html` (accessed March 2012)
[12] Piwko, K.: Hardware implementation of cryptographic algorithms in programmable logic devices. Dissertation for M.Sc. degree, Wrocław University of Technology, Faculty of Electronics (2010)
[13] RSA Laboratories, DES Challenges (1997-1999), `http://www.rsa.com`
[14] Sugier, J.: Low-cost hardware implementation of Serpent cipher in programmable devices. In: Monographs of System Dependability Technical Approach to Dependability, vol. 3, pp. 159–172. Publishing House of Wrocław University of Technology (2010)
[15] Sugier, J.: Implementing Serpent cipher in field programmable gate arrays. In: The 5th International Conference on Information Technology, ICIT 2011, Amman, Jordan, May 11-13, pp. 91–96 (2011)
[16] Wójcik, M.: Effective implementation of Serpent algorithm. Dissertation for M.Sc. degree, Faculty of Electronics and Information Technology, Warsaw University of Technology (2007)

Dependable Strategies for Job-Flows Dispatching and Scheduling in Virtual Organizations of Distributed Computing Environments

Victor Toporkov, Alexey Tselishchev, Dmitry Yemelyanov, and Alexander Bobchenkov

Abstract. This work presents dispatching strategies based on methods of job-flow and application-level scheduling in virtual organizations of distributed computational environments with non-dedicated resources. Dependable job-flow management is implemented with the set of specific rules for resource usage. Strategies are based on economic scheduling models and diverse administration policies inside resource domains. Job management structures and economic mechanisms for load balancing in distributed environments are considered. Scheduling methods composing priority algorithms for parallel applications and job batch scheduling in distributed computing with non-dedicated resources are proposed.

1 Introduction

Distributed computational environments such as Grid have been known for significant efficiency increase in shared computational resource usage and provision of scientific and enterprise communities with solutions for complex computational tasks. However, those who are responsible for setting up Grid infrastructure and economy encounter difficulties while defining policies and strategies for efficient resource management and job scheduling. The problem of establishing an optimal or at least good strategy based on current environment condition remains actual and prominent at the moment in the domain of distributed computing.

Victor Toporkov · Dmitry Yemelyanov · Alexander Bobchenkov
National Research University "MPEI", ul. Krasnokazarmennaya 14, Moscow, 111250 Russia
e-mail: ToporkovVV@mpei.ru,
{groddenator,yemelyanov.dmitry}@gmail.com

Alexey Tselishchev
CERN (European Organization for Nuclear Research),
CERN CH-1211 Genève 23 Switzerland
e-mail: Alexey.Tselishchev@cern.ch

W. Zamojski et al. (Eds.): Complex Systems and Dependability, AISC 170, pp. 289–304.
springerlink.com

Heterogeneity, changing composition, different owners of different nodes whose computing time is partially shared by users turn the organization of a distributed computational environment into an especially difficult task. Utility grid [1], multi-agent systems [2] and cloud computing [3] are types of distributed environments where usage of economic mechanisms is seen as promising. Those economic mechanisms are designed to solve tasks like resource management and scheduling of user jobs in a transparent and efficient way. Within the context of any used economic model the interests of different participants of a distributed computing environment (such as end-users or node owners) are often contradictory. Since the resources of distributed environment such as Grid are non-dedicated, it is assumed that node owners may have local job flows (their own tasks) and global job flow (which is formed by external user jobs) competing for limited computational resources of the node. Elaboration of pricing rules which are used to calculate a fee for node computing time usage and take into account user-required quality of service (QoS) is also a very serious problem [1-3]. An overview of various approaches to this problem is given in [4]. Heuristic algorithms for resource selection based on user-given utility function are described in [5]. Some resource management models offer simple search and selection of resources required by a user [6] and do not support any optimization. Others do not take into account features related to global and local job competition, the competition among users and other characteristics of distributed environments with non-dedicated computational resources [7]. A resource broker model [1-5] dynamically employs various economic policies which perform resource management which is decentralized and application-specific and have two parties: node owners and brokers representing users. Another common trend is related to virtual organizations [7-9] with central schedulers providing job-flow level scheduling and optimization. While former type of resource management is well-scalable, the simultaneous satisfaction of various application optimization criteria submitted by independent users is unreachable in essence and also can deteriorate such integral quality of service rates as total execution time of a sequence of jobs or overall resource utilization. The latter type, virtual organizations naturally restrict the scalability. However, scheduling based on uniform and controlled rules for allocation and consumption of resources makes it possible to improve the efficiency of resource usage and find a tradeoff between contradictory interests of different participants.

In this work, we propose two-level model of resource management system which is functioning within a virtual organization (VO). Resource management is implemented with a hierarchical structure consisting of a metascheduler and subordinate job schedulers that are controlled by the metascheduler and in turn interact with resource managers (e.g., with batch job processing systems). The application-level optimization begins when job-flow level optimization is finished. Such a flexible structure coupled with complex metascheduling approach enables multiaspect resource management and makes possible to control dynamic priority of job execution, resource selection and provide multicriterial optimization both on the job-flow scale and for specific job, according to its submitter requirements and optimization criteria. Hence, we may speak not only of a scheduling algorithm but rather of a scheduling strategy that is a combination of various methods of external and local scheduling. Such a mechanism allows finer control and higher overall resource management efficiency in a distributed computing environment. *Resource* is defined as an abstract computational entity, which can be used for execution of one and only one *task*. The complex set of connected interrelated tasks form a *job*. In some applications jobs require co-scheduling and resource co- allocation on several resources [10-13]. In this case resource allocation has a number of substantial specific features caused by autonomy, heterogeneity, dynamic content changes, and node failures [6-9]. In our model jobs are submitted to the system by end-users. The proposing approach is more or less the same as used in gLite Workload Management System, where Condor is used as a scheduling module [14]. But the significant difference between the approach proposed in this work and well-known scheduling solutions for distributed environments such as the Grid [1, 3-7] is the fact that the execution strategy is formed on a basis of formalized efficiency criteria, which efficiently allows to reflect economic principles of resource allocation by using relevant cost functions and solving a load balance problem for heterogeneous processor nodes. At the same time the inner structure of the job is taken into account when the resulting schedule is formed. Thus, two approaches are uniquely combined in a proposed two-tier model.

This work is organized as follows. Section 2 overviews model components and metascheduling workflow. In section 3 a strategy search is formalized. Section 4 contains simulation results. Section 5 summarizes the work and describes further research topics.

2 Basic Notions and Informal Model Components Description

Let us define basic model components presented in this work.

- VO, that defines resource co-allocation dispatching strategies, pricing policies and resource load-balancing mechanisms.
- Heterogeneous hierarchical computational environment that contains computational resources (Grid nodes, CPUs or others) with different performance indices. Each resource is considered as non-dedicated (i.e. it can have its own internal schedule and these schedules are sent to application-level schedulers upon request).
- Metascheduler, which implements resource management strategies and policies of the virtual organization.
- Application-level schedulers that analyze internal job structure and schedule single tasks.

The VO in our model of distributed computational environment includes three independent parties with their own interests.

- End-users of services provided within the VO such as computation services. End-users take steps to make resource requests to the environment, according to resource performance, time and budget estimations needed for running custom user jobs.

- VO administrators that set up resource usage policies to optimize scheduling and improve load balance. The administrators control metascheduler process running in the environment which is in fact the part of VO infrastructure software. Thus they are directly responsible for managing the parameters of higher level resource management.
- Owners of computational nodes that comprise the environment network and hardware base of the distributed computing environment. The owners offer part of their nodes computing time to VO for a fee. Computational nodes provide the only type of distributed resources used in our model.

Each computational node of the heterogeneous environment is mapped to a computational *resource line* in the metascheduler resource management routine. Several resource lines are combined into a virtual resource domain. Each resource line has two static attributes which are its performance P and its base price tag F for a computing time unit. The performance is an inherent parameter of a node and the base price tag is assigned by its owner. The dynamic characteristic of a node is represented with its local schedule which is a list of slots available for reservation. This list is sent to metascheduler by request. A slot is a continuous interval of time and is described with three parameters: its start time, its length and its fee [10-12]. The fee is calculated when the metascheduler applies its pricing policies taking in account resource type, slot length etc.

A resource request is a set of a few constraints determined by a user which correspond to the properties of the respective user job. They include:

a) minimal performance requirement for computational nodes, P_{min};

b) maximal price tag for a single timeslot, F_{max};

c) number n of simultaneously reserved timeslots;

d) minimal slot length;

e) the internal structure of a job as a directed acyclic graph (DAG), where vertices represent single tasks and edges represent data dependencies [13];

f) deadline for the job execution.

A job may require more than one timeslot if it includes several segments that can be executed in parallel way, for instance. Then the user specifies the number of reserved timeslots and minimal performance requirement that applies for them all. The whole job budget is determined by the timeslot number and the maximum price per timeslot. The minimal timeslot length requires an additional explanation. This is the minimal time estimated by the user which is required to complete job execution given the performance of the nodes meet the minimal requirement P_{min}. Hence, the metascheduler and the user share the responsibility since the probability of being run successfully for a job equally depends on primary user estimates and overall scheduling quality.

The hierarchical model of the computational environment implies two-tier scheduling (Fig. 1). On the job-flow level the set of independent jobs is distributed between resource domains according to dispatching strategies and economic criteria. Schedule on this level is defined by the metascheduler as a slot set for each job, which is optimal in terms of a whole job set. Application-level schedulers receive the list of resources which were meant to execute the job on and a strategy, which defines the rule used to execute tasks of a concrete job. On this level an optimal slot and specific resource are defined for each single task in a job, thus, making it possible to take internal job structure into account. On the job-flow level all end-user jobs are initially submitted into the global queue. The metascheduler can manage one or more job-flows which become sub-queues of the global queue. The mechanism of distribution of jobs between job-flows can be random or based on current load and actual efficiency of scheduling in certain job-flows. Scheduling process in each job-flow is performed by identical scheduling instance. We consider a single job-flow case.

The metascheduler works in cycles which are quanta of its process. For each cycle it has following information.

1. Information about distributed computing environment as a set of resource lines.
2. The global job queue.

What it needs then is a batch of jobs which is a ranked job list and a subset of available slots for a specific virtual resource domain and a certain timeframe which is called a scheduling interval. The length of the batch and the scheduling interval are parameterized by VO administrators.

Jobs are fetched into the batch accordingly to several variables, such as the maximum price tag, deadline, and the number of failed scheduling attempts for a job. These variables being weighted and added up determine job rank according to which it takes a position closer to head or tail of a batch.

The preparation phase ends and the actual scheduling process is executed as follows (see Fig. 1).

1. The metascheduler analyzes available slots and finds an optimal slot combination to accommodate every job in a batch using economic criteria. The budget and the deadline defined by the end-user are considered during this step. The algorithms for this step were detailed in [10-12].
2. After the domain is determined metascheduler defines the strategy for each job. For example as shown on Fig. 1, the user, who has sent the job i has the higher budget than the one who has sent the job k. The strategy for i may be expressed as "*execute as soon as possible*" while the strategy for k may be expressed as "*execute as late as possible within the defined deadline*". These jobs are later sent to application-level schedulers and the application-level scheduling begins.
3. Application-level schedulers query internal schedules for all the resources which were selected during step 2 for each job, analyze the job DAG and form a resulting schedule for every task according to the strategy from step 2. These schedules must support interruptions and delays and should be optimal in terms of the defined criteria (i.e. cost or resource load). The criterion for the job i would

be to minimize execution cost within the defined budget, criterion for the job k would be to maximize average resource load while meeting the defined deadline. As shown on Fig. 1, jobs i and k are scheduled to be executed on the same set of resources at once.

4. Application-level schedulers are guaranteeing that there are no collisions between the tasks which were scheduled during step 3 and local tasks, which may have priority over the job-flow from step 1.

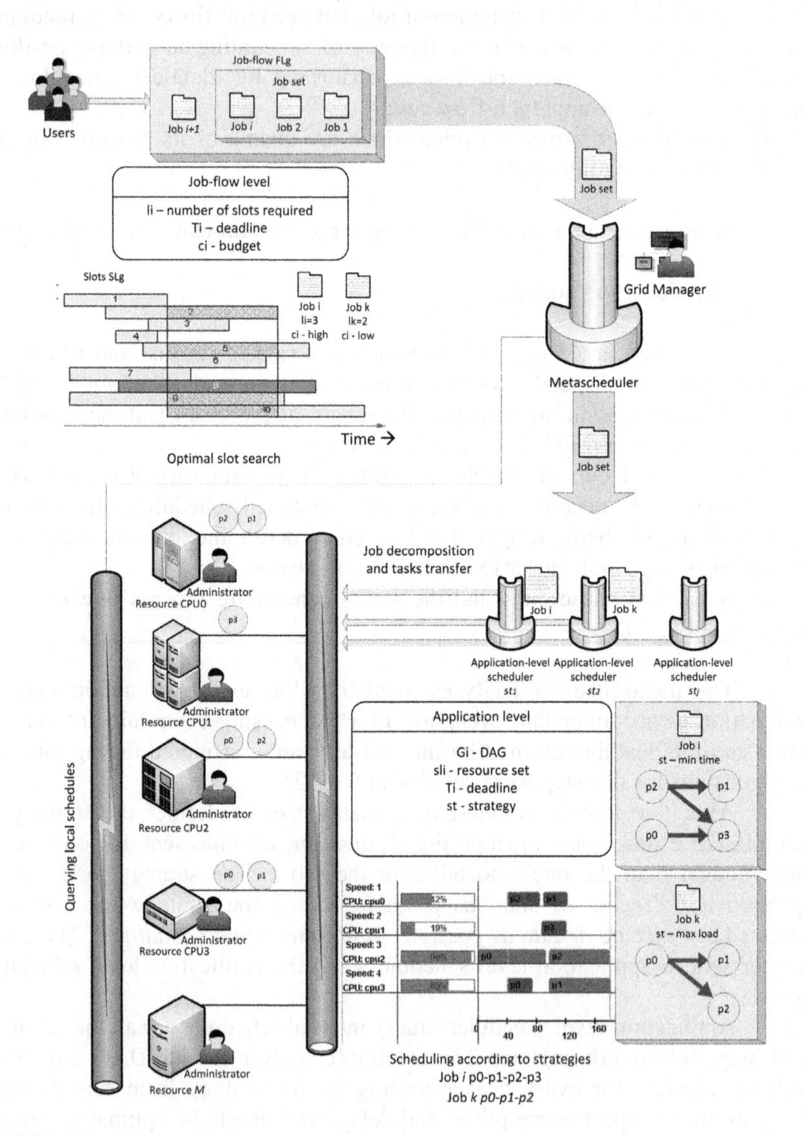

Fig. 1 Model components

3 Formalization of Scheduling

Let us note a global resource set $R_g = \{r_p, p = 1,..,M\}$,, which includes all resources. A global job-flow is a set of jobs received by the metascheduler in time: $FL_g = \{l_i, c_i, T_i, G_i, i = 1,..,I\}$, where the job i is represented as l_i – the amount of resource slots required, c_i - the maximal budget end-user is ready to allocate for execution of the job, T_i – deadline, G_i – the job DAG. Metascheduler at any time moment may query each resource, receive its local schedule and build a set of slots S_{gt} – idle time intervals.

Let us introduce a set of strategies $ST = \{st_l, l = 1,..,L\}$, which are based on economic criteria and are defined by Grid-managers and developers. Let SL be a set of K slots suitable to execute a subset of jobs $FL_p \subseteq Fl_g$. A slot set is considered as suitable for the job i if the execution is possible in terms of the resource number, the budget c_i and the deadline T_i . It is assumed that for every job there is at least one suitable slot set $sl_i \in SL, sl_i = k, k \in \{1,..,K\}$.

On a job-flow level for each job the metascheduler aims at finding a slot set sl_i and a strategy st_i for which the value of the function $g_i(sl_i)$, that defines whether the slot set is being effective for the job i, would be optimal [11]. The internal job structure G_i is not taken into account at this time. The mechanism to define $g_i(sl_i)$ which was developed in the previous works [10-12] is now improved. According to the resource request it is required to find a “window” with the following description: n concurrent time-slots providing resource performance rate at least P and maximal resource price not higher than $F_{\max}$ should be reserved for a time span T_i (the resource request type was described in more detail above). The length of each slot in the window is determined by the performance rate of the node on which it is allocated. Thus as a result we have a window with a “rough right edge” (Fig. 2). In addition, the criterion of selecting the most suitable set of slots could be specified. This could be the minimum cost, the minimum runtime or, for example, the minimum power consumption criterion. The window search is performed on the list of all available system slots sorted by their start time in ascending order (this condition is necessary to examine every slot in the list and for operation of search algorithms of linear complexity [10-12]).

The scheme of a search for a window that meets the requirements and effective by the given criterion can be represented as follows.

1°. From the list of available system slots the next suitable slot s_k is extracted and examined. Slot s_k suits, if following conditions are met:

a) resource performance rate $P(s_k) \geq P$ for slot s_k ;

b) slot length (time span) is enough (depending on the actual performance of the slot's resource) $L(s_k) \geq T_i * P(s_k) / P$.

If conditions **a)** and **b)** are met, the slot s_k is successfully added to the window list.

2°. A current window start time is a set equal to the start time of the last added slot.

3°. Slots whose length has expired considering new window start time T_{last} are removed from the list. The expiration means that remaining slot length $L'(s_k)$, calculated like shown in **step 1°b**, is not enough assuming the k -th slot start is equal to the last added slot start: $L'(s_k) < (T_i + (T_{last} - T(s_k)))P(s_k)/P$, where $T(s_k)$ is the slot's start time. Any combination of the remaining slots can form a window of necessary length.

4°. If the number of slots *m* in the current window is greater or equal to n, it is required to select n slots, effective on the specified criteria and at the same time satisfying the total cost and deadline restrictions. Suppose the window W of size n with a target criterion value equal to crW was selected. The problem of selecting efficient window consisting of n slots in the case of $m > n$ will be described below.

5°. The target criterion value crW of window *W* is compared with the cr' – the current best target criterion value for all previously found windows. If $crW < cr'$ (in case of a minimization problem) the window W announced as a new window-candidate and crW becomes the new best criteria value: $cr' = crW$. Go to **step 1°**.

6°.The algorithm ends after the last available slot is processed. The result of the algorithm is the window-candidate with the best target criteria value.

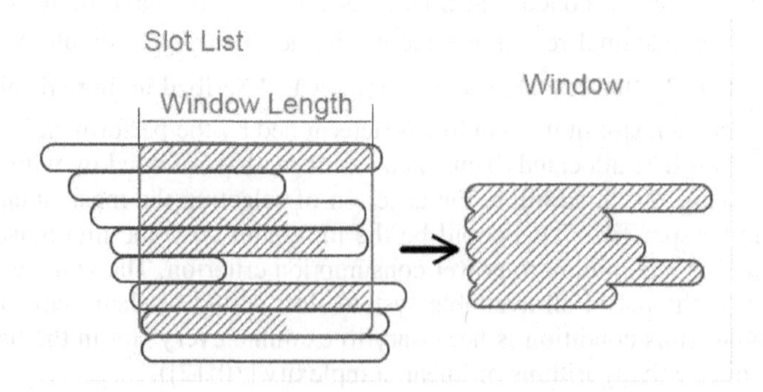

Fig. 2 Window with a "rough right edge"

The described algorithm can be compared to the algorithm of maximum/minimum value search in an array of flat values. The expanded window of size *m* "moves" through the ordered list of available system slots. At each step any combination of n slots inside it (in case when $n \le m$) can form a window that meets all the requirements to run the job. The effective on the specified criteria window of size n is selected from this m slots and compared with the results in the previous steps. By the end of the slot list the only solution with the best criteria value will be selected. Consider the problem of selecting a window of size n with a total cost no more than S from the list of $m > n$ slots (in case when $m = n$ the selection is trivial). The maximal budget is counted as $S = Ft_s n$, where t_s is a time span to reserve and n is the necessary number of slots. The current extended window consists of m slots $s_1, s_2, ..., s_m$. The cost of using each of the slots according to their required length is: $c_1, c_2, ..., c_m$. Each slot has a numeric characteristic z_i the total value of which should be minimized in the resulting window.

Then the problem could be formulated as follows:

$$a_1 z_1 + a_2 z_2 + ... + a_m z_m \rightarrow \min, \quad a_1 c_1 + a_2 c_2 + ... + a_m c_m \leq S,$$

$$a_1 + a_2 + ... + a_m = n, \; a_r \in \{0,1\}, \; r = 1, ..., m.$$

Additional restrictions can be added, for example, considering the specified value of deadline. Finding the coefficients $a_1, a_2, ..., a_m$ each of which takes integer values 0 or 1 (and the total number of '1' values is equal to n), determine the window with the specified criteria extreme value. Job-flow level scheduling ends here.

Application-level schedulers receive following input data.

- The optimal slot set sl and the description of all corresponding resources:

$$R = \{r_j, j = 1,.., J\} \subseteq R_g .$$

- The directed acyclic information graph $G = \{V, E\}$, where $V = \{v_i, i = 1,..n\}$ is a set of vertices that correspond to job tasks, for each of those execution time estimates τ_{ij}^0 on each of resources in R are provided, E – is a set of edges that define data dependencies between tasks and data transfer time intervals.
- The dispatching strategy st, which defines the criterion for a schedule expected
- The deadline T_i or the maximal budget c_i for the job (depends on a dispatching strategy and $g_i(sl_i)$.

The schedule which is being defined on an application level is presented as follows: $Sh = \{[s_i, f_i], \alpha_i, i = 1,.., n\}$, where $[s_i, f_i]$ is a time frame for a task i of a job and α_i - defines the selected resource. Sh is selected in the way that the criterion function $C = f(Sh)$ achieves an optimum value. The *critical jobs method* [13] which is used to find the optimal schedule and to define f consists of three main steps.

- Forming and ranging a set of critical jobs (longest sets of connected tasks) in the DAG.
- Consecutive planning of each critical job using dynamic programming methods.
- Resolution of possible collisions.
  Detailed algorithm description is presented in [13].

4 Simulation Results

The two-tier model described in the sections 2 and 3 was implemented in a simulation environment on two different and separated levels: on the job-flow level, where job-flows are optimally distributed between resource domains and on the application level, where jobs are decomposed and each task is executed in an optimal way on a selected resource.

4.1 Job-Flow Level Scheduling Simulation Results

Job-flow level metascheduling was simulated in a specially implemented and configured software that was written to test the features of the two-tier resource management.

An experiment was designed to compare the performance of our job-flow level metascheduling method with other approaches such as FCFS and backfilling. Let us remind that our scheduling method detailed in works [10] and [11] involves two stages that backfilling does not have at all, namely, slot set alternative generation and further elaboration of specific slots combination to optimize either time or cost characteristic for an entire job batch. Backfilling simply assigns "slot set" found to execute a job without an additional optimization phase. This behavior was simulated within our domain with random selection from an alternative slot, each job having one or more of them. So two modes were tested: with optimization ("OPT") and without optimization ("NO OPT").

The experiment was conducted as follows. Each mode was simulated in 5000 independent scheduling cycles. A job batch and environment condition was regenerated in every cycle in order to minimize other factor influence. A job batch contained 30 jobs. Slot selection was consistent throughout the experiment. If a job resource request could not be satisfied with actual resources available in the environment, then it was simply discarded.

For optimization mode as well as for no-optimization mode four optimization criteria or problems were used:

1. Maximize total budget, limit slot usage.
2. Minimize slot usage, limit total budget.
3. Minimize total budget, limit slot usage.
4. Maximize slot usage, limit slot budget.

Results presented in Table 1 apply for the problem 1. As one can see optimization mode, which is using additional optimization phase after slot set generation wins against random slot selection with about 13% gain in the problem 1 whose concern is about maximizing total slot budget thus raising total economical output per cycle and owners' profits.

Table 1 Experimental results for the problem 1: Total budget maximization with limited slot usage

Mode	Average jobs being processed per cycle (max 30)	Average total slot cost per cycle, *cost units*	Average total slot usage per cycle, *time units*	Average slot usage limit per cycle, *time units*
OPT	20.0	11945.98	421.22	471.14
NO OPT	20.0	10588.53	459.36	471.85

Comparable results were obtained for other problems which are summarized in Table 2. Optimized values are outlined in light grey.

Table 2 Experimental results for the problems 2-4

Mode	Average jobs being processed per cycle (max 30)	Average total budget (slot cost) per cycle, *cost units*	Average total slot usage per cycle, *time units*	GAIN, %
Problem 1: Maximize total budget, limit slot usage				
OPT	20.0	11945.9	421.2	+12.8
NO OPT	20.0	10588.5	459.4	
Problem 2: Minimize slot usage, limit total budget				
OPT	12.4	7980.4	300.9	+10.6
NO OPT	12.4	7830.9	332.8	
Problem 3: Minimize total budget, limit slot usage				
OPT	15.1	9242.4	410.057	+6.2
NO OPT	15.3	9813.9	406.612	
Problem 4: Maximize slot usage, limit total budget				
OPT	15.28	9870.8	416.835	+3.0
NO OPT	15.4	9718.1	404.8	

These results are showing the advantage of the metascheduling on the job-flow level. The next section describes the experiments on the application level.

4.2 Application Level Scheduling Simulation Results

The experiment results presented in Table 3 shows the advantage of the critical jobs method usage in a two-tier scheduling model compared to consecutive application-level scheduling. Here k=0.75 means that each job is sent to be scheduled after 75% of the time allocated for the previous one: while the scheduling cost for a job is more or less the same, 1000 jobs are planned 25% faster.

Consider another experiment: while changing the length of the scheduling interval, we will estimate the proportion of successfully distributed jobs. The length of the scheduling interval is equal to $L = l * h, h = 1.0,..,2.6$, with step 0.2, where l is the length of the longest critical path of tasks in the job and h is a distribution interval magnification factor. There were carried 200 experiments for each h (bold points on Fig. 3). Analysis of the Fig. 3 shows that increasing the scheduling interval (relatively to the execution time of the longest critical path on the nodes with the highest performance) is accompanied by a significant increase in the number of successfully distributed jobs. The detailed study of this dependence can give a priori estimates of an individual job successful distribution probability.

Table 3 Two-tier model vs consecutive application-level scheduling

Parameter	Application-level scheduling	Two-tier model (k=0.75)
Jobs number	1000	1000
Execution time	531089 time units	399465 time units
Optimal schedules	687	703
Mean collision count	3.85	4.41
Mean load (forecast)	0.1843	0.1836
Mean load (fact)	0.1841	0.1830
Mean job cost	14.51 units	14.47 units

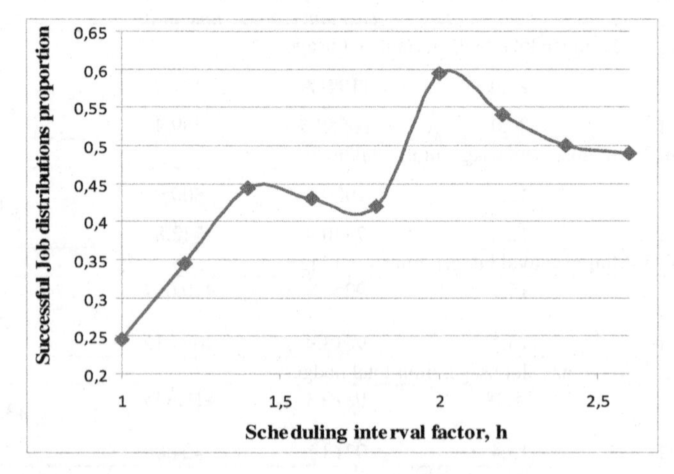

Fig. 3 Dependence of the proportion of the successful job distributions on the length of the distribution interval

In the next experiment we will consider the dependence of successful distributions number and the number of collisions per experiment on the level of resource instances availability. The experiments were performed in conditions of limited resources using the specific instances of the resources. The number of resources J in each experiment was determined as $J = j * N$, where j – factor (x-axis) and N – number of tiers in the graph. Fig. 4 shows results of the experiments with different j values and $N = 3,5,7$.

The obtained dependencies (Fig. 4) suggest that the collisions number depends on the resources availability. The lower the number of resource instances and the greater the number of tiers in the graph – the more collisions occurred during the

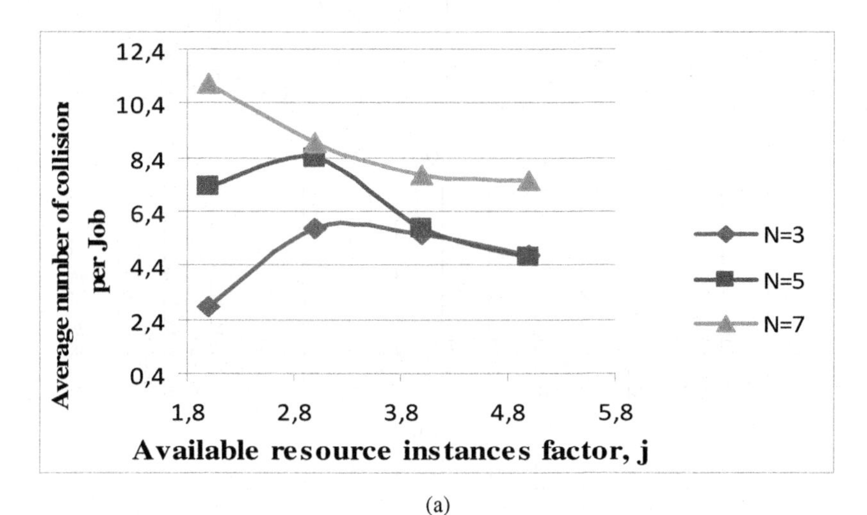

(a)

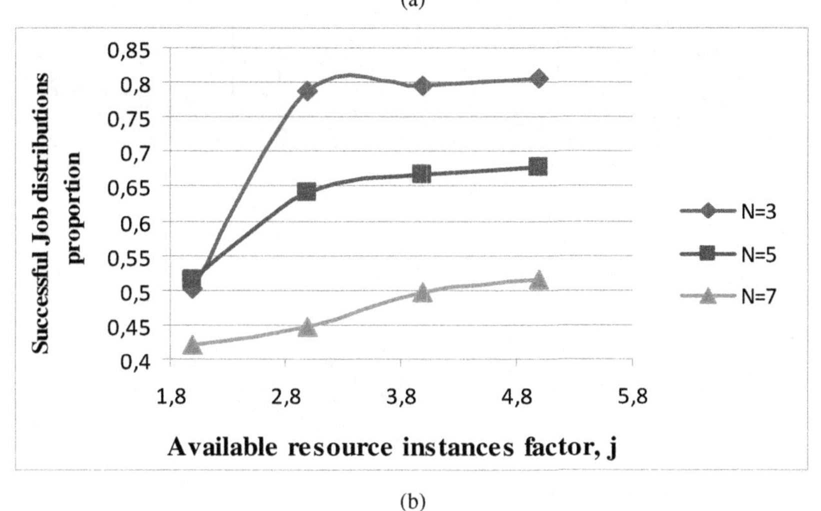

(b)

Fig. 4 Simulation results: resource dependencies of collisions number (a) and successful job distribution proportion (b) scheduling. At the same time the number of resource instances affects the successful distribution probability. With a value of $j > 4$ (that is, when the number of available resource instances is more than 4 times greater than the number of tiers in the graph) all cases provide the maximum value of successful distribution probability. These results are subject of future research of refined strategies on a job-flow level.

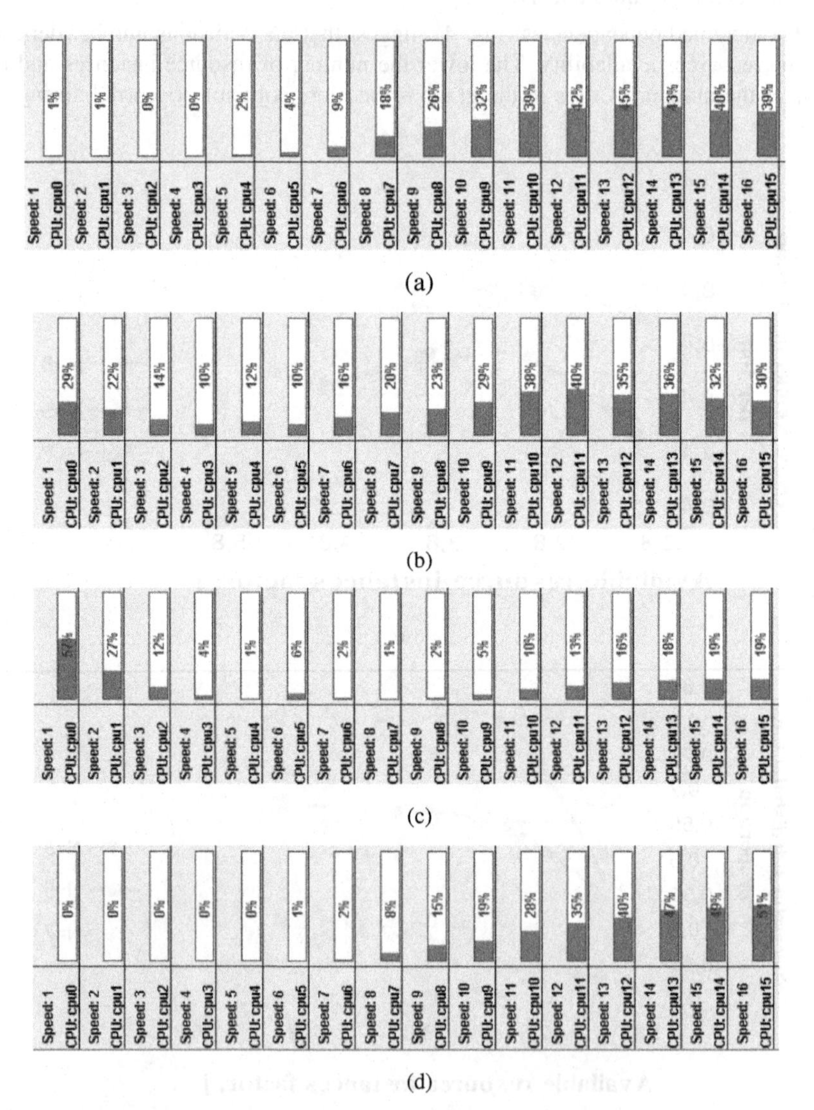

Fig. 5 Resource utilization level balancing: utilization maximization with $h = 1.66$ (a) and $h = 1.2$ (b), utilization minimization and distribution cost maximization (c), distribution cost minimization (d)

The next series of experiments aimed at identifying the priorities of selecting certain resource instances with different optimization criteria and various restrictions. Figures 5 (a-d) show resource utilization levels in the following problems: resource load balancing, distribution cost minimization and maximization. The scheduling interval is defined as $L = l * h, h = 1.66$ and 1.2, where l is execution time of the longest critical path and h is a scheduling interval factor. Processors with greater number have relatively lower cost and performance level. To maximize average resource utilization the priority is given to processors with relatively low performance (Fig. 5 (a)). In case of a shorter scheduling interval ($h = 1.2$) there is need to use resources with higher performance (Fig 5 (b)). During the resource utilization minimization and distribution cost maximization the priority is given to the nodes with higher performance and usage cost (Fig. 5 (c)). During the distribution cost minimization the priority is given to processor with low performance level and correspondingly low cost. These experiments show how strategies defined on a job-flow level are implemented on an application level, how flexible the strategies can be and how can resource load be controlled by the metascheduler.

5 Conclusions and Future Work

In this work, we address the problem of independent job-flow scheduling in heterogeneous environment with non-dedicated resources.

Each job consists of a number of interrelated tasks with data dependencies. Using the combination of existing methods with a number of original algorithms the resulting schedules are computed. These schedules meet the defined deadlines and budget expectations, provide optimal load-balance for all the resources and follows virtual organization's strategies, thus, allowing to achieve unprecedented quality of service and economic competitiveness for distributed systems such as Grid. The experiments which were conducted are showing the efficiency of methods developed for both job-flow and application level scheduling. The model proposed is showing the way these methods and advantages can be converged in one place making it possible to achieve the main goal.

Future research will include the simulation of connected job-flow and application levels and experiments on real Grid-jobs in order to get finer view on advantages of the approach proposed.

Acknowledgements. This work was partially supported by the Council on Grants of the President of the Russian Federation for State Support of Leading Scientific Schools (SS-316.2012.9), the Russian Foundation for Basic Research (grant no. 12-07-00042), and by the Federal Target Program "Research and scientific-pedagogical cadres of innovative Russia" (State contracts 16.740.11.0038 and 16.740.11.0516).

References

[1] Garg, S.K., Buyya, R., Siegel, H.J.: Scheduling parallel applications on utility Grids: time and cost trade-off management. In: Proc. of ACSC 2009, Wellington, New Zealand, pp. 151–159 (2009)

[2] Tesauro, G., Bredin, J.L.: Strategic sequential bidding in auctions using dynamic programming. In: Proc of the First International Joint Conference on Autonomous Agents and Multiagent Systems: part 2, pp. 591–598. ACM, New York (2002)

[3] Garg, S.K., Yeo, C.S., Anandasivam, A., Buyya, R.: Environment-conscious scheduling of HPC applications on distributed cloud-oriented data centers. J. of Parallel and Distributed Computing 71(6), 732–749 (2011)

[4] Buyya, R., Abramson, D., Giddy, J.: Economic models for resource management and scheduling in Grid computing. J. of Concurrency and Computation: Practice and Experience 14(5), 1507–1542 (2002)

[5] Ernemann, C., Hamscher, V., Yahyapour, R.: Economic Scheduling in Grid Computing. In: Feitelson, D.G., Rudolph, L., Schwiegelshohn, U. (eds.) JSSPP 2002. LNCS, vol. 2537, pp. 128–152. Springer, Heidelberg (2002)

[6] Voevodin, V.: The Solution of Large Problems in Distributed Computational Media. Automation and Remote Control. Pleiades Publishing, Inc. 68(5), 773–786 (2007)

[7] Kurowski, K., Nabrzyski, J., Oleksiak, A., et al.: Multicriteria aspects of Grid resource management. In: Nabr-zyski, J., Schopf, J.M., Weglarz, J. (eds.) Grid Resource Management. State of the Art and Future Trends, pp. 271–293. Kluwer Acad. Publ. (2003)

[8] Toporkov, V.: Application-Level and Job-Flow Scheduling: An Approach for Achieving Quality of Service in Distributed Computing. In: Malyshkin, V. (ed.) PaCT 2009. LNCS, vol. 5698, pp. 350–359. Springer, Heidelberg (2009)

[9] Toporkov, V.V.: Job and application-level scheduling in distributed computing. Ubiquitous Comput. Commun. J. 4, 559–570 (2009)

[10] Toporkov, V., Toporkova, A., Bobchenkov, A., Yemelyanov, D.: Resource selection al-gorithms for economic scheduling in distributed systems. Procedia Computer Science 4, 2267–2276 (2011)

[11] Toporkov, V., Yemelyanov, D., Toporkova, A., Bobchenkov, A.: Resource Co-allocation Algorithms for Job Batch Scheduling in Dependable Distributed Computing. In: Zamojski, W., Kacprzyk, J., Mazurkiewicz, J., Sugier, J., Walkowiak, T. (eds.) Dependable Computer Systems. AISC, vol. 97, pp. 243–256. Springer, Heidelberg (2011)

[12] Toporkov, V., Bobchenkov, A., Toporkova, A., Tselishchev, A., Yemelyanov, D.: Slot Selection and Co-allocation for Economic Scheduling in Distributed Computing. In: Malyshkin, V. (ed.) PaCT 2011. LNCS, vol. 6873, pp. 368–383. Springer, Heidelberg (2011)

[13] Toporkov, V.V., Tselishchev, A.S.: Safety scheduling strategies in distributed computing. Intern. J. of Critical Computer-Based Systems 1(1/2/3), 41–58 (2010)

[14] Cecchi, M., Capannini, F., Dorigo, A., et al.: The gLite Workload Management System. Journal of Physics: Conference Series 219(6), 062039 (2010)

Controlling Complex Lighting Systems

Igor Wojnicki and Leszek Kotulski

Abstract. Designing and controlling lighting systems become more and more complex. Focusing on the control problem a rule-based control system is proposed. The system allows to control such lighting systems with high level control logic constituting so-called profiles. The profiles express lighting system behavior under certain conditions defined with rules. The light point and sensor distribution in the grid are given as graph structures. Control command sequences are inferred based on current system state, profiles, light points topology and sensor input. System's architecture and a case study are also presented.

1 Motivation

Contemporary lighting systems become more and more complex. The complexity is a direct result of available capabilities of emerging new technologies such as LED. Light points based on LED can be more and more spatially distributed. The distribution includes not only location but also direction and light cone angle. Furthermore, each of the points can be very precisely controlled having multiple power states which correspond both to power consumption and light intensity. These provide more fine-grained control over lighting conditions. With an availability of such control of lighting conditions there are certain issues which need to be addressed. These are: design, and control.

Designing a lighting system which should provide certain lighting conditions become more and more complex. Different spatial distributions for a similar or the same effect can be chosen. Some CAD support seems to be necessary. At the design stage certain optimizations can be carried out. There could be many optimization

Igor Wojnicki · Leszek Kotulski
AGH University of Science and Technology, Faculty of Electrical Engineering, Automatics, Computer Science and Electronics, Department of Automatics, al. Mickiewicza 30, 30-059 Kraków, Poland
e-mail: wojnicki@agh.edu.pl, kotulski@agh.edu.pl

W. Zamojski et al. (Eds.): Complex Systems and Dependability, AISC 170, pp. 305–317.
springerlink.com

criteria such as: cost (number of light points reduction), robustness (adaptability to lighting conditions not covered by the original design), durability (providing similar or the same lighting conditions in case of some points failing) etc. So the design process involves multi-criteria optimization which also should be supported by a CAD software (there is an ongoing research at AGH-UST targeting this area: a Large-scale Intelligent Lighting – LaSIL [13]).

Controlling such a complex system also poses a challenge. First of all there are many independent light points. To obtain particular lighting conditions some of them should be activated. Furthermore, a light point is not just a simple on-off device. Upon activation the power level is to be specified. The main task of a control system would be fulfilling requirements for lighting conditions by activating certain points. The activation pattern (which indicates points, at what power levels, in what sequence) is also subject to optimization according to some criteria. These can be: power consumption, light point utilization, aesthetic etc. Switching from one lighting condition to another presents yet another challenge: how the light points should be switched not causing unpleasant effects (blinking, strobe, etc.).

Furthermore the successful control should be based on a feedback from sensors. It allows the control system to verify if the lighting conditions are met. It would also allow to asses if there is a need to provide other conditions (i.e. illuminating or dimming certain areas). Depending on particular application there should be different sensors considered such as: time, light, presence, movement etc.

This paper focuses on the control issue. The light points topology and their capabilities as well as the lighting conditions and sensor output need to be defined in a formal way. To provide a successful control a pattern matching between the sensor output and the lighting conditions is to be provided. Upon successful match a sequence of control commands is generated. The commands are directed to proper light points based on the topology. Generating such sequences requires planning and optimization. Alternative control sequences, regarding particular optimization criteria, should also be considered.

To provide pattern matching, planning, plan optimization and to cope with the complexity, applying certain Artificial Intelligence methods is proposed. In particular, using rules and implementing a Rule-based System capable of inferring appropriate control is further discussed.

2 Related Research

Rule based systems are fully capable of providing means for control applications. Gensym G2 and Tiger systems should be mentioned here. Gensym G2 is a general expert system which can be applied to variety of problems ranging from Business Rules to Real-Time Control [8]. Among other applications it has been successfully applied to control Unmanned Aerial Vehicles (UAV) or Power Systems management. Tiger (created by Intelligent Applications Ltd, presently Turbine Services

Ltd) is a real-time rule-based system designed to diagnose gas turbines [16]. It is estimated that the system alone has brought huge savings thanks to early detection of possible failures. The project initially was researched at LAAS in France [2]. There are also languages and run-time environments dedicated to handling rules, such as CLIPS, Jess or Prolog.

3 Expressing Topology and States

Both designing and controlling a lighting system in the open spaces need considering very large number of states. Furthermore each of these states depends on significant number of parameters. These parameters either can dynamically change their values (i.e. current value of sensor data) or they are permanent and depend on the formal government regulations (i.e.[4]). Moreover, such the selection of parameters to consider can also dynamically change if a detected situation changes. For example the parameters applied for assurance of the personal safety should be changed to the parameters applied for assurance of the safe movement, if such a movement is detected (see [15] for details). The illumination also changes in dependence on the atmospheric conditions, thus some additional intermediate states should be computed. Complexity of the problem causes that we have to introduce a formal notation that enables a formal description of the lighting problem in a way that will by computable and scalable. Graph structures seams to be a good choice in this case. In [13] the hypergraph representation for building, streets, squares and its elements has been introduced in the following way.

Definition 1. *A* ***cartographic hypergraph*** *is a tuple of the form*

$$G = (N, H, att_N, att_H, lab_N, lab_H),$$

where N is a set of nodes, $H \subset \bigcup_{i>1} P_i(N)$ is a set of hyperedges, $att_N : N \longrightarrow \mathscr{A}_N$ and $att_H : H \longrightarrow \mathscr{A}_H$ are node and hyperedge attributing functions respectively, $lab_N : N \longrightarrow \mathscr{L}_N$ and $lab_H : H \longrightarrow \mathscr{L}_H$ are node and hyperedge labeling functions. $\mathscr{A}_N$ and $\mathscr{A}_H$ denote sets of node and hyperedge attributes, $\mathscr{L}_N$ and $\mathscr{L}_H$ are sets of node and hyperedge labels. The family of cartographic hypergraphs is denoted as $\mathscr{H}_{Cart}$.

Elements of the node set N correspond to such physical objects as streets, paths, squares and so on. Elements of H correspond to physical junction points of streets, paths and so on.

In [14] the idea of introduction to this notation elements representing sensors and agents that gather some knowledge while moving via the "virtual streets" is presented.

We will consider the permanent part of graph i.e. representation of the buildings, street, squares (that will not move and their parameters will not change) and temporary part of graph that represent the fragment of description of the specific illumination behavior (called profile). The representation of the set of profiles is input for the proposed Intelligent Control System.

In Figure 1 the sample map and the corresponding hypergraph for the permanent representation are shown. The part of map consists of streets sections $s_1, \ldots, s_5$, and the square p which are represented by the cartographic hypergraph vertices. Their junction points marked with bolded points correspond to its hyperedges.

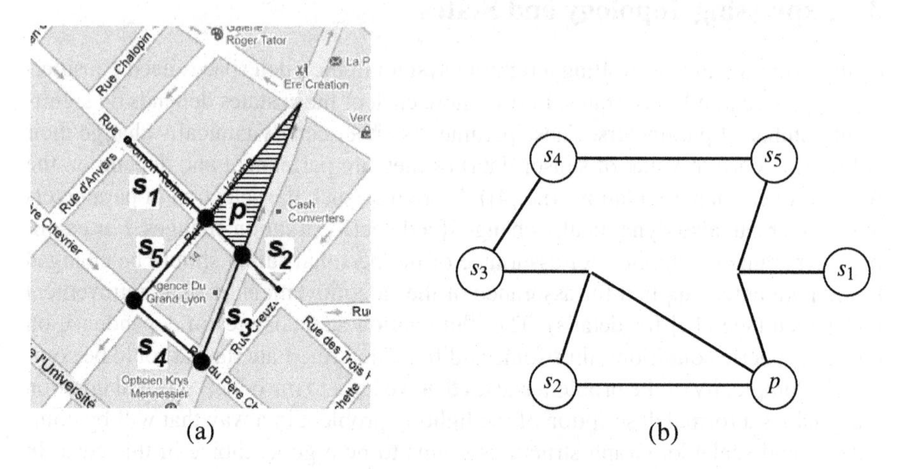

Fig. 1 (a) The sample map (b) Hypergraph representing the map

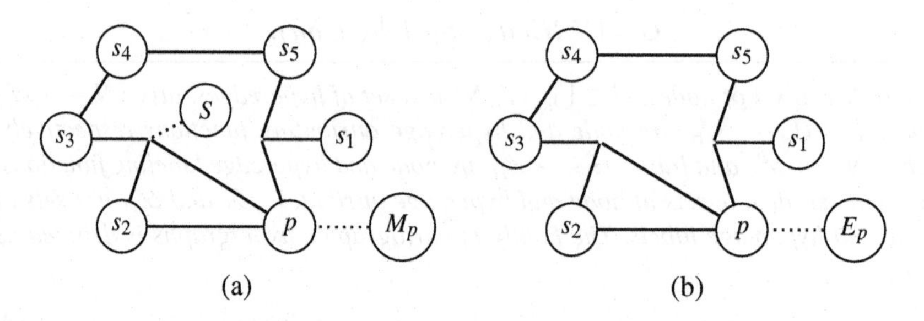

Fig. 2 (a) Movement and (b) Standby profiles represented with hypergraphs

With the square p we can join the temporary part of graph such as economic lighting scheme, see Figure 2 (for simplicity represented here as a single node labeled by Ep – in a normal situation, so called Standby profile, it is a complex subgraph representing lamps and their attributes) and illumination while some movement is detected, so called Movement profile, labeled by Mp.

Both situations are presented in Figure 2 where a) represents the Movement profile and b) the Standby profile. Let us note that Mp profile can be added to the graph only if some sensor (labeled by S), that is associated with the square p, has detected a movement. In the graph this means that p is connected with S by some hyperedge, and p is also connected with the graph representing the profile (here Ep).

4 Complex Control Requirements

The following section discusses requirements for handling complex lighting systems. It introduces a concept of profile, proposes an architecture and outlines system behavior.

4.1 Profiles

A lighting grid forms a complex system of light points. The points are spatially distributed, possibly non uniformly. Each point has certain parameters, in addition to its location, such as available output power levels, energy consumption, direction etc. Each point can be managed, controlled separately to make optimization of the entire grid according to chosen criteria possible.

Lighting control should be subject to profiles. A profile defines a mode of operation for a lighting grid for particular purpose. It provides behavioral model under certain circumstances, precisely defining lighting conditions. Choosing a profile can depend on such factors as natural lighting, weather (snow, rain, fog), time (time of day, working days), social events, energy shortage, traffic etc.

Since some profiles might depend or subsume each other, they can form a hierarchy. In general, this hierarchy can be expressed as a tree, graph or a hypergraph. An example set of profiles for public access areas are presented in Figure 3. There are two major profiles *Normal* and *Emergency* defining lighting conditions for regular and emergency operations. Within regular operations there are *Tracking* and *Standby* profiles defined. The *Standby* describes a profile which provides minimum lighting, while *Tracking* illuminates regions in which some activity can be detected i.e. people presence.

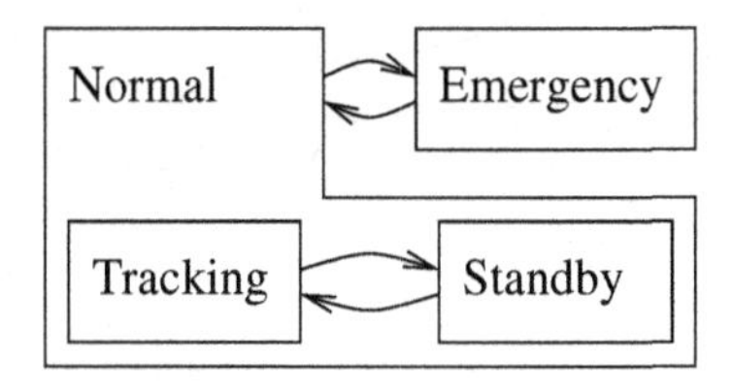

Fig. 3 Environment Profiles

4.2 *Massive Input Processing*

General architecture of proposed control system is given in Figure 4. Input for such lighting control system is threefold. First, there is static input. It is topology related information, regarding spatial distribution of light points in the grid, their parameters such as: direction, yield, power consumption characteristic etc., and sensors, including: type, location, etc. It rarely changes during operation, unless some light points become uncontrollable or inoperational. Second, there is profile definitions being also static. Third, there is dynamic input. It includes all information coming from sensors and it is subject to change often. All above inputs are represented by graph structures thus they are called input graphs.

To control in such a case is to detect certain patters defined by profiles matched against light points and sensor input. From a formal point of view it is detecting patterns in input graphs or subgraphs derived from them.

4.3 *Reactive and Deductive Behavior*

A successful control system should be able to process incoming data, and generate appropriate response sending control commands to light points. It constitutes stimuli-response reactive behavior.

Furthermore the control system should be capable of more complex actions. Based on incoming data, it should be able to anticipate where and what to

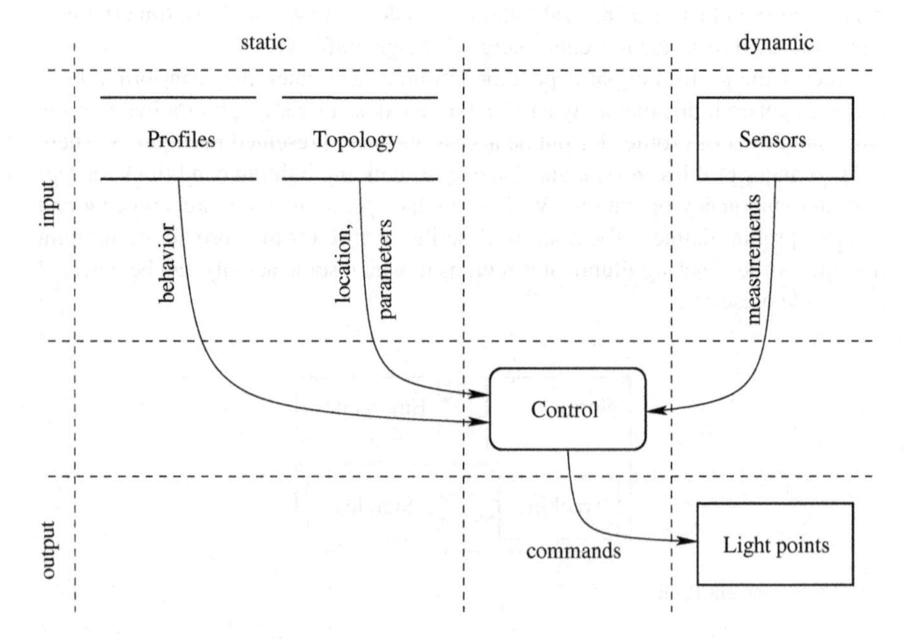

Fig. 4 General Architecure, DFD

control, predicting which light points should be enabled, when and with what intensity. Such a behavior allows for more sparse sensor grid, reducing number of sensors deployed, decreasing cost of entire system. It would also make the system work in case of communication problems, covering for missing data, or in case of missing, destroyed, or disconnected sensors (i.e. in case of natural disasters or riots).

Anticipation requires planning. Having different control plans identified, it is possible to evaluate them according to chosen criteria and pick the optimal plan. Different criteria are possible, including safety maximization, power consumption reduction, or human comfort maximization. The criteria themselves can be part of the profile data.

5 Proposed Solution: A Rule-Based System

A straightforward way of defining profiles is to use rules. A set of rules defines both conditions and actions which should be met and carried out, respectively, within the profile. Using rules simplifies profile design process employing knowledge engineering approach and tools. Furthermore formal analysis is possible including checking for subsumption and completeness.

To have a working control system the rules need to be interpreted. The interpretation is provided by an inference engine. This approach constitutes a rule-based system (RBS). Such a system consists of a knowledge base and an inference engine[5, 3, 12]. The knowledge base is the actual control logic. It is a dynamic aspect of the system, being subject to change. A change happens if the requirements for the behavior of the system change, in other words when the system needs to be reprogrammed, or if new facts become known or available, i.e. new sensor data. The inference engine provides general means for interpreting it. It works according to established inference algorithm which remains fixed.

In order to store knowledge, RBS use various knowledge representation methods tailored to particular needs such as: expert systems, decision support, or control and monitoring systems. Quite often logical representations are used i.e. propositional logic or predicates.

In the presented case the knowledge base consists of rules describing the profiles, and facts identifying topology, sensor data, control commands and system state (see Figure 5). The *Profiles* and *Topology*, provided by designer, implement system behavior and define spatial distribution and properties of the light points and sensors. The *State* identifies what state the system is in, including information about active profiles and which light points are activated. It is given as a graph. The *Control* and *Input* are facts representing control sequences and input data, from sensors, respectively. The run-time system consists of the *Inference Engine*, input/output subsystem (*I/O System*) and the *Explanatory Facility*. The *I/O System* provides a separate software module handling input/output operations[19]. It feeds data from *Sensors* into the *Knowledge Base* and makes sure that the *control sequences* are delivered to *Actuators*. The *Explanatory Facility* is optional. It provides feedback to the user

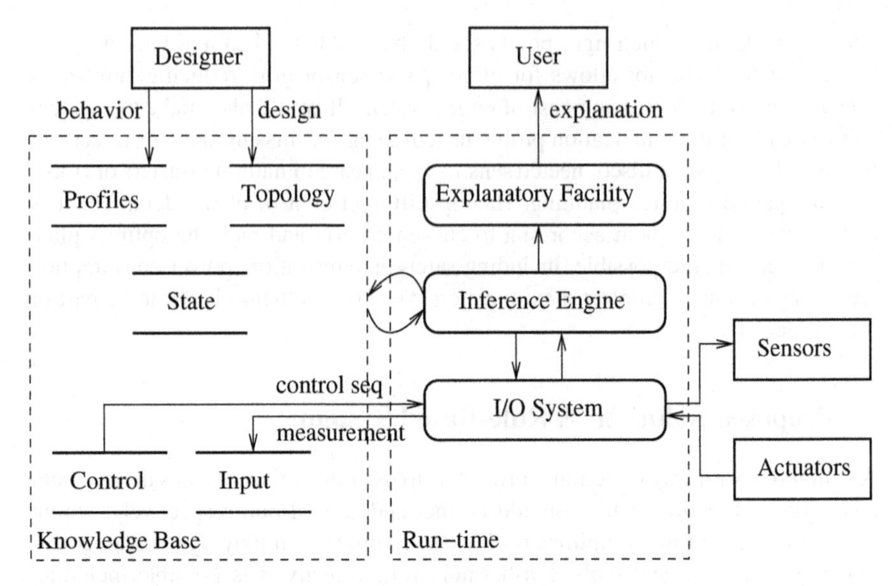

Fig. 5 Lighting Control Rule-based System

regarding system decisions. It explains why certain control sequences are chosen. It plays both diagnostic and informational roles.

5.1 Backward and Forward Chaining

Since both deductive and reactive behavior for the control system is requested, the rule-based system must be capable of handling both features. There are two approaches to the inference process that can be used here, these are forward and backward chainings.

The forward chaining is suitable for modeling reactive systems. The inference process takes into account available data (facts) and generates new data based on rules. This process continues until a goal is reached, which is an appropriate control command sequence. In the particular case presented here, the available data is defined by input graphs. The ever changing dynamic input keeps delivering facts, while the inference engine, taking into account the static input, keeps sending control commands.

To cover the deductive characteristics, mentioned before, a backward chaining strategy can be used. It starts with hypothesis and works toward facts, looking for these which support it. The backward chaining is suitable for synthesizing plans, as well as finding alternative solutions. Looking for optimal solutions can be applied throughout the process, at each inference step employing the given criteria. Alternatively, having alternative solutions synthesized, at the end of the inference process, they can be easily evaluated and the optimal one can be selected then.

5.2 Contexts

Implementing the profiles is provided straightforwardly by gathering rules in subsets – one for each profile. The inference engine acts, interpreting the facts based on the rules, only if a profile is activated. It applies rules from a given subset corresponding to the active profile only.

Assuming that some dependencies or subsumption of profiles can take place more advanced implementation and execution schema is considered. Handling multiple, hierarchical profiles is provided by the Context Based Reasoning (CxBR) [1]. The profiles would correspond to the contexts. The contexts gather rules. They form a hierarchy which can be described by a hypergraph. Furthermore applying CxBR concept the control system is capable of switching among profiles (contexts) easily. It is accomplished by introducing additional context switching rules present in each of the contexts.

5.3 Design and Redesign

Using rules for describing the profiles has also some impact on the design process. The process of synthesizing rules, while being natural, has a strong background which comes from the Knowledge Engineering field. There are several design methodologies and patterns that can be used here [6]. Especially structured rule-bases which were introduced to support control systems [7, 11]. Rule design methodologies such as ARD+ and DPD, can be also applied here. They were developed at AGH-UST [10] with full support of CASE tools [9].

Rules can be easily structured and visualized in different ways, exposing different aspects using: decision tables, trees, and recently developed contextual networked decision tables (CoNDeT) [17]. For control system purposes particularly the CoNDeT approach is suitable. An ongoing research indicates that it can handle a wide range of applications from control systems to general purpose software [18]. It makes the rule-base more readable, easier to construct or alter, simplifying the redesign process if needed.

5.4 Formal Analysis

Rules are formally defined. They are often represented with propositional or predicate, usually first order, logic. It allows for their formal analysis. In the proposed solution rules are used.

From the control system point of view the most important property of such rule-based system is completeness. It is essential that the system reacts properly for any possible input. Completeness detection can be applied not only during run-time but also in the design stage. It allows for early detection of missing rules, typos, wrong values etc.

Furthermore, considering knowledge representation of the distribution of light points or sensors, their formal analysis and transformation is also available.

Applying graph transformations allows to verify certain properties or identify flows and compensate for them.

6 A Case Study

A case study presented here regards a lighting control system for a gas station. The station is equipped with several light points, motion sensors and smoke detectors. The profile definitions, from the RBS point of view, are provided in Figure 6. There are four profiles forming a hypergraph. Two major profiles: *Normal* and *Emergency*, and two minor ones: *Standby* and *Tracking*.

A profile consists of rules and, optionally, other profiles. To consider the control system active at least one profile has to be active. The default active profile (*Normal*) is indicated by an incoming arrow. A rule within a profile can trigger actions, generating control sequences, or switch to another profile. A single rule consists of conditions and conclusions separated with an arrow symbol (->). Conditions involve reading data from sensor regarding current time (*h*), motion sensors (*m*) and smoke detectors (*s*). They may also involve using logical operators and graph matching functions. In general case rule conditions involve graph transformations on graph structures representing current state, topology and sensor inputs.

The *Normal* profile defines two rules:

1. if there is smoke detected switch to the *Emergency* profile,
2. if there is no smoke detected activate the *Standby* profile.

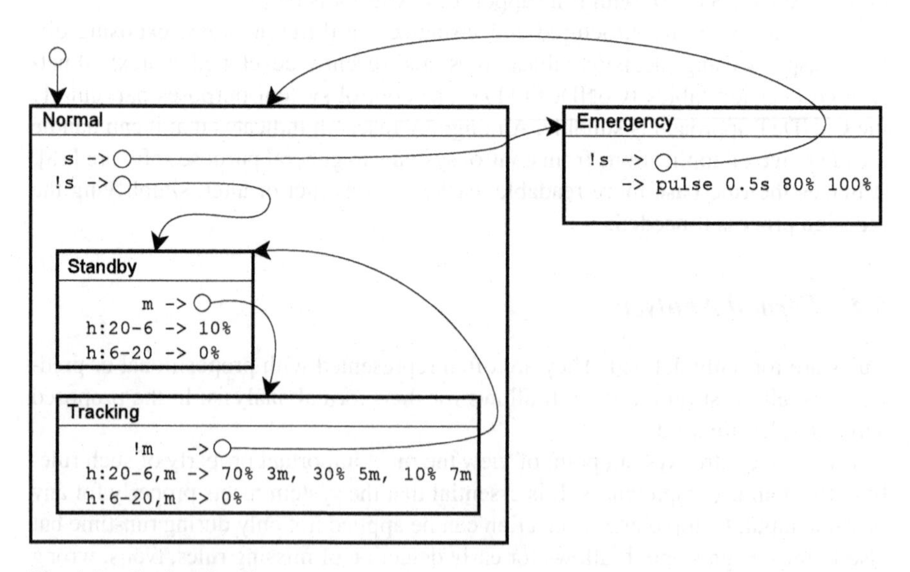

Fig. 6 Gas Station, Profiles

Activating a minor profile (subprofile) causes both minor and major profiles to be active. A minor profile is not allowed to be active if its major profile is not active. The *Standby* profile defines the following rules:

1. if any motion is detected (m) switch to the *Tracking* profile,
2. if current hour is between 20:00 and 6:00 set all lighting at 10%,
3. if current hour is between 6:00 and 20:00 set all lighting at 0%.

The *Tracking* profile rules can be read as:

1. if there is no motion detected switch to the *Standby* profile,
2. if current hour is between 20:00 and 6:00 and there is motion detected set lighting at 70% within 3 meter radius from the motion source, 30% within 5 meter radius, 10% within 7 meter radius,
3. if current hour is between 6:00 and 20:00 set lighting at 0% regardless of the detected motion.

The *Emergency* profile is switched to if smoke is detected. There are the following rules:

1. switch to the *Normal* profile if there is no smoke present,
2. unconditionally set lighting to pulse changing luminosity from 80% to 100% at 0.5 second interval.

In this example rule conditions regard data from simple, single sensors (smoke detector, motion sensor) or real-time clock. However in general case this data can come from complex sensor grids. In such a case it would be expressed with graphs. Rule conditions will regard results of graph transformations providing certain pattern matching: identification of subgraphs, graph properties etc. Similarly rule decisions involve graph structures – activating certain light points within the grid, and simultaneously changing system state, which is also given as a graph. An example of such a decision, which affects a grid of light points, is given in the second rule of *Tracking* profile ((`h:20-6,m -> 70% 3m, 30% 5m, 10% 7m`), see Figure 6). In general case a decision can use far more complex graph-related patterns as well.

It needs to be pointed out that transitions upon switching lighting conditions should be aesthetic and smooth. Blinking, strobe effects, sudden darkness should be avoided as causing disorientation. For example, firing the second rule in the *Tracking* profile (`h:20-6,m -> 70% 3m, 30% 5m, 10% 7m`) results in a certain lighting condition change. However the change is not abrupt. It is gradual from whatever state the lighting system is in to the one defined by the rule. Such a smooth transition is also computed by the proposed control system. The computation is based on gradual graph transformations, from the current state to the one switched to by the rule. I means that if AI rule decides that the system should move from the state described by graph G_i to the state described by the state G_{i+1} (which additionally might involve changing current profile as well), the graph transformation subsystem parses the graph state and designates sequence of elementary productions that transform the G_i to G_{i+1}. Applying the elementary productions generates a sequence of control commands to be delivered to actuators.

7 Summary

This paper tackles the problem of controlling complex lighting systems. The problem is approached with Artificial Intelligence techniques, particularly the rule-based paradigm.

Spatial distribution (topology) of light points is expressed as a graph. The main goal of the control system is to activate certain light points in given situations. The particular situations are defined with so-called profiles.

Profile activation is triggered by rules. There could exist complex relationships among profiles, thus they are expressed with graphs as well. Within each profile there is a set of rules precisely defining lighting conditions. This delivers a high abstract control layer which can successfully cover defining complex profiles. Simultaneously it provides a very intuitive and straightforward programming model.

Profile activation and rule firing, thus sending control sequences powering up light points, is provided by a rule-based system. The system uses a structured knowledge-base for storing data regarding the spatial distribution of light points, profiles, sensor measurements and control commands and current state. The inference engine is based on the context-based reasoning to cover the profiles and provide a tool for proper profile activation.

Further research focuses on graph transformations, extending deductive capabilities of the proposed rule base system and fine tuning the light transitions. The graph transformations will provide more operators to be used both in the rule condition and decision parts. Research regarding the deductive capabilities will provide automatic optimization methods for different criteria mentioned earlier. Finally fine-tuning the automatic transitions among lighting conditions will provide more aesthetic visual effects.

References

1. Gonzalez, A.J., Stensrud, B.S., Barrett, G.C.: Formalizing context-based reasoning: A modeling paradigm for representing tactical human behavior. Int. J. Intell. Syst. 23(7), 822–847 (2008)
2. Gouyon, J.P.: Kheops users's guide. Report of Laboratoire d'Automatique et d'Analyse des Systemes (92503) (1994)
3. Jackson, P.: Introduction to Expert Systems, 3rd edn. Addison–Wesley (1999) ISBN 0-201-87686-8
4. Kotulski, L., Strug, B.: Distributed Adaptive Design with Hierarchical Autonomous Graph Transformation Systems. In: Shi, Y., van Albada, G.D., Dongarra, J., Sloot, P.M.A. (eds.) ICCS 2007, Part II. LNCS, vol. 4488, pp. 880–887. Springer, Heidelberg (2007)
5. Liebowitz, J. (ed.): The Handbook of Applied Expert Systems. CRC Press, Boca Raton (1998)
6. Ligęza, A. (ed.): Logical Foundations for Rule-Based Systems. Springer, Heidelberg (2006)
7. Ligęza, A., Wojnicki, I., Nalepa, G.J.: Tab-Trees: A CASE Tool for the Design of Extended Tabular Systems. In: Mayr, H.C., Lazanský, J., Quirchmayr, G., Vogel, P. (eds.) DEXA 2001. LNCS, vol. 2113, pp. 422–431. Springer, Heidelberg (2001)

8. Moore, R., Lindenfilzer, P., Hawkinson, L., Matthews, B.: Process control with the g2 real-time expert system. In: Proceedings of the 1st International Conference on Industrial and Engineering Applications of Artificial Intelligence and Expert Systems, IEA/AIE 1988, pp. 492–497. ACM, New York (1988), doi: http://doi.acm.org/10.1145/51909.51965
9. Nalepa, G.J., Wojnicki, I.: VARDA Rule Design and Visualization Tool-Chain. In: Dengel, A.R., Berns, K., Breuel, T.M., Bomarius, F., Roth-Berghofer, T.R. (eds.) KI 2008. LNCS (LNAI), vol. 5243, pp. 395–396. Springer, Heidelberg (2008)
10. Nalepa, G.J., Wojnicki, I.: Ard+ a prototyping method for decision rules. method overview, tools, and the thermostat case study. Tech. Rep. CSLTR 01/2009, AGH University of Science and Technology (2009)
11. Nalepa, G.J., Wojnicki, I.: Visual Generalized Rule Programming Model for Prolog with Hybrid Operators. In: Seipel, D., Hanus, M., Wolf, A. (eds.) INAP 2007. LNCS (LNAI), vol. 5437, pp. 178–194. Springer, Heidelberg (2009)
12. Negnevitsky, M.: Artificial Intelligence. A Guide to Intelligent Systems. Addison-Wesley, Harlow (2002) ISBN 0-201-71159-1
13. Sędziwy, A.: Representation of objects in agent-based lighting design problem. In: DepCoS: Dependability and Complex Systems. DepCoS-RELCOMEX (2012)
14. Sędziwy, A., Kotulski, L.: Solving large-scale multipoint lighting design problem using multi-agent environment. In: Su, D., Xue, K., Zhu, S. (eds.) Key Engineering Materials. Advanced Design and Manufacture IV (2011)
15. Sędziwy, A., Kozień-Woźniak, M.: Computational methods supporting street lighting design. In: DepCoS: Dependability and Complex Systems. DepCos-RELCOMEX (2012)
16. Travé-Massuyès, L., Milne, R.: Gas-turbine condition monitoring using qualitative model-based diagnosis. IEEE Expert: Intelligent Systems and Their Applications 12, 22–31 (1997), doi: http://dx.doi.org/10.1109/64.590070
17. Wojnicki, I.: From tabular trees to networked decision tables: an evolution of modularized knowledge-base representations. Pomiary Automatyka Kontrola 12 (2011)
18. Wojnicki, I.: Implementing general purpose applications with the rule-based approach. In: RuleML 2011: Proceedings of the 5th International Conference on Rule-based Reasoning, Programming, and Applications, pp. 360–367. Springer, Heidelberg (2011)
19. Wojnicki, I.: Separating i/o from application logic for rule-based control systems. Decision Making in Manufacturing and Services (2011)

Service Renaming in Component Composition

Wlodek M. Zuberek

Abstract. In component-based systems, the behavior of components is usually described at component interfaces and the components are characterized as requester (active) and provider (reactive) components. Two interacting components are considered compatible if all possible sequences of services requested by one component can be provided by the other component. This concept of component compatibility can be extended to sets of interacting components, however, in the case of several requester components interacting with one or more provider components, as is typically the case of cleint–server applications, the requests from different components can be interleaved and then verifying component compatibility must take into account all possible interleavings of requests. Such interleaving of requests can lead to unexpected behavior of the composed system, e.g. a deadlock can occur. Service renaming is proposed as a method of systematic eliminating of such unexpected effects and streamlining component compositions.

1 Introduction

In component-based systems, components represent high-level software abstraction which must be generic enough to work in a variety of contexts and in cooperation with other components, but which also must be specific enough to provide easy reuse [9].

Primary reasons for using components are [8]: separability of components from their contexts; independent component development, testing and later reuse; upgrade and replacement in running systems. Component composability is often taken for granted, while it actually is influenced by a number of factors such as operating platforms, programming languages or the specific middleware technology in which the components are based. Ideally, the development, quality control, and deployment of software components should be automated similarly to other

Wlodek M. Zuberek
Department of Computer Science, Memorial University, St.John's, NL, Canada A1B 3X5
and
Department of Applied Informatics, University of Life Sciences, 02-787 Warszawa, Poland
e-mail: wlodek@mun.ca

W. Zamojski et al. (Eds.): Complex Systems and Dependability, AISC 170, pp. 319–330.
springerlink.com

engineering domains, which deal with the construction of large systems composed of well-understood elements with predictable properties and under acceptable budget and timing constraints [14]. For this to happen, automated component-based software engineering must resolve a number of issues, including efficient verification of component compatibility.

The behavior of components is usually described at component interfaces [16] and the components are characterized as requester (active) and provider (reactive) components. Although several approaches to checking component composability have been proposed [1] [4] [10] [12], further research is needed to make these ideas practical [8]. Usually two interacting components are considered compatible if all sequences of services requested by one component can be provides by the other component. In the case of several components interacting with a single provider, as is typically the case of internet applications (e.g., client-server systems), the requests from different components can be interleaved and then verifying component compatibility must check all possible interleavings of requests from all interacting components for possible conflicts. Such interleaving of service requests can lead to unexpected behavior of the composed system; e.g., a deadlock can occur. Service renaming is proposed as a systematic method of eliminating request conflicts and streamlining component composition.

The idea of service renaming for the elimination of conflicting requests can be illustrated by the following simple example. Let the languages of two requester components be specified by regular expressions "`(abc)*`" and "`(bac)*`" ("`a`", "`b`" and "`c`" are services requested by the components), and let the language of the corresponding provider component be "`((ab + ba)c)*`". It can be easily checked that both requester components are compatible with the provider [20], however, one of possible interleavings of the requests is "`(ab(bc + ac))*`", which - if allowed - results in a deadlock. Service renaming which eliminates this deadlock replaces, for example, "`a`" by "`A`" in the sequence "`bac`" and "`b`" by "`B`" in the sequence "`abc`", and then any interleaving of "`(aBc)*`" with "`(bAc)`" is compatible with the provider "`((aB + bA)c)*`".

The paper is continuation of previous work on component compatibility and substitutability [7] [19] [20] [21]. Using the same formal specification of component behavior in the form of component languages, the paper proposes an approach to identify component conflicts in component composition and systematic renaming of services as a conflict removal method.

Since component languages are usually infinite, their compact finite specification is needed for effective verification, comparisons and other operations. Labeled Petri nets are used as such specification.

Petri nets [13] [15] are formal models of systems which exhibit concurrent activities with constraints on frequency or orderings of these activities. In labeled Petri nets, labels, which represent services, are associated with elements of nets in order to identify interacting components. Well-developed mathematical theory of Petri nets provides a convenient formal foundation for analysis of systems modeled by Petri nets.

Section 2 recalls the concept of component languages as a characterization of component's behavior. Component languages are used in Section 3 to define component compatibility. Service renaming is described in Section 4 while Section 5 concludes the chapter.

2 Modeling Component Behavior

The behavior of a component, at its interface, can be represented by a cyclic labeled Petri net [6] [20]:

$$\mathscr{M}_i = (P_i, T_i, A_i, S_i, m_i, \ell_i, F_i),$$

where P_i and T_i are disjoint sets of places and transitions, respectively, A_i is the set of directed arcs, $A_i \subseteq P_i \times T_i \cup T_i \times P_i$, S_i is an alphabet representing the set of services that are associated with transitions by the labeling function $\ell_i : T_i \to S_i \cup \{\varepsilon\}$ (ε is the "empty" service; it labels transitions which do not represent services), m_i is the initial marking function $m_i : P_i \to \{0, 1, ...\}$, and F_i is the set of final markings (which are used to capture the cyclic nature of sequences of firings).

Sometimes it is convenient to separate net structure $\mathscr{N} = (P, T, A)$ from the initial marking function m.

In order to represent component interactions, the interfaces are divided into *provider* interfaces (or p-interfaces) and *requester* interfaces (or r-interfaces). In the context of a provider interface, a labeled transition can be thought of as a service provided by that component; in the context of a requester interface, a labeled transition is a request for a corresponding service. For example, the label can represent a conventional procedure or method invocation. It is assumed that if the p-interface requires parameters from the r-interface, then the appropriate number and types of parameters are delivered by the r-interface. Similarly, it is assumed that the p-interface provides an appropriate return value, if such a value is required. The equality of symbols representing component services (provided and requested) implies that all such requirements are satisfied.

For unambiguous interactions of requester and provider interfaces, it is required that in each p-interface there is exactly one labeled transition for each provided service:

$$\forall t_i, t_j \in T : \ell(t_i) = \ell(t_j) \neq \varepsilon \Rightarrow t_i = t_j.$$

Moreover, to express the reactive nature of provider components, all provider models are required to be ε–conflict–free, *i.e.*:

$$\forall t \in T \ \forall p \in Inp(t) : Out(p) \neq \{t\} \Rightarrow \ell(t) \neq \varepsilon$$

where $Out(p) = \{t \in T \mid (p,t) \in A\}$; the condition for ε–conflict–freeness could be used in a more relaxed form but this is not discussed here for simplicity of presentation.

Component behavior is determined by the set of all possible sequences of services (required or provided by a component) at a particular interface. Such a set of sequences is called the *interface language*.

Let $\mathscr{F}(\mathscr{M})$ denote the set of firing sequences in $\mathscr{M}$ such that the marking created by each firing sequence belongs to the set of final markings F of $\mathscr{M}$. The interface language $\mathscr{L}(\mathscr{M})$, of a component represented by a labeled Petri net $\mathscr{M}$, is the set of all labeled firing sequences of $\mathscr{M}$:

$$\mathscr{L}(\mathscr{M}) = \{\ell(\sigma) \mid \sigma \in \mathscr{F}(\mathscr{M})\},$$

where $\ell(t_{i_1}t_{i_2}...t_{i_k}) = \ell(t_{i_1})\ell(t_{i_2})...\ell(t_{i_k})$.

By using the concept of final markings, interface languages reflect the cyclic behavior of (requester as well as provider) components.

Interface languages defined by Petri nets include regular languages, some context–free and even context–sensitive languages [11]. Therefore, they are significantly more general than languages defined by finite automata [5], but their compatibility verification is also more difficult than in the case of regular languages.

3 Component Compatibility

Interface languages of interacting components can be used to define the compatibility of components; a requester component $\mathscr{M}_r$ is compatible with a provider component $\mathscr{M}_p$ if and only if all sequences of services requested by $\mathscr{M}_r$ can be provided by $\mathscr{M}_p$, i.e., if and only if:

$$\mathscr{L}(\mathscr{M}_r) \subseteq \mathscr{L}(\mathscr{M}_p).$$

Checking the inclusion relation between the requester and provider languages defined by Petri nets $\mathscr{M}_r$ and $\mathscr{M}_p$ can be performed by systematic checking if the services requested by one of the interacting nets can be provided by the other net at each stage of the interaction.

3.1 Bounded Case

In the case of bounded nets, checking compatibility of a single requester with a single provider components performs a breadth–first traversal of the reachability graph $\mathscr{G}(\mathscr{M}_r)$ verifying that for each transition in $\mathscr{G}(\mathscr{M}_r)$ there is a corresponding transition in $\mathscr{G}(\mathscr{M}_p)$, which is described in detail in [20]. For the case of several requester components $\mathscr{M}_i, i = 1,...,k$, interacting with a single provider component $\mathscr{M}_p$, first the compatibility of each requester with the provider is checked in [20]. Then the interleaving of requests are checked for progress, and is a deadlock is discovered, the set of interacting components cannot be compatible. For simplicity, the family of requester components is represented by a vector $\mathbf{N}_r$ with individual components $\mathbf{N}_r[1]$, $\mathbf{N}_r[2]$, ... $\mathbf{N}_r[k]$. Similarly, the markings for $\mathbf{N}_r$ are denoted be a vector $\mathbf{m}_r$ with individual marking functions $\mathbf{m}_r[1]$, $\mathbf{m}_r[2]$, ... $\mathbf{m}_r[k]$.

The following logical function *CheckProgressB* is used when all requester and provider languages are defined by bounded marked Petri nets $(\mathcal{N}_i, m_i)$, $i = 1,...,k$, and $(\mathcal{N}_p, m_p)$, respectively. The function performs exhaustive analysis of possible interleavings of requests, checking the progress of the composed model; if there is no progress (which means, a deadlock has been created), FALSE is returned. In the pseudocode below, *New* is a sequence (a queue) of markings to be checked, *head* and *tail* are operations on sequences that return the first element and remaining part of the sequence, respectively, $append(s,a)$ appends an element a to a sequence s, *Analyzed* is the set of markings that have been analyzed, $Enabled(\mathcal{N},m)$ returns the set of labels of transitions enabled in the net $\mathcal{N}$ by the marking m (including ε if the enabled transitions include transitions without labels), and $next(\mathcal{N},m,a)$ returns the marking obtained in the net $\mathcal{N}$ from the marking m by firing the transition labeled by x):

```
proc CheckProgressB(N_r, m_r, 𝒩_p, m_p);
begin
    New := (m_r, m_p);
    Analyzed := {};
    while New ≠ {} do
        (m, n) := head(New);
        New := tail(New);
        if (m, n) ∉ Analyzed then
            Analyzed := Analyzed ∪ {(m, n)};
            noprogress := true;
            for i := 1 to k do
                Symbols1 := Enabled(N_r[i], SkipEps(N_r[i], m[i]));
                Symbols2 := Enabled(𝒩_p, SkipEps(𝒩_p, n));
                if Symbols1 ∩ Symbols2 ≠ {} then
                    noprogress := false;
                    m' := m;
                    for each x in Symbols1 ∩ Symbols2 do
                        m'[i] := next(N_r[i], m[i], x)
                        append(New, (m', next(𝒩_p, n, x))
                    od
                fi
            od;
            if noprogress return FALSE fi
        fi
    od;
    return TRUE
end;
```

The function *SkipEps(m)* advances the marking function m through all transitions labeled by ε:

proc *SkipEps*($\mathcal{N}$, m);
begin
  while $\varepsilon \in Enabled(\mathcal{N}, m)$ **do** $m := next(\mathcal{N}, m, \varepsilon)$ **od**;
  return m
end;

where the ε parameter of the function *next* refers to any transition enabled by m that is labeled by ε.

Example. Fig.1 shows a simple configuration of two (cyclic) requester components and a single provider of three services named `a`, `b` and `c`.

In this case, the languages of the requesters are described by regular expressions "`(abc)*`" and "`(bac)*`" and the language of the provider by "`((ab+ba)c)*`". It can be easily checked that both requesters are compatible with the provider; the languages "`(abc)*`" and "`(bac)*`" are subsets of the language "`((ab+ba)c)*`".

As indicated in the introduction, the combined requests from both requester components are not compatible with the provider shown in Fig.1. For example, if the first request from requester-1 (i.e., "`a`") is followed by the first request from requester-2 (i.e., "`b`"), the composed system becomes deadlocked because further requests are "`b`" (from requester-1) and "`a`" (from requester-2) while the only provided service at this stage is "`c`".

The steps performed by the function *CheckProgressB* for the nets shown in Fig.1 are illustrated in a table, in which the first column, "*conf*", identifies the configuration of the model, while the last column, "ℓ", indicates the next configuration, reached in effect of the requested/provided service shown in column "x"; columns **m** and n show the markings of the requester and provider nets, respectively; i indicates

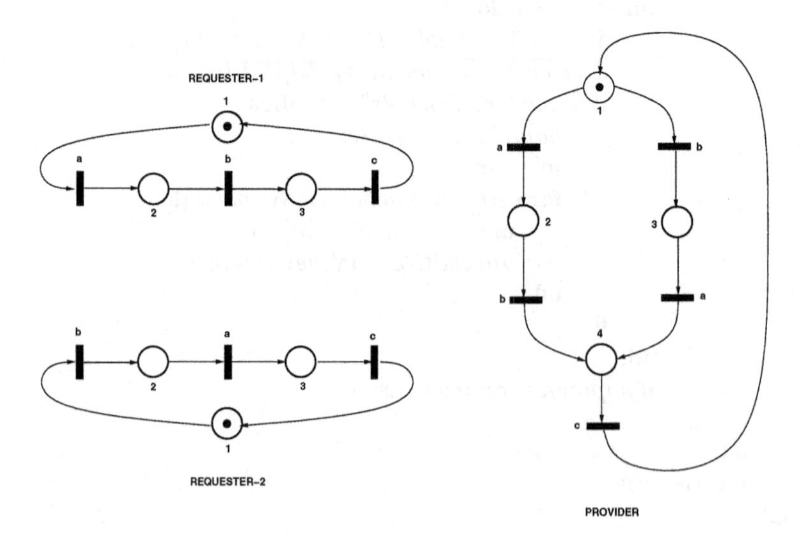

Fig. 1 Two requesters and a single provider.

the requester component (used for interleaving), as in function *CheckProgressB*, similarly to the remaining columns of the table:

$conf$	$\mathbf{m}$	n	i	$Symbols1$	$Symbols2$	x	$next(\mathbf{N}_r[i],\mathbf{m}[i],x)$	$next(\mathcal{N}_p,n,x)$	ℓ
0	[(1,0,0),(1,0,0)]	(1,0,0,0)	1	$\{a\}$	$\{a,b\}$	a	(0,1,0)	(0,1,0,0)	1
			2	$\{b\}$	$\{a,b\}$	b	(0,1,0)	(0,0,1,0)	2
1	[(0,1,0),(1,0,0)]	(0,1,0,0)	1	$\{b\}$	$\{b\}$	b	(0,0,1)	(0,0,0,1)	3
			2	$\{b\}$	$\{b\}$	b	(0,1,0)	(0,0,0,1)	4
2	[(1,0,0),(0,1,0)]	(0,0,1,0)	1	$\{a\}$	$\{a\}$	a	(0,1,0)	(0,0,0,1)	4
			2	$\{a\}$	$\{a\}$	a	(0,0,1)	(0,0,0,1)	5
3	[(0,0,1),(1,0,0)]	(0,0,0,1)	1	$\{c\}$	$\{c\}$	c	(1,0,0)	(1,0,0,0)	0
			2	$\{b\}$	$\{c\}$				–
4	[(0,1,0),(0,1,0)]	(0,0,0,1)	1	$\{b\}$	$\{c\}$				–
			2	$\{a\}$	$\{c\}$				–
5	[(1,0,0),(0,0,1)]	(0,0,0,1)	1	$\{a\}$	$\{c\}$				–
			2	$\{c\}$	$\{c\}$	c	(1,0,0)	(1,0,0,0)	0

No component can progress in configuration 4, so this is a deadlock configuration. Consequently the components shown in Fig.1 cannot be compatible. It can be observed that this deadlock configuration can be reached from configuration 0 by requesting service "a" (by requester-1) and then in configuration 1, requesting service "b" (by requester-2). Configuration 4 can also be reached from configuration 0 by first requesting service "b" (by requester-2) and then, in configuration 2, service "a" (by requester-2).

It should also be noted that in configurations 3 and 5, only one of the requester components does not progress, so these configurations are not deadlocks.

3.2 Unbounded Case

For the unbounded case, compatibility checking must include checking the unboundedness condition (a marked net $(\mathcal{N}, m_0)$ is unbounded if there exist markings m' and m'' reachable from m_0 such that m'' is reachable from m' and m'' is componentwise greater or equal to m'). This condition is checked for the requesters as well as for the provider nets by combining the markings together. More specifically, for each analyzed pair of markings $(\mathbf{m},n)$, an additional check is performed if the set *Analyzed* contains a pair of markings, which is componentwise smaller than $(\mathbf{m},n)$ and from which $(\mathbf{m},n)$ is reachable; if the set *Analyzed* contains such a pair, analysis of $(\mathbf{m},n)$ is discontinued. This additional check is performed by a logical function *Reachable*$((\mathbf{m},n),$*Analyzed*$)$, in which the first argument is a vector of marking functions (which - in this particular case - can be considered as a single marking function obtained by concatenation of all consecutive elements of $\mathbf{m}$). As in the bounded case, $\mathbf{N}_r$ is a vector of requester nets $[\mathcal{N}_1,\mathcal{N}_2,...,\mathcal{N}_k]$, also denoted $\mathbf{N}_r[i], i=1,...,k$, and $\mathbf{m}_r$ is a vector of marking functions for nets $\mathbf{N}_r[i], i=1,...,k$:

```
proc CheckProgressU(N_r, m_r, 𝒩_p, m_p);
begin
    New := (m_r, m_p);
    Analyzed := {};
    while New ≠ {} do
        (m, n) := head(New);
        New := tail(New);
        if (m, n) ∉ Analyzed then
            Analyzed := Analyzed ∪ {(m, n)};
            noprogress := true;
            if not Reachable((m, n), Analyzed) then
                for i := 1 to k do
                    Symbols1 := Enabled(N_r[i], SkipEps(N_r[i], m[i]));
                    Symbols2 := Enabled(𝒩_p, SkipEps(𝒩_p, n));
                    if Symbols1 ∩ Symbols2 ≠ {} then
                        noprogress := false;
                        m' := m;
                        for each x in Symbols1𝒮†⇕⌊≀⇕∫∈ do
                            m'[i] := next(N_r[i], m[i], x)
                            append(New, (next(m', next(𝒩_p, n, x))
                        do
                    fi
                od;
                if noprogress return FALSE fi
            fi
        fi
    od;
    return TRUE
end;
```

As in the bounded case, the function *CheckProgressU* returns FALSE if there is a sequence of service requests which cannot be satisfied by the provider component.

Example. Fig.2 shows another model composed of two requester components with languages "`(abc)*`" and "`(ab*c)*`" and an unbounded provider which accepts any sequence of requests of services "a", "b" and "c" such that any prefix of this sequence (including the whole sequence) contains not less requests for service "a" than for service "c"; the language of this provider is nonregular.

Both requester components are compatible with the provider as their languages are subsets of the provider's language.

The steps used by the function *CheckProgressU* for the components shown in Fig.2 are illustrated in the following table:

It can be easily checked that both requester components shown in Fig.2 are compatible with the provider. The steps used by *CheckProgressU* for the components shown in Fig.2 are illustrated by the following table:

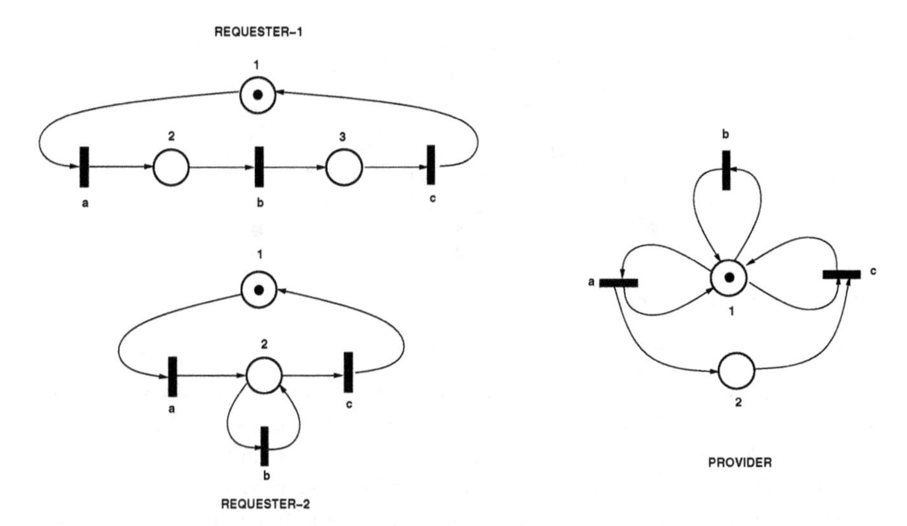

Fig. 2 Two requesters with an unbounded provider.

conf	**m**	*n*	*i*	*Symbols*1	*Symbols*2	*x*	$next(\mathbf{N}_r[i],\mathbf{m}[i],x)$	$next(\mathcal{N}_p,n,x)$	ℓ
0	[(1,0,0),(1,0)]	(1,0)	1	$\{a\}$	$\{a,b\}$	a	(0,1,0)	(1,1)	1
			2	$\{a\}$	$\{a,b\}$	a	(0,1)	(1,1)	2
1	[(0,1,0),(1,0)]	(1,1)	1	$\{b\}$	$\{a,b,c\}$	b	(0,0,1)	(1,1)	3
			2	$\{a\}$	$\{a,b,c\}$	a	(0,1)	(1,2)	4
2	[(1,0,0),(0,1)]	(1,1)	1	$\{a\}$	$\{a,b,c\}$	b	(0,1,0)	(1,2)	4
			2	$\{b,c\}$	$\{a,b,c\}$	b	(0,1)	(1,1)	2
						c	(1,0)	(1,0)	0
3	[(0,0,1),(1,0)]	(1,1)	1	$\{c\}$	$\{a,b,c\}$	c	(1,0,0)	(1,0)	3
			2	$\{a\}$	$\{a,b,c\}$	a	(0,1)	(1,2)	4
4	[(0,0,1),(0,1)]	(1,2)	1	$\{c\}$	$\{a,b,c\}$	c	(1,0,0)	(1,1)	2
			2	$\{b,c\}$	$\{a,b,c\}$	b	(0,1)	(1,2)	4
						c	(1,0)	(1,1)	3

Since there is no deadlock configuration, the components shown in Fig.2 are compatible.

4 Service Renaming

The progress of interactions for the model shown in Fig.1 can be illustrated by a "request graph" shown in Fig.3, in which the nodes are configurations of the model (from the column "*conf*" of the table following Fig.1 and the edges are labeled by the services requested/provided by the interacting components in the form "`x/i`" where "`x`" is the service and "`i`" is the index of the requester component. It should be observed that Fig.3 is a graphical representation of the table following Fig.1 with node 4 representing the deadlock.

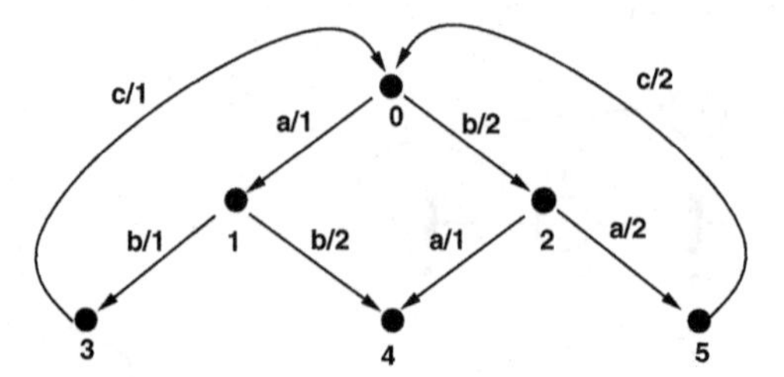

Fig. 3 Request graph for Fig.1.

There are two paths leading (from the initial node 0) to the deadlock node; one is "`(a/1,b/2)`" and the second is "`b/2,a/1)`", as discussed in the example in Section 3.1. Moreover, the cycles including nodes 0–1–3–0 and 0–2–5–0 represent the compatibility of single requester components with the provider.

For the renamed services, the steps performed by the function *CheckProgressB* are illustrated in the following table:

conf	**m**	*n*	*i*	*Symbols*1	*Symbols*2	*x*	$next(\mathbf{N}_r[i],\mathbf{m}[i],x)$	$next(\mathscr{N}_p,n,x)$	ℓ
0	[(1,0,0),(1,0,0)]	(1,0,0,0)	1	$\{a\}$	$\{a,b\}$	a	(0,1,0)	(0,1,0,0)	1
			2	$\{b\}$	$\{a,b\}$	b	(0,1,0)	(0,0,1,0)	2
1	[(0,1,0),(1,0,0)]	(0,1,0,0)	1	$\{B\}$	$\{B\}$	B	(0,0,1)	(0,0,0,1)	3
			2	$\{b\}$	$\{B\}$				–
2	[(1,0,0),(0,1,0)]	(0,0,1,0)	1	$\{a\}$	$\{A\}$				–
			2	$\{A\}$	$\{A\}$	A	(0,0,1)	(0,0,0,1)	4
3	[(0,0,1),(1,0,0)]	(0,0,0,1)	1	$\{c\}$	$\{c\}$	c	(1,0,0)	(1,0,0,0)	0
			2	$\{b\}$	$\{c\}$				–
4	[(1,0,0),(0,0,1)]	(0,0,0,1)	1	$\{a\}$	$\{c\}$				–
			2	$\{c\}$	$\{c\}$	c	(1,0,0)	(1,0,0,0)	0

The deadlock node from Fig.3 has been eliminated, so the components – after service renaming – are compatible and can be composed into a deadlock–free system.

It can be observed that there are several other service renamings which result in the same behavior of composed system, for example "`(Abc)*`", "`(Bac)*`" and "`(Ab+Ba)c)*`" as well as "`(ABc)*`", "`(bac)*`" and "`(AB+ba)c)*`", so some other criteria can be taken into account in using service renaming for eliminating conflicting requests.

5 Concluding Remarks

In component–based systems, when several requester components are interacting with one or more provider components, the requests from different components can be interleaved and then the properties of the composed system can differ significantly from the properties of components. As shown in Section 3.1, the compatibilities of pairs of interacting components are not sufficient – in general case – for the compatibility of composed system. Consequently, the compatibility of each composition must be verified independently of the compatibility of interacting pairs of components. Straightforward algorithms for such verifications are outlined in Section 3.

In the case of incompatibilities (represented by deadlocks in the composed system), service renaming has been proposed as a systematic approach to eliminating conflicting requests.

Practical service renaming can be performed by connectors [2] or component adaptors [3] [17].

The discussion (in Section 3) was restricted to systems of several requester components interacting with a single provider component because it can be shown that systems with several provider components can be decomposed into several systems, each with one provider component, and analyzed one after another.

It can be observed that the proposed service renaming can be used for restricting request interleaving that may be required for incremental composition [20].

Acknowledgements. The Natural Sciences and Engineering Research Council of Canada partially supported this research through grant RGPIN-8222.

References

1. Attiogbé, J.C., André, P., Ardourel, G.: Checking Component Composability. In: Löwe, W., Südholt, M. (eds.) SC 2006. LNCS, vol. 4089, pp. 18–33. Springer, Heidelberg (2006)
2. Baier, C., Klein, J., Klüppelholz, S.: Modeling and Verification of Components and Connectors. In: Bernardo, M., Issarny, V. (eds.) SFM 2011. LNCS, vol. 6659, pp. 114–147. Springer, Heidelberg (2011)
3. Bracciali, A., Brogi, A., Canal, C.: A formal approach to component adaptations. The Journal of Systems and Software 74(1), 45–54 (2005)
4. Broy, M.: A theory of system interaction: components, interfaces, and services. In: Interactive Computations: The New Paradigm, pp. 41–96. Springer, Heidelberg (2006)
5. Chaki, S., Clarke, S.M., Groce, A., Jha, S., Veith, H.: Modular verification of software components in C. IEEE Trans. on Software Engineering 30(6), 388–402 (2004)
6. Craig, D.C., Zuberek, W.M.: Compatibility of software components – modeling and verification. In: Proc. Int. Conf. on Dependability of Computer Systems, Szklarska Poreba, Poland, pp. 11–18 (2006)
7. Craig, D.C., Zuberek, W.M.: Petri nets in modeling component behavior and verifying component compatibility. In: Proc. Int. Workshop on Petri Nets and Software Engineering, Siedlce, Poland, pp. 160–174 (2007)
8. Crnkovic, I., Schmidt, H.W., Stafford, J., Wallnau, K.: Automated component-based software engineering. The Journal of Systems and Software 74(1), 1–3 (2005)
9. Garlan, D.: Formal Modeling and Analysis of Software Architecture: Components, Connectors, and Events. In: Bernardo, M., Inverardi, P. (eds.) SFM 2003. LNCS, vol. 2804, pp. 1–24. Springer, Heidelberg (2003)

10. Henrio, L., Kammüller, F., Khan, M.U.: A Framework for Reasoning on Component Composition. In: de Boer, F.S., Bonsangue, M.M., Hallerstede, S., Leuschel, M. (eds.) FMCO 2009. LNCS, vol. 6286, pp. 1–20. Springer, Heidelberg (2010)
11. Hopcroft, J.E., Motwani, R., Ullman, J.D.: Introduction to automata theory, languages, and computations, 2nd edn. Addison-Wesley (2001)
12. Leicher, A., Busse, S., Süß, J.G.: Analysis of Compositional Conflicts in Component-Based Systems. In: Gschwind, T., Aßmann, U., Wang, J. (eds.) SC 2005. LNCS, vol. 3628, pp. 67–82. Springer, Heidelberg (2005)
13. Murata, T.: Petri nets: properties, analysis, and applications. Proceedings of the IEEE 77(4), 541–580 (1989)
14. Nierstrasz, O., Meijler, T.: Research directions on software composition. ACM Computing Surveys 27(2), 262–264 (1995)
15. Reisig, W.: Petri nets – an introduction. EATCS Monographs on Theoretical Computer Science, vol. 4. Springer (1985)
16. Szyperski, C.: Component software: beyond object-oriented programming, 2nd edn. Addison–Wesley Professional (2002)
17. Yellin, D.M., Strom, R.E.: Protocol specifications and component adaptors. ACM Trans. on Programming Languages and Systems 19(2), 292–333 (1997)
18. Zaremski, A.M., Wang, J.M.: Specification matching of software components. ACM Trans. on Software Engineering and Methodology 6(4), 333–369 (1997)
19. Zuberek, W.M.: Checking compatibility and substitutability of software components. In: Models and Methodology of System Dependability, ch. 14, pp. 175–186. Oficyna Wydawnicza Politechniki Wroclawskiej, Wroclaw (2010)
20. Zuberek, W.M.: Incremental Composition of Software Components. In: Zamojski, W., Kacprzyk, J., Mazurkiewicz, J., Sugier, J., Walkowiak, T. (eds.) Dependable Computer Systems. AISC, vol. 97, pp. 301–311. Springer, Heidelberg (2011)
21. Zuberek, W.M., Bluemke, I., Craig, D.C.: Modeling and performance analysis of component-based systems. Int. Journal of Critical Computer-Based Systems 1(1-3), 191–207 (2010)

Author Index